WIRELESS SECURITY AND CRYPTOGRAPHY

Specifications and Implementations

WIRELESS SECURITY AND CRYPTOGRAPHY

Specifications and Implementations

Edited by
NICOLAS SKLAVOS
XINMIAO ZHANG

CRC Press is an imprint of the
Taylor & Francis Group, an informa business

CRC Press
Taylor & Francis Group
6000 Broken Sound Parkway NW, Suite 300
Boca Raton, FL 33487-2742

Printed in the United States of America on acid-free paper
10 9 8 7 6 5 4 3 2 1

International Standard Book Number-10: 0-8493-8771-X (Hardcover)
International Standard Book Number-13: 978-0-8493-8771-5 (Hardcover)

Library of Congress Cataloging-in-Publication Data

Wireless security and cryptography : specifications and implementations / edited by Nicolas Sklavos and Xinmiao Zhang.
p. cm.
Includes bibliographical references and index.
ISBN-13: 978-0-8493-8771-5
ISBN-10: 0-8493-8771-X
1. Wireless communication systems--Security measures. 2. Cryptography. I. Sklavos, Nicolas. II. Zhang, Xinmiao. III. Title.

TK5103.2.W57415 2007
005.8--dc22 2006100063

Visit the Taylor & Francis Web site at
http://www.taylorandfrancis.com

and the CRC Press Web site at
http://www.crcpress.com

Dedication

To my family and to all my teachers

— Dr. Nicolas Sklavos

To Ruoyu and my parents

— Dr. Xinmiao Zhang

Editors

Nicolas Sklavos received a Ph.D. in electrical and computer engineering and a diploma in electrical and computer engineering in 2004 and 2000, respectively, both from the Electrical and Computer Engineering Department, University of Patras, Greece. In 2005, he joined the Telecommunications Systems and Networks Department of the Technological Educational Institute of Messolonghi, Nafpaktos, Greece, where he works as an assistant professor. His research interests include security and privacy, wireless communications security, and mobile networks. He holds an award for his Ph.D. thesis on "VLSI Designs of Wireless Communications Security Systems," from IFIP VLSI SOC, Germany (2003). He has also contributed to international journals and participated in the organization of conferences, as program committee and guest editor. Dr. Sklavos is a member of the IEEE, IEE, the Technical Chamber of Greece, and the Greek Electrical Engineering Society. He has authored and coauthored up to 90 scientific articles, books and book chapters, reviews, and technical reports in the areas of his research. He can be contacted at nsklavos@ieee.org.

Xinmiao Zhang received B.S. and M.S. degrees in electrical engineering from Tianjin University, Tianjin, China, in 1997 and 2000, respectively. She received a Ph.D. in electrical engineering from the University of Minnesota–Twin Cities, in 2005. Since then, she has been with Case Western Reserve University, where she is currently a Timothy E. and Allison L. Schroeder Assistant Professor in the Department of Electrical Engineering and Computer Science. Her research interests include efficient VLSI architecture design for communications, cryptosystems, and digital signal processing. Dr. Zhang is the recipient of the Best Paper Award at ACM Great Lake Symposium on VLSI 2004. She also won the first prize in the Student Paper Contest at the Asilomar Conference on Signals, Systems, and Computers 2004. She is a member of the IEEE.

Contributors

Apostolos P. Fournaris
Electrical and Computer
Engineering Department
University of Patras
Patras, Greece

Panu Hämäläinen
Nokia Technology Platforms
Tampere, Finland

and

Institute of Digital and Computer
Systems
Tampere University of Technology
Tampere, Finland

Timo D. Hämäläinen
Institute of Digital and Computer
Systems
Tampere University of Technology
Tampere, Finland

Marko Hännikäinen
Institute of Digital and Computer
Systems
Tampere University of Technology
Tampere, Finland

Vesna Hassler
European Patent Office
Sub-office Vienna
Vienna, Austria

Paris Kitsos
School of Science and Technology
Hellenic Open University
Patras, Greece

Çetin Kaya Koç
School of Electrical Engineering
and Computer Science
Oregon State University
Corvallis, Oregon

and

Istanbul Commerce University
Istanbul, Turkey

O. Koufopavlou
Electrical and Computer
Engineering Department
University of Patras
Patras, Greece

Martin Manninger
Austria Card GmbH
Vienna, Austria

John V. McCanny
The Institute of Electronics,
Communications and Information
Technology (ECIT)
Queen's University Belfast
Belfast, Northern Ireland

Maire McLoone
The Institute of Electronics,
Communications and Information
Technology (ECIT)
Queen's University Belfast
Belfast, Northern Ireland

Siddika Berna Örs
Department of Electronics and
Communication Engineering
Istanbul Technical University
Istanbul, Turkey

Norbert Pramstaller
Institute for Applied Information Processing and Communications (IAIK)—Krypto Group
Graz University of Technology
Graz, Austria

Bart Preneel
Department of Electrical Engineering
Catholic University of Leuven
SCD/COSIC, Belgium

Vincent Rijmen
Institute for Applied Information Processing and Communications (IAIK)–Krypto Group
Graz University of Technology
Graz, Austria

Palash Sarkar
Applied Statistics Unit
Indian Statistical Institute
Kolkata, India

Erkay Savas
Faculty of Engineering and Natural Sciences
Sabanci University
Istanbul, Turkey

Nicolas Sklavos
Telecommunication Systems and Networks Department
Technological Educational Institute of Messolonghi
Nafpaktos, Greece

Neil Smyth
Conescant Systems Inc.
Belfast, Northern Ireland

Lo'ai A. Tawalbeh
Computer Engineering Department
Jordan University of Science and Technology (JUST)
Irbid, Jordan

Ingrid Verbauwhede
Department of Electrical Engineering
Catholic University of Leuven
SCD/COSIC, Belgium

Xinmiao Zhang
Department of Electrical Engineering and Computer Science
Case Western Reserve University
Cleveland, Ohio

Introduction

Wireless communications have become a very attractive and interesting sector for the provision of electronic services. Mobile networks are available almost anytime and anywhere, and the popularity of wireless handheld devices is high. The services offered are strongly increasing because of the wide range of the users' needs. They vary from simple communication services to applications for special and sensitive purposes such as electronic commerce and digital cash.

As wireless devices are used in offices and houses, the need for strong and secure transport protocols seems to be one of the most important issues in mobile standards. It is obvious that in future wireless protocols and communication environments (networks), security will play a key role in transmitted information operations. From e-mail services to cellular-provided applications and from secure internet possibilities to banking operations, cryptography is an essential part of today's users' needs. Recent and future mobile communication systems have special needs for cryptography. They must support the three basic types of cryptography: bulk encryption, message authentication, and data integrity. Most of the widely used wireless systems support all the three different types of encryption. Additionally, some systems offer users the choice to select from two or three alternative ciphers for each encryption operation. The user can select the best-suited algorithm for the needs of the application. In most of the cases, implementation of the same encryption system supports all the three different types of cryptography.

The standards for mobile applications and services are maturing, and new specifications in security systems are defined. This leads to a huge set of possible technologies that a service provider can choose. Although organizations and forums seem to agree with the increasing need for secure and strong systems cryptography is still troublesome for wireless networks because of the difficulties in implementation. The security layers of many wireless protocols use outdated encryption algorithms, which have proved unsuitable for hardware implementations, especially for wireless handheld devices. In general, the ciphers use large arithmetic and algebraic modifications, which are not appropriate for hardware implementations. That is why cipher implementations allocate many of the system resources, in hardware terms, to be used as components. Therefore, in many cases, software applications have been developed to support the needs of security and cryptography. However, the software solution is not acceptable in the case of handheld devices and mobile communications with high-speed and low-power consumption requirements.

This book summarizes key issues that should be solved to achieve the desirable performance in security implementations and to focus on alternative integration approaches for wireless communication security. It gives an overview of the current security layer of wireless protocols and presents the performance characteristics of implementations in both software and hardware.

This book also proposes efficient and novel methods to implement security schemes in wireless protocols with high performance. The purpose of this book is to provide the state-of-the-art research trends in implementations of wireless protocol security for current and future wireless communications.

This book contains 13 chapters in total.

The introduction is by Nicolas Sklavos and Xinmiao Zhang. The basic security primitives relevant to all communication protocols are dealt with in Chapter 1, by Palash Sarkar. The main scope of this chapter is to explain the underlying ideas of the described complete solutions, which are given in the subsequent chapters of this book.

Chapter 2, by Vesna Hassler, addresses the basic communication security concepts. It first explains the threats that are encountered in a communication network of any type, such as a LAN, wireless local area networking (WLAN), or Universal Mobile Telecommunication System (UMTS), and then presents the security services that protect against those threats as well as the security mechanisms and techniques to implement the services.

In Chapter 3, Xinmiao Zhang addresses various algorithmic and architectural optimization approaches for efficient hardware implementation of the advanced encryption standard (AES) algorithm. Three architectural-level optimization techniques, as well as the speedup factor and area consumption of each technique, are presented in this chapter. In addition, various algorithmic modifications of the AES algorithm are introduced. Finally, resource sharing between encryptors and decryptors is explored.

Chapter 4 is dedicated to hardware design issues in elliptic curve cryptography for wireless systems. Design problems of elliptic curve cryptosystems (ECCs) are presented. The authors Apostolos P. Fournaris and O. Koufopavlou deal with it along with algorithms and methods of solving such problems.

Chapter 5, by Lo'ai A. Tawalbeh and Çetin Kaya Koç, presents an efficient elliptic curve cryptographic hardware design for wireless security. It is based on a new algorithm called unified division/multiplication algorithm (UDMA). The scalability feature of the proposed cryptoprocessor allows the adjustment of the word size used in the datapath to meet area and performance requirements.

Vincent Rijmen and Norbert Pramstaller, in Chapter 6, discusses cryptographic primitives and the security services they can deliver and argues that by using only a block cipher it is possible to deliver a wide range of security services. In this chapter, the implementation of the AES, which is used for symmetric encryption and authentication, and Whirlpool, which is a dedicated hash function standardized in ISO/IEC 10118-3, is also presented.

In Chapter 7, the authors Siddika Berna Örs et al. deal with side-channel analysis attacks on hardware implementations. The chapter introduces the passive attacks that the authors have conducted on the hardware implementations of an ECC over GF(p), the AES, and the data encryption standard (DES). The chapter also summarizes the previous work on these side-channel attacks.

Panu Hämäläinen et al., in Chapter 8, present a novel enhanced security layer (ESL) for Bluetooth. As ESL is placed on top of the standard controller interface, it can be integrated into any standard Bluetooth implementation. A full-scale embedded prototype implementation of ESL is also presented. AES and its operation modes are implemented in hardware for high performance. The easy-to-use programming interface supports straightforward application development.

In Chapter 9, Neil Smyth et al. discuss two contrasting approaches that may be taken in the design of a hardware accelerator targeted at IEEE 802.11i. The first approach is a programmable design that comprises the authors' own primitive reduced instruction set computer (RISC) processor design and two hardware accelerators, which perform AES and RC4 encryptions. The WLAN processor has been designed specifically to perform the frame processing requirements of WEP, TKIP, WRAP, and CCMP, as specified in Draft 3.0 of the IEEE 802.11i standard. The second approach evaluates the performance of a fixed-functionality WLAN security design.

Paris Kitsos and Nicolas Sklavos, in Chapter 10, propose a hardware implementation of the UMTS security mechanism. The proposed system supports the authentication and key agreement (AKA) procedure and the data confidentiality and integrity protection procedures. The AKA procedure is based on AES. The data confidentiality and integrity protection procedures are based on the Kasumi block cipher.

In Chapter 11, by Nicolas Sklavos, a security processor for the wireless application protocol (WAP) is presented. Wireless transport layer security (WTLS) is dedicated to the security of WAP. In this chapter, an efficient architecture and the implementation of WTLS are introduced. The proposed processor supports privacy, authentication, and data integrity.

In Chapter 12 of this handbook, Erkay Savas proposes different algorithms for GF(p). Their performances from the perspectives of both software and hardware implementations are discussed. Inversion algorithms for GF(2^n) are also presented.

Last but not the least, in Chapter 13, Martin Manninger describes smart card technology. To achieve better security on the technical level, secure hardware such as smart cards can be employed. The chapter explains the basics of smart card technology, and it further shows how smart cards can help in establishing end-to-end transaction security in wireless environments.

We would like to thank Allison Taub of CRC Press/Taylor & Francis for her personal interest in this book and for her help. We also wish to thank everyone connected with the CRC Press/Taylor & Francis team, including project coordinators Theresa Delforn and Marsha Pronin, project editor Richard Tressider, and Suryakala Arulprakasam of SPi for their help in the production of this book.

Xinmiao Zhang would like to thank Keshab Parhi for encouraging her to undertake the implementation of the AES algorithm during her Ph.D. study. She would also like to give special thanks to her husband, parents, and grandparents for their love and support. Last, by no means the least, she would like to thank the coeditor, Nicolas Sklavos, for this enjoyable and productive collaboration.

We also thank Maja Matijasevic for her interest in this project and for volunteering to support. Many thanks to the anonymous reviewers for their comments on and suggestions for this publication. Their efforts helped us to improve the quality of this work.

Special thanks to the authors of this book. We expect that their ideas introduced here would contribute to the research community in great measure, to go a step forward not only in science, research, and engineering, but also to a more secured world.

Nicolas Sklavos and Xinmiao Zhang

Table of Contents

1 Overview of Cryptographic Primitives for Secure Communication

Palash Sarkar

CONTENTS

1.1 INTRODUCTION

Cryptography is essentially the art of secret writing. To most people, this is confined to the pages of a detective story (remember the "dancing figures" faced by Sherlock Holmes) or is something that is relevant in the context of military communication. In a war, messages need to be exchanged between units of the same army to coordinate joint maneuvers. Since such messages can easily fall into enemy hands, it should be ensured that none but the intended recipient can read the message. In fact, a system of exchanging secret messages was practiced in the time of Julius Caesar, and the system is called "Caesar shift" after him.

The subject of cryptology has an ancient history. Interested persons can read the encyclopedic book by D. Kahn called *The Codebreakers*. The book covers cryptology from its initial use by the Egyptians some 4000 years ago to the twentieth century where it played an important role in the outcome of both the World Wars. Another equally fascinating book is the *Code Book* by Simon Singh, which covers the development of modern cryptology.

In the present day, secure communication is no longer confined to the pages of a story book or to military communication. In the modern business world, vital information needs to be exchanged between parties for the successful completion of a transaction. Moreover, current business practices are dependent on extensive use of computers and the Internet. In fact, in e-commerce applications, whole business transactions are completed over the Internet. This possibility gives rise to various kinds of subtle security problems.

This chapter attempts to provide an overview of some of the fundamental primitives used in modern cryptography. In symmetric key cryptography, we cover block and stream ciphers as well as hash functions. In public key cryptography, we cover key agreement, public key encryption (PKE), digital signatures, and the current research topic of identity-based encryption (IBE).

We believe the above primitives to be of fundamental importance to modern cryptography. While discussing these topics we also discuss related topics. For example, in the discussion on block ciphers we deal with message authentication code (MAC) and various modes of operations.

For each of the topics, we present the basic concept, sketch some construction methods, and describe the formal model and security notions. None of the constructions and protocols described here are meant to be used directly in practice. They are presented for illustrating the underlying ideas rather than for providing complete description of ready-to-use protocols. The latter is not the goal of this chapter. This chapter is intended to serve as an introduction to the main ideas of cryptography and should be accessible to a general engineering audience. Lastly, we must add that our selection of topics, constructions, and formalism is based on our knowledge and belief of what is important in cryptography. We make no claims of providing a complete and comprehensive treatment of cryptography. The subject is too vast to be condensed to a few pages. There are several books on cryptology [1,2,3], which may be consulted for further reading. Though a little old, the handbook of applied cryptography [4] is an excellent source of reference.

1.2 BLOCK CIPHERS

In general terms, a block cipher is a map $E: \mathcal{K} \times \mathcal{M} \rightarrow \mathcal{M}$, where for each $K \in \mathcal{K}$, the map $E_K: \mathcal{M} \rightarrow \mathcal{M}$, defined by $E_K(M) = E(K,M)$, is a bijection. In other words, $E_K()$ is a permutation of $\mathcal{M}$. The set $\mathcal{K}$ is called the key space and the set $\mathcal{M}$ is called the message space. The output of $E_K()$ lies in the cipher space and in our definition, the cipher space is the same as the message space. The inverse of E_K is a map $D: \mathcal{K} \times \mathcal{M} \rightarrow \mathcal{M}$ and we write $D_K(M) = D(K,M)$. By the inverse property we have $M = D_K(E_K(M))$. Practical block ciphers have $\mathcal{M} = \{0,1\}^n$ and $\mathcal{K} = \{0,1\}^k$. The values of n and k need not be equal, but both of them must be large enough such that exhaustive search requiring 2^n and 2^k operations is infeasible. Typical values of k and n are 128, 192, and 256.

In basic terms, a sender and a receiver share an element K of $\mathcal{K}$. This K is known to both of them and is not known to anybody else, that is, K is a secret key shared by the sender and the receiver. This key is shared between the two parties using a secure channel. To encrypt a message (or plaintext) $M \in \mathcal{M}$, the sender computes $C = E_K(M)$ and transmits C to the receiver over a public channel. The receiver decrypts by computing $D_K(C)$.

The security of a block cipher has been defined precisely in the literature. We discuss it a little later. At this point, let us try to intuitively understand what it means for a block cipher to be secure. It is usually assumed that the adversary who is trying to crack (or break) the cipher knows the particular block cipher that is used, though he does not know the value of K. Further, the adversary has access to the public channel and hence knows C. The target of the adversary is to find K or M. Thus, for security, it must be infeasible to find K or M from C. The same K may be used to encrypt many messages, say, $M_1, \ldots, M_t$ and the adversary knows the corresponding ciphertexts $C_1, \ldots, C_t$. Knowing more than one ciphertexts may possibly provide the adversary with more information about the key. However, for a secure block cipher, it should still be infeasible for the adversary to find K or any of the M_i.

The scenario just described assumes that the adversary gets to know only the ciphertexts. This is called a ciphertext only attack. A stronger attack is when the adversary knows a few plaintext–ciphertext pairs, that is, it knows a few pairs of inputs and outputs of $E_K()$. This is called a known plaintext attack. Since the adversary has access to more information, the attack is stronger than a ciphertext only attack. An even stronger attack assumes that the adversary is able to choose (as opposed to simply knowing) a few plaintexts and gets to know the corresponding ciphertexts. This scenario is called a chosen plaintext attack. The goal of the adversary in both the known and chosen plaintext attacks is to either find K or to find the plaintext corresponding to a ciphertext it has not seen earlier.

The design of practical block ciphers has a long history. Many ciphers have been proposed and analyzed in the literature. In the process, certain design principles have become accepted. The basic structure of almost all proposed block ciphers can be described in the following manner. The encryption process consists of several rounds that are applied to the plaintext one after another. The key K is expanded using a key schedule algorithm into a set of round keys $K_1, \ldots, K_r$. Each round takes the round key as input and the output of the previous round and produces an output. For a fixed round key, the round function is a bijective map. For a plaintext M, let M_0, $M_1, \ldots,$ M_{r-1} denote the inputs to the r rounds. The input M_0 to the first round is M itself and let $M_r = C$ be the final output of $E_K()$. If we denote the ith round function by R_i, then we have $M_i = R_i(K_i, M_{i-1})$.

This reduces the task of designing a block cipher to the task of designing a key-scheduling algorithm and that of designing the round functions. Usually for a cipher, the round functions are same or very similar. Here, we briefly describe two methods for designing round functions.

1.2.1 Feistel Structure

In a Feistel structure, the input M_{i-1} to the ith round is divided into two equal halves L_{i-1} and R_{i-1}, that is, $M_{i-1} = L_{i-1} \,\|\, R_{i-1}$. The output $M_i = L_i \,\|\, R_i$ is defined as follows:

$$
\begin{aligned}
L_i &= R_{i-1}, \\
R_i &= L_{i-1} \oplus f(R_{i-1}, K_i).
\end{aligned}
$$

This defines an invertible map, that is, from $L_i \,\|\, R_i$, it is possible (and easy) to obtain $L_{i-1} \,\|\, R_{i-1}$. Invertibility does not depend on the function $f(.,.)$. In fact, it is this function f that one has to design to obtain a specific algorithm and the security of the algorithm depends on the design of $f(.,.)$ (and the key-scheduling algorithm). Among several properties, f must be a nonlinear map. The data encryption standard (DES) is the most famous example of a block cipher based on the Feistel structure.

1.2.2 Substitution–Permutation Network

In a substitution–permutation network (SPN), each round function consists of a few alternating layers called substitution and permutation layers. The input to a substitution layer is divided into small blocks of bits, say blocks of 8 bits each. An S-box (or substitution box) is applied to each block. The S-box substitutes its input bits by an equal number of bits. Each S-box is a bijective map, so that the entire substitution layer is also a bijective map. (In general, an S-box replaces t_1 bits by t_2 bits where t_1 and t_2 can be unequal. Hence, an S-box is not necessarily a bijection.) The effect of a substitution layer is local in the sense that an output bit in a particular position depends only on a few of the input bits in its nearby positions. This local effect is compensated with a permutation layer, which performs a permutation of its input bits. The round key is usually incorporated in between a substitution and a permutation layer. The advanced encryption standard (AES) employs the SPN style of design with the following modification. In the permutation layer, instead of applying a bit permutation it applies a carefully designed affine transformation. See [5] for a detailed description of the algorithm.

1.2.3 Modes of Operations

As mentioned earlier, a block cipher is a fundamental primitive in cryptography. A block cipher by itself can encrypt only fixed-length strings of length n. Applications in general require encryption of long and arbitrary-length strings. A mode of operation of a block cipher is used to extend the domain of applicability from fixed-length strings to long and variable-length strings. The four classical modes of operations are as follows. Let the long message consist of n-bit blocks denoted by $M_1, \ldots, M_m$. The ciphertext blocks $C_1, \ldots, C_m$ in the different modes of operations are obtained as follows. Some of the modes of operations require the use of an initialization vector (IV).

Electronic codebook (ECB) mode: $C_i = E_K(M_i)$.
Cipher block chaining (CBC) mode: Let $C_0 = \text{IV}$ and for $i \geq 1$, define $C_i = E_K(C_{i-1} \oplus M_i)$.
Cipher feedback (CFB) mode: Let $C_0 = \text{IV}$ and for $i \geq 1$, define $Z_i = E_K(C_{i-1})$; $C_i = M_i \oplus Z_i$.
Output feedback (OFB) mode: Let $Z_0 = \text{IV}$ and for $i \geq 1$, define $Z_i = E_K(Z_{i-1})$; $C_i = M_i \oplus Z_i$.

A mode of operation must be secure in the sense that one should be able to prove that the only way of attacking a mode of operation is to attack the underlying block cipher. The ECB is not a secure mode of operation. This is because if two of the M_is are equal the corresponding ciphertext blocks are also equal. This is undesirable from a security standpoint.

There are several different goals of a mode of operation. The basic goal is privacy or confidentiality of the message. Another equally important goal is to provide authentication. This means that instead of encrypting a message, we produce a tag (which is a fixed-length string), such that if the message is tampered, then the tag of the tampered message will not equal the original tag. Such a feature allows tamper detection and is important in many practical applications. The tag is also called a MAC.

Very often, applications require both privacy and authentication. A mode of operation providing both is called authenticated encryption (AE). The problem of designing a secure AE mode of operation has been a topic of intense research. A simple way to achieve AE is to use a two-pass algorithm. In the first pass, the message is encrypted and the ciphertext is produced. The second pass computes a tag of the ciphertext and the final output is the ciphertext followed by the tag. Using two passes makes the scheme inefficient. Jutla [6] was the first to point out that both encryption and authentication can be achieved by a one-pass algorithm. Other one-pass algorithms include a design named offset codebook (OCB) by Rogaway [7]. Unfortunately, all previous one-pass algorithms have pending patent applications, which severely restrict their widespread adoption. Very recently, several new one-pass algorithms have been proposed [52] without fresh patent claims.

There are several other interesting modes of operations. Consider the application of disk encryption. This capability is built into the disk controller. All data kept on the disk are encrypted. The atomicity of encryption is at the sector level, that is, a sector is considered to be a single message and encrypted. The same key is, however, used to encrypt all the sectors. The basic goal of such a mode of operation is to provide privacy. A secondary (but also important) goal is to achieve tamper resistance or nonmalleability. An adversary may change a few bits of an encrypted sector in such a manner that a decryption of the tampered sector leads to a valid but different data from what was originally encrypted. If this is possible, then the mode of operation is malleable. One way to achieve nonmalleability is to use a MAC as described earlier. The problem is that we will need to store the tag on the disk and hence waste disk space. Another option is to design a mode of encryption, such that decrypting a tampered sector provides a message that looks entirely random (it will be computationally indistinguishable from a random message). This also provides a limited form of authentication and achieves nonmalleability. In some sense, this is the maximum authentication one can hope to achieve without storing a tag. Work on this problem has led to several interesting designs [8].

1.2.4 Formal Security Model

The formal model of security for a block cipher is a pseudorandom permutation (PRP) [9,10]. This notion is defined in terms of an adversarial game. The adversary interacts with an oracle, that is, the adversary provides an input and is provided with an output corresponding to the input. The queries can be made in an adaptive manner, that is, a particular query can depend on the previous queries and its outputs. At the end of the interaction, the adversary outputs a bit. By instantiating the oracle in two ways, we obtain two games. In the real game, a random secret key is chosen and the oracle is instantiated with $E_K()$, whereas in the random game the oracle is instantiated with a random permutation. Let p_0 (resp. p_1) be the probability that the adversary outputs 1 in the real (resp. random) game. The difference $|p_0 - p_1|$ is the adversary's advantage in distinguishing $E_K()$ from a random permutation.

We say that $E_K()$ is a PRP, if this advantage is negligible. A stronger notion is that of a strong pseudorandom permutation (SPRP). In this notion, the adversary interacts with two oracles—the encryption and the decryption oracles in the real game; and a random permutation and its inverse in the random game. The advantage is defined as given previously and the block cipher is said to be an SPRP if this advantage is negligible.

At this point, we should remark on the utility of the formal model. None of the practical block ciphers (including AES) can be actually proved to be a PRP or an SPRP. On the other hand, one usually constructs protocols where the block cipher is a component. For example, a mode of operation can be considered to be a protocol to encrypt long messages using a block cipher. Such protocols have their own appropriate notion of security. To show that a particular protocol satisfies this notion of security one requires the underlying block cipher (and other components) to be a PRP or an SPRP. Another way of viewing this situation is to consider a PRP or an SPRP to be an idealization of practical block ciphers.

1.3 STREAM CIPHERS

Stream ciphers are the second basic cryptographic primitives for encryption. They are used widely for both defense communications and industrial applications. The basic principle behind stream cipher encryption is quite simple. Assume that for $t \geq 0$, $z^{(t)}$ is a random-bit sequence, which is known both to the sender and the receiver. Suppose the sender wants to transmit a message-bit sequence $m^{(t)}$. The cipher-bit sequence is computed as $c^{(t)} = m^{(t)} \oplus z^{(t)}$, which is then transmitted. Since the receiver knows $z^{(t)}$, it is possible for him to compute $m^{(t)}$ as $m^{(t)} = c^{(t)} \oplus z^{(t)}$. This simple scheme satisfies a strongest possible notion of secrecy called perfect secrecy [11]. In other words, access to the cipher-bit sequence provides no information about the message-bit sequence. This property arises because the masking sequence $z^{(t)}$ (also called key sequence) is a true random sequence. Since it is a random sequence, it cannot be reused and hence this scheme is also called a one-time pad.

The main problem with the one-time pad is that the key sequence, which is a true random sequence, is as long as the message sequence. Since the key sequence is required at both the sender and the receiver ends, the entire key sequence must be transmitted securely before its use in encryption and decryption. Since the key sequence has to be transmitted through a secure channel, the problem of securely transmitting a long sequence remains. Note that the main issue here is the fact that a true random sequence cannot be produced by a deterministic method. In fact, extracting true random bits from electronic devices is a difficult problem.

One way of getting around the above problem is to use a pseudorandom generator (PRG) as a key sequence. (PRG is different from a PRP discussed earlier.) A PRG is a deterministic algorithm, which extends a short fixed-length bit string (called a seed) into a long sequence of bits. The seed is the secret key, which is shared between the sender and the receiver. Consequently, both the sender and the receiver can generate the same key sequence.

The security of the system depends on the security of the PRG. There are several ways of defining a PRG. Here we consider the notion of computational security. Informally, a PRG is said to be secure if the knowledge of a segment of the key sequence does not allow an adversary with practical computational resources to guess the next bit with probability significantly more than half. Alternatively, it should not be possible to computationally distinguish the output of a PRG from a true random sequence. Both these notions have been formalized and shown to be equivalent [12,13].

Practical stream ciphers have been around for a very long time and certainly before the notion of computational pseudorandomness came to be formalized. The goal of practical stream ciphers is essentially to construct a secure PRG. As in the case of block ciphers, it is not possible to prove any practical stream cipher to be a secure PRG. Thus, the theoretical concept must be seen as an idealization of practical stream ciphers. We, however, note that there are certain constructions [14], which can be proved to be a secure PRG assuming the hardness of certain computational problem such as determining quadratic residues. Though interesting from a theoretical point of view, such designs are usually too slow to meet the application requirements.

1.3.1 Linear Feedback Shift Register

One of the most important structures used in the construction of practical stream ciphers is that of a linear feedback shift register (LFSR). This is essentially a register consisting of k bits. At each clock, the register changes state. The next state is determined from the current state using a simple linear transformation. Let $a^{(i)} = (a_{k-1}^{(i)}, \ldots, a_0^{(i)})$ be a sequence of k-bit vectors providing the successive states of an LFSR. The linear mapping is given by

$$\left.\begin{aligned} a_j^{(i+1)} &= a_{j+1}^{(i)} && \text{for } 0 \le j \le k-2; \\ a_{k-1}^{(i+1)} &= t_1 a_{k-1}^{(i)} \oplus t_2 a_{k-2}^{(i)} \oplus \cdots \oplus t_{k-1} a_1^{(i)} \oplus t_k a_0^{(i)}. && \end{aligned}\right\} \quad (1.1)$$

Let $p(x) = t_k x^k \oplus t_{k-1} x^{k-1} \oplus t_1 x \oplus 1$. The polynomial $p(x)$ is called the connection polynomial and completely determines the next state function. The output of an LFSR is usually taken to be the least significant bit of each $a^{(i)}$. Of special interest is the case when $p(x)$ is a primitive polynomial. If $a^{(0)}$ is not the zero vector, then the sequence $a^{(i)}$ has a period $2^k - 1$. In this case, the output also has a period $2^k - 1$ and is called an m-sequence. There is an extensive literature on LFSRs [15] and other linear finite state machines. Since sequences produced by LFSRs satisfy linear recurrences, these cannot be directly used for cryptographic purposes. They are used as building blocks of secure stream ciphers.

There are two classical models of stream ciphers—the nonlinear-filter model and the nonlinear-combiner model. Both the models are built using LFSRs and Boolean functions. In the nonlinear-combiner model, exactly one bit sequence is extracted from each LFSR and all the bit sequences are combined using a Boolean function to generate the key sequence. In the nonlinear-filter model, several bit sequences are generated from a single LFSR and these are then combined using a Boolean function to generate the key sequence. See [4] for more details on these models and other classical stream ciphers. Extensive research on these models has shown that the Boolean functions used must have certain necessary properties. Construction methods and bounds for suitable functions are known [16].

LFSRs are also used in several different ways to design stream ciphers. Examples are the shrinking generator and the A5 stream cipher. The LFSRs described earlier are also called bit-oriented LFSRs. Such LFSRs are well suited for hardware implementation, but their software implementation is not efficient. For efficient software implementation one usually uses a word- or block-oriented LFSR [17,18].

Another important design principle for software-efficient stream cipher is the exchange-shuffle paradigm. This is based on the following idea. Consider an array of length 2^k, such that the array contains all possible k-bit strings. For example, $[0, \ldots, 255]$ is such an array where $k = 8$. We now repeatedly perform the following operation on the array. Choose two random locations of the array and exchange the elements contained in those positions. If we perform this operation sufficiently large number of times (usually a small multiple of 2^k times) then we obtain an array, which is a random permutation of the k-bit strings. From this point onward, it is possible to extract a k-bit string at each step by the following principle: Select two positions, swap their contents, and extract one k-bit string. To make this idea more concrete, we need to specify the method of choosing the positions to swap and the position from which to extract the k-bit string. RC4 is a stream cipher designed by Rivest and is the first cipher that is based on this principle.

Most modern stream ciphers use an IV. The role of the IV is not to increase security but to provide variability. In this case, the PRG is seeded by the (key, IV) pair rather than only by the key itself. While the key is secret and not known to the adversary, the IV is not secret and the adversary gets to know it. The same key may be used with distinct IVs and the constraint on the protocol usage is that a (key, IV) pair should not be repeated.

At present, stream ciphers have a similar structure, which can be described as follows. A stream cipher has an internal state that evolves under a state update map. An output function is applied to the current internal state to extract a fixed number of pseudorandom bits. The cipher goes through an initialization or key setup phase before the actual extraction of pseudorandom bits begins. In this phase, the (key, IV) pair is placed into the internal state and an initialization function is applied to the state without extracting any output. This initialization function may consist of applying the state update function a fixed number of times or it may be a different function. The aim of the initialization phase is to ensure that the internal state from which the key extraction starts becomes a complex nonlinear function of the initial internal state. On the other hand, this phase should not be too long, since during this phase no key stream is produced and there can be no encryption.

Currently, there are many stream cipher proposals as part of the Ecrypt call for stream cipher primitives [19]. Most of the proposals follow the methodology described earlier; an exception is Salsa 20, which uses a different principle. The home page contains a great deal of information and is a must-read for anybody who is seriously interested in the design and analysis of stream ciphers.

1.3.2 Self-Synchronizing Stream Cipher (SSSC)

Consider the use of a stream cipher in an error-prone channel. The channel errors may result in bit flips or in bit inserts and bit slips. The latter two errors are more serious since they destroy alignment and result in loss of synchronization between sender and receiver. In a bit-oriented stream cipher, a bit flip due to channel error causes a single bit of the received sequence to be erroneous. On the other hand, a bit slip or a bit insert causes all subsequent bits to be erroneous until the alignment is restored by a complementary error.

Channels with noisy characteristics are quite common in defense applications. Moreover, such channels usually have low bandwidth so that the employment of error-correcting codes is not feasible owing to the redundancy introduced by such codes. Yet we require secure communication on such channels. The solution is to design a cipher satisfying the following requirement. Starting from any point in the ciphertext, if a fixed number of bits are properly received, then all subsequent bits can be properly decrypted. This allows automatic synchronization between the sender and the receiver without them sharing a common clock. Hence, such ciphers are also called asynchronous stream ciphers. Apart from recovery from errors, other possible uses of self-synchronizing stream cipher (SSSC) are

1. The receiver can switch at any time into an ongoing enciphered message without knowing the current bit position in the message and decrypt from within a few bits of the time of their joining.
2. Users can join a broadcast at any point of time and be able to decrypt from within a few bits of the time of their joining.

Currently, the only known secure SSSC is to use a block cipher in a 1-bit CFB mode (see [4]). This method is inefficient since it requires a block cipher call per bit of encryption. There have been other direct proposals of SSSC. Unfortunately, all such proposals have turned out to be insecure.

1.4 HASH FUNCTIONS

A hash function maps a long message to a fixed-length bit string. The domain of a hash function is the set of all binary strings. (Actually, the domain is the set of all binary strings of a maximum possible length, such as the set of all binary strings of length less than 2^{64}.) The range consists of all binary strings of a fixed length. For example, the range can be the set of all binary strings of length equal to 128. The output of a hash function on a particular message is often called the digest of the message or simply the message digest.

Hash functions are extensively used in cryptographic protocols. One of the main uses of hash functions is in digital signature protocols, where the message digest produced by the hash function is signed. Because of the central importance of hash functions in cryptography, there has been a lot of work in this area. See [20] for a slightly outdated survey.

For a hash function H to be used in cryptographic protocols, it must satisfy certain well-known necessary properties. In a recent paper [21], Stinson provides a comprehensive discussion of these properties and also relations among them. Depending on a particular application, a secure hash function must satisfy some or all of the following properties:

1. Preimage Resistance: Finding a preimage of a given message digest must be computationally infeasible. In other words, given z it should be computationally infeasible to find x such that $H(x)=z$. A function satisfying this property is also called a one-way function. Such functions are of central importance in cryptography and were introduced by Diffie–Hellman in their seminal paper on modern cryptology [22].
2. Second Preimage Resistance: Finding a second preimage of a digest given one preimage of the same digest must be computationally infeasible. In other words, given x and z such that $H(x)=z$, it should be computationally infeasible to find y such that $x \neq y$ and $H(y)=z$. The notion of second preimage resistance was introduced by Merkle in [23].
3. Collision Resistance: Finding a collision must be computationally infeasible. In other words, it should be computationally infeasible to find x, y such that $x \neq y$ but $H(x)=H(y)$. This property was first formally defined by Damgård in [24].

It is clear that if it is possible to find a second preimage, then it is possible to find collisions. Hence, it is usually sufficient to study collision resistance. However, as pointed out in [21], there is no satisfactory reduction from collision resistance to preimage resistance or vice versa. Therefore, the goal of a practical hash function should be to achieve both preimage and collision resistance.

A generic attack for finding collisions uses the so-called birthday paradox. Suppose the hash function $H()$ produces digests of length m. In this method, one randomly chooses k distinct elements $x_1, \dots, x_k$ from the domain of $H()$ and computes the corresponding digests $y_1, \dots, y_k$. If $y_i = y_j$ for some $i \neq j$, then we have a collision. The birthday paradox states that if $k \approx 2^{m/2}$, then the probability of finding a collision using this method is around $1/2$. To prevent such an attack, we must have m to be such that it is not computationally feasible to compute $2^{m/2}$ digests in a reasonable amount of time. Consequently, message digests are at least 128 bits long and preferably 160, 256, or 512 bits long.

It is possible to construct hash functions where one can prove that finding collisions is equivalent to solving certain known difficult problems (see, for example, [25]). However, from a practical point of view such hash functions are unacceptably slow. Hence, practical hash functions are constructed from simple arithmetic/logical operations so that they are fast. The trade-off is that for such hash functions it is not possible to relate the difficulty of finding collisions to known hard problems.

Research in the design of hash functions has evolved certain principles for designing secure and practical hash functions. One of the important papers in this area is by Damgård [26]. An important point made in [26] is that it is easier to design a secure hash function with a short fixed domain than a hash function with a very large (or infinite) domain. However, for a hash function to be useful it must be possible to hash arbitrary long messages. Hence, one must look for techniques that can extend the domain of a hash function while preserving the relevant security properties.

An important construction for securely extending the domain of a secure hash function has been described by Merkle [23] and Damgård [26]. The construction is called the Merkle–Damgård (MD) construction. The MD construction is a sequential construction and provides a basic guideline for designing practical hash functions. Many of the practical hash functions such as SHA-256, SHA-512, and RIPEMD-160 are based on the MD method. We provide a simplified description of this method here.

Let h be a function that maps an n-bit string to an m-bit string and $n > m$. Such a function is usually called a compression function. This function is assumed to be collision resistant. The MD algorithm uses h to construct a hash function H, which maps long strings to the m-bit digest. Let IV be an m-bit IV.

This can be chosen randomly, but then it becomes fixed and part of the specification of $H()$.

Let x be the message to be hashed. Format x into substrings $x_1, x_2, \ldots, x_{t-1}, x_t$, where $|x_i| = n - m$. If the length of x is not a multiple of $(n - m)$, then x_t consists of the broken block padded with 1 followed by a required number of zeros to make the length equal to $(n - m)$. Let x_{t+1} be the $(n - m)$-bit binary representation of the length of x. We now define variables $z_0, z_1, \ldots, z_{t+1}$ in the following manner:

$$
\begin{aligned}
z_0 &= \mathrm{IV}, \\
z_i &= h(x_i, z_{i-1}) \quad \text{for } 1 \leq i \leq t+1.
\end{aligned}
$$

The final digest of x under $H()$ is defined to be z_{t+1}. It is simple to prove by backward induction that if it is possible to find a collision for $H()$ then it is also possible to find a collision for $h()$. Thus, we have $H()$ to be collision resistant under the assumption that the compression function $h()$ is collision resistant. The hash function families MD, SHA, and RIPEMD follow a variant of this strategy.

The cryptographic literature contains some very successful attacks on practical hash functions. The attack by Dobbertin [27] on MD4 in the mid-1990s was extremely powerful. He could show a collision for two meaningful messages. Partial attacks on MD5 were also reported. In the recent past, there have been some powerful attacks on MD5, RIPEMD, SHA, and other hash functions by Wang and others [28,29]. The hash functions RIPEMD-160 and SHA-256 survive these attacks. However, the development of the new attacks has resulted in a serious rethinking on the design strategy of practical hash functions.

Another old theme for designing hash functions is to use block ciphers. The MD-family of hash function proposals was developed by Rivest in the early 1990s. Concurrently, there has been active research on designing secure hash functions based on secure block ciphers. A basic motivation for basing hash functions on block ciphers is that one can then put his entire trust on a single well-studied primitive such as a block cipher. The disadvantage is that hash functions designed from block ciphers are generally slower than hash functions built from scratch.

The first systematic study of block cipher-based hash functions was made by Preneel, Govaerts and Vandewalle (PGV) in [30]. This study considered 64 possible constructions and suggested that some of these are secure while others are not. A formal treatment of the 64 PGV constructions was made in [31]. They proved that some of the PGV constructions are collision resistant using either the MD paradigm or otherwise. The study in [32] develops the area by proving some more bounds and corresponding attacks. A more recent topic on hash function is the multicollision attack by Joux [33] and the work on designing hash functions to avoid such attacks.

1.5 KEY AGREEMENT

Let us consider the basic problem of secure information exchange. Consider the scenario where n persons want to communicate with each other and the communication between any two persons should not be intelligible to the others. Such a situation may arise in the stock market, where any pair of brokers may want to exchange information without any of the other brokers knowing what is exchanged. Suppose a block or a stream cipher is used to protect the communication between any two parties. Each person maintains a list of $n-1$ secret keys, which are used for communication with the other $n-1$ persons. When person i wants to send a message to person j, he chooses from his list the secret key corresponding to j and uses it to construct the cipher, which he then sends to person j. When person j gets the message from i, he uses the key corresponding to i (which is the same key that person i has corresponding to j) to decipher the message.

In this scenario, for each pair of communication one needs a secret key and thus this gives rise to a total of $\binom{n}{2}$ keys for the whole system. Therefore, if there are 1000 brokers in a stock market each one of them will have a list of 999 secret keys and the system will have a total of $\binom{1000}{2}$ secret keys overall. Clearly maintaining and managing the secrecy of so many keys is a difficult administrative problem. In addition, a broker might need to communicate with some other broker very infrequently (or not at all). Thus, it is not very sensible to maintain a secret key with such a person. Moreover, if a new broker enters the market, this person has to establish a secret key with all the existing brokers, which is a time-consuming and costly affair.

A brilliant solution to this problem was proposed by Diffie and Hellman in 1976 [22]; they introduced the concept of public key cryptography. Their solution is to allow any two parties to dynamically agree on a secret key by public discussion. First, each of the two parties chooses a random secret that is not known to anybody else. Then the parties exchange information using a previously agreed on protocol and also perform some private computations. The information exchange is done over a public channel and this information is available to an adversary. Finally, the two parties agree on a common secret key, which is known only to two of them and not to anybody else. A protocol that achieves this is called a two-party key agreement protocol. Clearly, this notion can be generalized to the case of more than two parties and it is then called multiparty key agreement.

We next describe the two-party key agreement protocol developed by Diffie–Hellmann. Let G be a cyclic group whose order is a large prime p having a generator g. The generator g and the prime p are publicly known. Suppose Alice and Bob wish to agree on a common secret key. They follow the protocol in Table 1.1.

The public information consists of $p, g, g_1 = g^r$, and $g_2 = g^s$. From this, the adversary has to compute g^{rs}. This is believed to be a computationally

TABLE 1.1
Diffie–Hellman Key Agreement Protocol

Alice-Phase 1	**Bob-Phase 1**
Choose r randomly from $\{0, \ldots, p-1\}$	Choose s randomly from $\{0, \ldots, p-1\}$
Compute $g_1 = g^r$	Compute $g_2 = g^s$
Transmit g_1 to Bob	Transmit g_2 to Alice
Alice-Phase 2	**Bob-Phase 2**
Compute $h = g_2^r = (g^s)^r = g^{rs}$	Compute $h = g_1^s = (g^r)^s = g^{rs}$

infeasible task and is called the Diffie–Hellman assumption. The DH problem (DHP) is related to the discrete log problem (DLP), which is to find the value of a given a pair (g, g^a). If the DLP can be solved in G, then the DHP can also be solved in G. The converse, however, is not known to be either true or false. Currently, the DHP is believed to be hard for properly chosen group G.

The DH key agreement protocol can be extended to a multiparty key agreement using a tree-based structure [34]. This requires several rounds of interaction among the involved parties. A very interesting key agreement protocol was proposed by Burmester–Desmedt [35]. In this protocol, any number of parties can agree on a common secret key in just two rounds.

The protocols discussed so far are unauthenticated. The adversary is assumed to be passive, that is, the adversary listens to what is flowing across the public channel but does not attempt to change or alter it. A more powerful adversary is an active adversary, who can alter or stop the flow of information across the public channel. The DH protocol is insecure against such an adversary because of a man-in-the-middle attack. In this attack, the adversary establishes separate common keys with Alice and Bob without Alice and Bob realizing it. As a result, the adversary can read (and forward) any message that Alice sends to Bob, or vice versa. Key agreement protocols that remove this problem include some kind of authentication measure. This allows Alice and Bob to verify that they are indeed interacting with each other and not with a third party. Authenticated key agreement protocols have appeared in the literature. Perhaps the most important example is a generic conversion of the Burmester–Desmedt protocol into an authenticated protocol [36].

1.6 PUBLIC KEY ENCRYPTION

The notion of PKE was introduced by Diffie–Hellman in [22]. The novel idea is for each user to have exactly two keys—an encryption key and a decryption key. The encryption key is made public, that is, it is made known to everybody and the decryption key is kept secret.

Going back to our stock market example, each broker has an encryption key and a decryption key. The encryption keys are published in a global (broker) directory and the decryption keys are kept secret by the respective brokers. Again suppose that broker A wants to send a message x to broker B. Broker A chooses the encryption key e_B of broker B from the global directory and uses the publicly known encryption method to encrypt x to obtain a message y, that is, $y = E(e_B, x)$, where $E(.,.)$ is the encryption function and the key e_B and x are parameters to this function. This y is transmitted to broker B. On receiving y, broker B uses the secret decryption key d_B and the publicly known decryption method to decrypt y and obtain x, that is, $x = D(d_B, y) = D(d_B, E(e_B, x))$. A little reflection will convince the reader that such a scheme removes the difficulties explained in the previous section.

In a PKE protocol, the encryption and decryption keys are different and hence they are sometimes called asymmetric key cryptosystems, whereas secret key cryptosystems, where the encryption and decryption keys are equal, are called symmetric key systems.

Let us now consider what the security requirements on such a system are. The functions $E(.,.)$ and $D(.,.)$, the encryption key e_B, and the cipher y are known. From these it would be infeasible to obtain either the message x or the secret decryption key d_B. Viewed another way, it should be easy to obtain y from x, but without the knowledge of d_B it should be difficult to obtain x from y, that is, computation in one direction is easy, whereas it is hard in the reverse direction. As mentioned earlier in connection with hash function, functions satisfying such a criterion are called one-way functions. However, the encryption function used here is not exactly a one-way function, since knowledge of d_B makes it easy to go back. Therefore, d_B can be considered a sort of trapdoor that allows easy inversion. Hence, the function $E(.,.)$ is actually a trapdoor one-way function.

To implement a public key cryptosystem one has to design a trapdoor one-way function. The most popular and widely used system employing a trapdoor one-way function is the system proposed by Rivest, Shamir, and Adleman [37] and called the RSA system after them.

To set up the RSA system each user chooses two large primes p and q and forms the product $N = pq$. From N, find $\phi(N) = \phi(pq) = \phi(p)\phi(q) = (p-1)(q-1)$. (Here $\phi(N)$ is the number of integers between 1 and $(N-1)$, which are coprime to N.) Next two positive integers e and d are chosen using the extended Euclidean algorithm such that $1 < e, d < \phi(N)$ and $ed \equiv 1 \mod \phi(N)$. Once e and d are obtained, it is no longer required to preserve the individual values of p, q, or $\phi(N)$. The public key is declared to be the pair (e, N) and the private key that is kept secret is the pair (d, N). In fact, only d is kept secret.

To encrypt a nonnegative integer x less than N one uses the public key (e, N) and forms $y = x^e \mod N$. This y is the cipher corresponding to x and is transmitted. To decrypt all that is required is to form $z \equiv y^d \mod N$. This z is equal to x and hence the original message has been recovered ($z \equiv x^{ed} \bmod N \equiv x^{1+k\phi(N)} \bmod N \equiv x \bmod N$. Note $x^{1+k\phi(N)} \equiv x \bmod N$ if and only if $N \mid x(x^{k\phi(N)} - 1)$. Now use the fact that either $p \mid x$ or $q \mid x$ or $\gcd(N, x) = 1$).

Let us now briefly try to understand the security of the system. The secret key is (d, N), which a cryptanalyst will try to recover. If from N one can obtain the factors p and q of N, then it is easy to find $\phi(N)$ and since e is known, one can also find d using the Euclidean algorithm. It is believed that if N is a large composite number it is difficult to obtain the factors of N. Thus, trying to break RSA by factoring N will be difficult. Therefore, one might try to obtain d in other ways. However, it can be shown that if one can obtain d or $\phi(N)$ from N, then one can also find p and q, that is, factorize N. Since all known attacks on RSA ultimately boil down to the problem of factoring N, it is generally believed (but not proved) that breaking the RSA system is as hard as factoring N. See [38] for a survey of attacks on the RSA cryptosystem.

An alternative method of PKE was proposed by ElGamal [39] and is based on the Diffie–Hellman key agreement protocol. Next, we describe the basic ElGamal protocol. There are many variants to this protocol, but the underlying idea remains the same.

As in the case of DH key agreement protocol let G be a cyclic group of large prime order p with g as a generator. The secret key of a user, Bob, is a random integer $a \in \{0, \ldots, p-1\}$ and the corresponding public key is $h = g^a$. Suppose Alice wants to send a message x to Bob. She chooses a random k from $\{0, \ldots, p-1\}$ and computes $g_1 = g^k$ and $y = h^k \times x$. She sends (g_1, y) to Bob. To decrypt, Bob computes $g_2 = g_1^a = g^{ka}$ and then $x = g_2^{-1} y$. The quantity $h^k = g^{ka}$ is used to mask the message x and the auxiliary information g_1 is provided to Bob to enable him to compute the mask using his secret key a.

The main advantage of the ElGamal protocol is that it works over any cyclic group for which the DHP is difficult. A cornerstone of modern cryptography is the discovery that certain groups obtained from elliptic curves can be used for building ElGamal protocols [40,41]. For properly chosen elliptic curve groups, the only known method for solving DLP (and DHP) is to employ a generic attack such as Pollard's rho method [42], which is an exponential algorithm. On the other hand, development of the number and function field sieve algorithms has resulted in subexponential algorithms for factoring and DLP in finite fields. The consequence of all this is that for elliptic curves one can use smaller size parameters, leading to lesser storage space and more efficient protocols. See [42] for more on elliptic curve cryptography.

1.6.1 Hybrid Encryption

Public key algorithms are significantly slower than secret key algorithms. Thus, encrypting large messages using a PKE protocol is inherently inefficient.

One way of solving this problem is to use hybrid encryption, which couples together a secret key and a public key algorithm. Let us illustrate this with a simple example based on the ElGamal protocol described earlier. Recall that $g_1 = g^k$ is the auxiliary information (also called the ephemeral key) and the masking of the message x is done using $h^k = g^{ak}$. Suppose that instead of masking x directly, we consider h^k to be the secret key of a symmetric encryption algorithm. (The value h^k may be hashed to obtain the secret key.) The actual encryption of the message x is done using the symmetric encryption algorithm. Even if the message x is long, the encryption will be reasonably efficient. During decryption, Bob computes $g_1^a = g^{ak}$ and uses this to obtain the secret key employed to encrypt x. He can then use the corresponding symmetric decryption algorithm and recover the message x. The above is a simplified description, intended to convey the basic idea. It should not be used as described since there are several subtleties that have not been discussed. For practical hybrid encryption algorithms, one may consult [43].

1.6.2 Formal Model

Formally, an asymmetric encryption scheme asym is a tuple asym $=$ $(\mathcal{M}, \mathcal{C}, \mathcal{SK}, \mathcal{PK},$ keygen, enc, dec), where $\mathcal{M}$ and $\mathcal{C}$ are, respectively, the message and cipher spaces; $\mathcal{SK}$ and $\mathcal{PK}$ are, respectively, the secret and public key spaces; enc(pk, M) is the encryption algorithm, which takes a key pk $\in \mathcal{PK}$ and a message $M \in \mathcal{M}$ as input and produces a cipher $C \in \mathcal{C}$; dec(sk, C) is the decryption algorithm, which takes a key sk $\in \mathcal{SK}$ and a cipher $C \in \mathcal{C}$ as input and either returns bad or produces a message $M \in \mathcal{M}$ such that $\texttt{dec}(\texttt{sk}, \texttt{enc}(\texttt{pk}, M)) = M$.

All the above algorithms are probabilistic algorithms, which run in time upper bounded by a polynomial in the security parameter. The security parameter specifies the level of security to be attained by the protocol. A matching pair of private–public keys (sk, pk) is produced by invoking the key generation algorithm keygen on the security parameter.

The notion of security for asymmetric encryption is as follows. The adversary is considered to run in two stages—the find stage followed by the guess stage. In both stages, the adversary has access to a decryption oracle, which is the decryption algorithm instantiated by a randomly chosen secret (i.e., unknown to the adversary) key. In both stages, the adversary can query the decryption oracle with ciphertexts and receive either bad or the corresponding messages. At the end of the find stage, the adversary outputs two messages (x_0, x_1). A bit $b \in \{0,1\}$ is selected at random and x_b is encrypted using the encryption oracle. The adversary then starts the guess stage. In the guess stage, the adversary is not allowed to query the decryption oracle on the target y. At the end of the guess stage, it outputs a bit b'. The adversary's advantage in breaking the system is defined to be $2\,|\Pr[b = b'] - 1/2|$.

The formal security model is useful for designing and proving protocols. The best-known example of a secure PKE protocol is the Cramer–Shoup protocol [43]. This protocol is proved to be secure assuming the hardness of a variant of the Diffie–Hellman problem. Another example of a secure PKE protocol is the RSA-OAEP [44]. However, this protocol (like many others) uses several hash functions and assumes that the hash functions are random functions. Thus, the proof holds under the random oracle assumption or in the random oracle model.

1.7 DIGITAL SIGNATURES

The notion of digital signatures is almost as old as the notion of PKE itself. The basic idea of a digital signature is that one person can sign a message, whereas anybody can verify the correctness of the signature. Thus, a message can be authenticated by a user and the authentication can be publicly verified. It may be recalled that MAC also is a method of authentication. The main difference between an MAC and a digital signature is that in an MAC algorithm, verification can only be done by somebody who possesses a secret key, whereas in a digital signature protocol, the verification can be done publicly.

A digital signature protocol consists of three probabilistic algorithms—setup, sign, and verify. The setup algorithm generates the secret signing key and the public parameters of a user. The signing algorithm takes the signing key, the public parameters, and a message as input and produces a signature on the message as output. The verification algorithm takes the message, the signature, and the public parameters as input. It outputs true if the (message, signature) pair is valid, else it outputs false.

A method for signing messages was given by the inventors of RSA [37]. The idea is to use the public key algorithm in reverse. Let $N = pq$ and e and d be generated by the setup of the RSA algorithm. The pair (e, N) is the public key, whereas d is the secret signing key. To sign a message x, a user computes the signature $\sigma = x^d \bmod N$. The pair (x, σ) constitutes a message–signature pair. Verification can be done by computing $\sigma^e \bmod N$ and comparing with x. Note that verification can be done using only the public parameters. By itself, this protocol cannot be proved to be secure, but it illustrates the basic idea of obtaining a digital signature protocol from a PKE protocol.

We describe a simplified version of the ElGamal signature protocol. The cryptosystem is setup as follows. Choose p to be a prime and α to be a generator of Z_p^*. Let $\beta = \alpha^a$ for some $a \in \{1, \dots, p-1\}$. The tuple (p, α, β) is made public, whereas a is kept secret. A message x is an integer $1 \leq x \leq p-1$. Signing is done in the following manner. Choose a secret $k \in Z_{p-1}^*$. The signature is $\sigma = (\gamma, \delta)$, where $\gamma = \alpha^k \bmod p$ and $\delta = (x - a\gamma)k^{-1} \bmod (p-1)$. Note that signing requires the use of the secret a. A message–signature pair (x, σ) with $\sigma = (\gamma, \delta)$ is declared to be valid if and only if $\beta^\gamma \gamma^\delta \equiv \alpha^x \bmod p$.

This verification can be done publicly. Perhaps the most widely used digital signature protocol today is the elliptic curve digital signature algorithm (ECDSA), which is based on a variant of the ElGamal signature protocol.

Among all the modern concepts of cryptography, digital signatures have arguably the most number of variants. There are one-time, blind, group, ring, unique, and proxy signatures to name a few. These concepts arise in connection with the different subtle requirements of modern business. Unfortunately, there does not exist a good survey or textbook discussion of the various signature protocols. This makes it very difficult for a newcomer to grasp the different concepts, tools, and proofs used for constructing and proving the security of the multitude of signature protocols.

1.7.1 Public Key Infrastructure

The widespread deployment of PKE technology requires an infrastructure that is often called public key infrastructure (PKI). The main component of such an infrastructure is a certifying authority (CA). The basic role of a CA in a PKI is to issue digital certificates to individual users. A CA itself has a public and a private key. An individual user, Alice, can approach a CA for a certificate. The first step of the CA is to perform an extensive physical validation of Alice's identity. Once satisfied, the CA generates a (public key, private key) pair for Alice. It provides Alice with the private key using a secure channel. Alternatively, and in practice, Alice will generate her own (public key, private key) pair, provide the CA with the public key and keep the private key to herself. The CA uses its own private key to digitally sign a message consisting of Alice's identity and her public key. It next prepares a certificate for Alice consisting of her identity, her public key, and the CA's signature on these two. This certificate is provided to Alice.

When Alice wants to communicate with Bob, she first presents the certificate she obtained from the CA to Bob. Bob verifies the CA's signature on the certificate by using the public key of the CA. Alice performs a similar verification of Bob's certificate. Once both are verified, Alice and Bob can communicate with each other using their public keys. It may happen that Alice and Bob have obtained their certificates from two different CAs. In this situation, Alice and Bob will trust each other if their CAs trust each other. The existence of many CAs leads to the notion of a web of trust and complicates the implementation of PKI.

There is another problem that complicates PKI implementation. A CA issues certificates. For certain reasons, a CA may later decide to revoke the certificate. Since a certificate has already been issued, there is no way of taking it back. Instead, the CA publishes a certificate revocation list (CRL), which specifies the certificates that have been revoked by the CA. When Bob authenticates Alice's certificate, he must take care to ensure that Alice's certificate is not in the CRL published by the corresponding CA. This situation becomes more complicated when Alice and Bob have certificates issued by separate CAs.

1.8 IDENTITY-BASED ENCRYPTION (IBE)

IBE was proposed by Shamir [45]. An IBE is a public key protocol in which the public key can be any binary string. There is a trusted authority called a private key generator (PKG), which provides the private key corresponding to an identity. In other words, the public key of Bob can be his email address such as bob@crypto1234.com. To obtain a private key for this identity, Bob approaches the PKG and is supplied with a corresponding private key through a secure channel. The role of the PKG in an IBE is somewhat different from the role of a CA in a PKI. This can potentially simplify the implementation of PKI. An IBE also has other applications [46].

Since its introduction, there have been a few proposals for IBE, but these were more of a theoretical nature. The first practical solutions were based on the notion of cryptographic bilinear maps [47,46]. A proper security model for IBE was given by Boneh and Franklin [46] and they proved their protocol to be secure in the model using the random oracle assumption.

1.8.1 Cryptographic Bilinear Map

Let G_1 and G_2 be cyclic groups of the same prime order p and $G_1 = \langle P \rangle$, where we write G_1 additively and G_2 multiplicatively. A mapping e: $G_1 \times G_1 \longrightarrow G_2$ is called a cryptographic bilinear map if it satisfies the following properties:

- Bilinearity: $e(aP, bQ) = e(P, Q)^{ab}$ for all P, $Q \in G_1$ and a, $b \in Z_p$.
- Nondegeneracy: If $G_1 = \langle P \rangle$, then $G_2 = \langle e(P, P) \rangle$.
- Computability: There exists an efficient algorithm to compute $e(P, Q)$ for all P, $Q \in G_1$.

Since $e(aP, bP) = e(P, P)^{ab} = e(bP, aP)$, $e()$ also satisfies the symmetry property. Modified Weil pairing [46] and Tate pairing [48,49] are examples of cryptographic bilinear maps. These examples have G_1 to be an elliptic curve group and G_2 to be a subgroup of a multiplicative group of a finite field.

1.8.2 Hardness Assumption

The main hardness assumption for bilinear maps is a variant of the DH assumption and is called the decision bilinear Diffie–Hellman (DBDH) assumption. The DBDH problem [46] in $\langle G_1, G_2, e \rangle$ is as follows:

> Given a tuple $\langle P, aP, bP, cP, Z \rangle$, where Z $\in G_2$, decide whether $Z = e(P, P)^{abc}$, which we denote as Z is real or Z is random.

1.8.3 Identity-Based Encryption Protocol

Following [46], an IBE scheme is specified by four probabilistic algorithms: setup, key generation, encryption, and decryption.

Setup: It takes a security parameter as input and returns the system parameters together with the master key. The system parameters include a description of the message space, the ciphertext space, and the identity space. They are publicly known, whereas the master key is known only to the PKG.

Key Generation: It takes an identity v as input and returns a private key d_v, using the master key. The identity v is used as the public key whereas d_v is the corresponding private key.

Encryption: It takes the identity v, the public parameters of the PKG, and a message from the message space as input. The output is a ciphertext in the cipher space.

Decryption: It takes the ciphertext, the public parameters of the PKG, the identity v, and the private key d_v corresponding to v as input and returns the message or bad if the ciphertext is not valid.

1.8.4 Security Model

Security of an IBE protocol is defined using an adversarial game. An adversary $\mathcal{A}$ is allowed to query two oracles—a decryption oracle and a key-extraction oracle. At the initiation, it is provided with the system public parameters. There are two query phases with a challenge phase in between.

Query Phase 1: Adversary $\mathcal{A}$ makes a finite number of queries and each query is addressed either to the decryption oracle or to the key-extraction oracle. In a query to the decryption oracle, it provides the ciphertext as well as the identity under which it wants the decryption. Similarly, in a query to the key-extraction oracle, it asks for the private key of the identity it provides. Further, $\mathcal{A}$ is allowed to make these queries adaptively, that is, any query may depend on the previous queries as well as their answers.

Challenge: At this stage, $\mathcal{A}$ fixes an identity v* and two equal length messages M_0, M_1 under the (obvious) constraint that it has not asked for the private key of v* and gets a ciphertext C^* corresponding to M_b, where b is a random bit.

Query Phase 2: $\mathcal{A}$ now issues additional queries just as in Phase 1, with the (obvious) restriction that it cannot ask the decryption oracle for the decryption of C^* under v* nor the key-extraction oracle for the private key of v*.

Guess: $\mathcal{A}$ outputs a guess b' of b.

The advantage of $\mathcal{A}$ in attacking the scheme is defined as $\mathtt{Adv}_{\mathcal{A}}^{\mathtt{IBE}} = 2\ |\Pr[(b = b')] - 1/2|$. The quantity $\mathtt{Adv}^{\mathtt{IBE}}(t, q_{\mathrm{ID}}, q_{\mathrm{C}})$ denotes the maximum of $\mathtt{Adv}_{\mathcal{A}}^{\mathtt{IBE}}$, where the maximum is taken over all adversaries running in time at most t and making at most q_{C} queries to the decryption oracle and $q_{\mathtt{ID}}$ queries to the key-extraction oracle. Any IBE scheme secure against such an adversary is said to be secure against chosen ciphertext attack (CCA).

We next describe the basic Boneh–Franklin IBE [46].

Setup: Let $\langle G_1, G_2, e\rangle$ define the cryptographic bilinear map $e(,)$, where $G_1 = \langle P\rangle$ and the order of both G_1 and G_2 is a prime p. The DBDH assumption holds for $\langle G_1, G_2, e\rangle$. The master secret of the PKG is an integer s chosen randomly from $\{0, \ldots, p-1\}$. Let $Q = sP$. The public parameters of the PKG consist of $\langle P, Q\rangle$ and two hash functions H_1: $\{0,1\}^* \longrightarrow G_1$ and $H_2 : G_2 \longrightarrow \{0,1\}^n$. The function H_1 maps an arbitrary string to an element of G_1, while H_2 maps an element of G_2 into a binary string of length n. The message space consists of all binary strings of length n, whereas the identity space consists of all binary strings.

Key Generation by PKG: Let v be an identity. The private key corresponding to v is defined to be $Q_{\mathrm{v}} = sH_1(\mathrm{v})$. The PKG knows s and hence can generate this identity.

Encryption: Let M be the message to be encrypted. Choose a random integer $r \in \{0, \ldots, p-1\}$. The ciphertext is $C = \langle rP, M \oplus H_2(e(Q, H_1(\mathrm{v}))^r)\rangle$.

Decryption: Let $C = \langle C_0, C_1\rangle$ be a ciphertext corresponding to an identity v. Compute $M = C_1 \oplus H_2(e(C_0, Q_{\mathrm{v}}))$:

The decryption succeeds due to the following equalities:

$$e(Q, H_1(\mathrm{v}))^r = e(sP, H_1(\mathrm{v}))^r = e(rP, sH_1(\mathrm{v})) = e(C_0, Q_{\mathrm{v}}).$$

The above computation uses the bilinearity property of $e(,)$. This scheme by itself cannot be proved to be secure. It is combined with the Fujisaki–Okamoto transformation to obtain a protocol that can be proved to be secure. The proof of security assumes that $H_1()$ and $H_2()$ are random functions, that is, the proof is obtained under the random oracle assumption. Later works [50,51] have shown how to construct efficient IBE protocols that can be proved to be secure without using the random oracle assumption.

1.9 CONCLUSION

In this chapter, we have provided a brief description of some of the most important topics in modern cryptography. There are other topics like secret sharing, commitment protocols, multiparty computation, and so on that have not been covered. Even in the topics that have been discussed, we have only sketched the basic ideas. Practical and ready-to-use algorithms are out of scope of this paper and can be found in the references. In summary, we have attempted to provide a quick and gentle introduction to several important aspects of modern cryptography and will be satisfied if the reader finds the material useful.

ACKNOWLEDGMENT

We thank Rana Barua and Kishan Chand Gupta for reading an earlier version of this article and providing several suggestions for improvement.

REFERENCES

1. Douglas R. Stinson. *Cryptography: Theory and Practice, 3rd ed.* Discrete Mathematics and Its Applications. CRC Press, 2005.
2. Bruce Schneier. *Applied Cryptography: Protocols, Algorithms, and Source Code in C, 2nd ed.* Wiley, 1995.
3. Oded Goldreich. *Foundations of Cryptography*. Cambridge University Press, 2001.
4. Alfred J. Menezes, Paul C. van Oorschot, and Scott A. Vanstone. *Handbook of Applied Cryptography*. CRC Press, 1996.
5. Vincent Rijmen Joan Daemen. *The Design of Rijndael: AES—The Advanced Encryption Standard.* Information Security and Cryptography. Springer, 2002.
6. Charanjit S. Jutla. Encryption modes with almost free message integrity. In Birgit Pfitzmann, editor, *EUROCRYPT*, vol. 2045 of *Lecture Notes in Computer Science*, pp. 529–544. Springer, 2001.
7. Phillip Rogaway. Efficient instantiations of tweakable blockciphers and refinements to modes OCB and PMAC. In Pil Joong Lee, editor, *ASIACRYPT*, vol. 3329 of *Lecture Notes in Computer Science*, pp. 16–31. Springer, 2004.
8. Debrup Chakraborty and Palash Sarkar. HCH: A new tweakable enciphering scheme using the Hash-Encrypt-Hash approach. In Rana Barua and Tanja Lange, editors, *Indocrypt 2006*, vol. 4329 of *Lecture Notes in Computer Science*, pp. 287–302, Springer, 2006.
9. Michael Luby and Charles Rackoff. How to construct pseudorandom permutations from pseudorandom functions. *SIAM J. Comput.*, 17(2):373–386, 1988.
10. Moni Naor and Omer Reingold. On the construction of pseudorandom permutations: Luby–Rackoff revisited. *J. Cryptol.*, 12(1):29–66, 1999.
11. Claude E. Shannon. Communication theory of secrecy systems. *Bell System Tech. J.*, 28:656–715, 1949.
12. Andrew Chi-Chih Yao. Theory and applications of trapdoor functions (extended abstract). In *FOCS*, pp. 80–91. IEEE, 1982.
13. Manuel Blum and Silvio Micali. How to generate cryptographically strong sequences of pseudo-random bits. *SIAM J. Comput.*, 13(4):850–864, 1984.
14. Lenore Blum, Manuel Blum, and Mike Shub. A simple unpredictable pseudo-random number generator. *SIAM J. Comput.*, 15(2):364–383, 1986.

15. Harald Niederreiter and Rudolf Lidl. *Introduction to Finite Fields and Their Applications*. Cambridge University Press, 1994.
16. Palash Sarkar and Subhamoy Maitra. Nonlinearity bounds and constructions of resilient Boolean functions. In Mihir Bellare, editor, *CRYPTO*, vol. 1880 of *Lecture Notes in Computer Science*, pp. 515–532. Springer, 2000.
17. Patrik Ekdahl and Thomas Johansson. A new version of the stream cipher SNOW. In Kaisa Nyberg and Howard M. Heys, editors, *Selected Areas in Cryptography*, vol. 2595 of *Lecture Notes in Computer Science*, pp. 47–61. Springer, 2002.
18. Gregory G. Rose and Philip Hawkes. Turing: A fast stream cipher. In Thomas Johansson, editor, *FSE*, vol. 2887 of *Lecture Notes in Computer Science*, pp. 290–306. Springer, 2003.
19. http://www.ecrypt.eu.org/stream/.
20. Bart Preneel. The state of cryptographic hash functions. In Ivan Damgård, editor, *Lectures on Data Security*, vol. 1561 of *Lecture Notes in Computer Science*, pp. 158–182. Springer, 1998.
21. Douglas R. Stinson. Some observations on the theory of cryptographic hash functions. *Des., Codes, Cryptography*, 38:259–277, 2006.
22. Whitfield Diffie and Martin Hellman. New directions in cryptography. *IEEE Trans. Inform. Theory*, 22(6):644–654, 1976.
23. Ralph C. Merkle. One way hash functions and DES. Gilles Brassard, editor. *Advances in Cryptology—CRYPTO'89, 9th Annual International Cryptology Conference*, Santa Barbara, California, USA, August 20–24, 1989, *Proceedings*, vol. 435 of *Lecture Notes in Computer Science*, pp. 428–446. Springer, 1990.
24. Ivan Damgård. Collision free hash functions and public key signature schemes. In *EUROCRYPT*, pp. 203–216, 1987.
25. David Chaum, Eugène van Heijst, and Birgit Pfitzmann. Cryptographically strong undeniable signatures, unconditionally secure for the signer. In Joan Feigenbaum, editor, *CRYPTO*, vol. 576 of *Lecture Notes in Computer Science*, pp. 470–484. Springer, 1991.
26. Ivan Damgård. A design principle for hash functions. Gilles Brassard, editor. *Advances in Cryptology—CRYPTO'89, 9th Annual International Cryptology Conference*, Santa Barbara, California, USA, August 20–24, 1989, *Proceedings*, vol. 435 of *Lecture Notes in Computer Science*, pp. 416–427. Springer, 1990.
27. Hans Dobbertin. Cryptanalysis of MD4. *J. Cryptology*, 11(4):253–271, 1998.
28. Xiaoyun Wang and Hongbo Yu. How to break MD5 and other hash functions. Ronald Cramer, editor. *Advances in Cryptology—EUROCRYPT 2005, 24th Annual International Conference on the Theory and Applications of Cryptographic Techniques*, Aarhus, Denmark, May 22–26, 2005, *Proceedings*, vol. 3494 of *Lecture Notes in Computer Science*, pp. 19–35. Springer, 2005.
29. Xiaoyun Wang, Yiqun Lisa Yin, and Hongbo Yu. Finding collisions in the full SHA-1. In Victor Shoup, editor, *CRYPTO*, vol. 3621 of *Lecture Notes in Computer Science*, pp. 17–36. Springer, 2005.
30. Bart Preneel, René Govaerts, and Joos Vandewalle. Hash functions based on block ciphers: A synthetic approach. In Douglas R. Stinson, editor, *CRYPTO*, vol. 773 of *Lecture Notes in Computer Science*, pp. 368–378. Springer, 1993.
31. John Black, Phillip Rogaway, and Thomas Shrimpton. Black-box analysis of the blockcipher-based hash-function constructions from PGV. In Moti Yung, editor, *Advances in Cryptology—CRYPTO 2002, 22nd Annual International Cryptology*

Conference, Santa Barbara, California, USA, August 18–22, 2002, *Proceedings*, vol. 2442 of *Lecture Notes in Computer Science*, pp. 320–335. Springer, 2002.

32. Wonil Lee, Mridul Nandi, Palash Sarkar, Donghoon Chang, Sangjin Lee, and Kouichi Sakurai. PGV-style block-cipher-based hash families and black-box analysis. *IEICE Transactions*, 88-A(1):39–48, 2005.
33. Antoine Joux. Multicollisions in iterated hash functions. Application to cascaded constructions. In Matthew K. Franklin, editor, *CRYPTO*, vol. 3152 of *Lecture Notes in Computer Science*, pp. 306–316. Springer, 2004.
34. Yongdae Kim, Adrian Perrig, and Gene Tsudik. Tree-based group key agreement. *ACM Trans. Inf. Syst. Secur.*, 7(1):60–96, 2004.
35. Mike Burmester and Yvo Desmedt. A secure and efficient conference key distribution system (extended abstract). In *EUROCRYPT*, pp. 275–286, 1994.
36. Jonathan Katz and Moti Yung. Scalable protocols for authenticated group key exchange. Dan Boneh, editor. *Advances in Cryptology—CRYPTO 2003, 23rd Annual International Cryptology Conference*, Santa Barbara, California, USA, August 17–21, 2003, *Proceedings*, vol. 2729 of *Lecture Notes in Computer Science*, pp. 110–125. Springer, 2003.
37. Ronald L. Rivest, Adi Shamir, and Leonard M. Adleman. A method for obtaining digital signatures and public-key cryptosystems. *Commun. ACM*, 21(2):120–126, 1978.
38. Dan Boneh. Twenty years of attacks on the RSA cryptosystem. *Notices Am. Math. Soc.*, 46(2):203–213, 1999.
39. T. ElGamal. A public key cryptosystem and a signature scheme based on discrete logarithms. *IEEE Trans. Inform. Theory*, 31(4):469–472, 1985.
40. Neal Koblitz. Elliptic curve cryptosystems. *Math. Comput.*, 48:203–209, 1987.
41. Victor S. Miller. Use of elliptic curves in cryptography. In Hugh C. Williams, editor, *CRYPTO*, vol. 218 of *Lecture Notes in Computer Science*, pp. 417–426. Springer, 1985.
42. Darrel Hankerson, Alfred J. Menezes, and Scott A. Vanstone. *Guide to Elliptic Curve Cryptography*. Springer, 2005.
43. Ronald Cramer and Victor Shoup. Design and analysis of practical public-key encryption schemes secure against adaptive chosen ciphertext attack. *SIAM J. Comput.*, 33:167–226, 2003.
44. Eiichiro Fujisaki, Tatsuaki Okamoto, David Pointcheval, and Jacques Stern. RSA-OAEP is secure under the RSA assumption. *J. Cryptol.*, 17(2):81–104, 2004.
45. Adi Shamir. Identity-based cryptosystems and signature schemes. In *Advances in Cryptology—CRYPTO*, pp. 47–53, 1984.
46. Dan Boneh and Matthew K. Franklin. Identity-based encryption from the Weil pairing. *SIAM J. Comput.*, 32(3):586–615, 2003.
47. R. Sakai, K. Ohgishi, and M. Kasahara. Cryptosystems based on pairing. In *The 2000 Symposium on Cryptography and Information Security*, pp. 417–426, 2000. Okinawa, Japan, January 2000.
48. Paulo S.L.M. Barreto, Hae Yong Kim, Ben Lynn, and Michael Scott. Efficient algorithms for pairing-based cryptosystems. In Moti Yung, editor, *Advances in Cryptology—CRYPTO 2002, 22nd Annual International Cryptology Conference*, Santa Barbara, California, USA, August 18–22, 2002, *Proceedings*, vol. 2442 of *Lecture Notes in Computer Science*, pp. 354–368. Springer, 2002.

49. Steven D. Galbraith, Keith Harrison, and David Soldera. Implementing the Tate pairing. In Claus Fieker and David R. Kohel, editors, *ANTS*, vol. 2369 of *Lecture Notes in Computer Science*, pp. 324–337. Springer, 2002.
50. Brent Waters. Efficient identity-based encryption without random oracles. Ronald Cramer, editor. *Advances in Cryptology—EUROCRYPT 2005, 24th Annual International Conference on the Theory and Applications of Cryptographic Techniques*, Aarhus, Denmark, May 22–26, 2005, *Proceedings*, vol. 3494 of *Lecture Notes in Computer Science*, pp. 114–127. Springer, 2005.
51. Sanjit Chatterjee and Palash Sarkar. Trading time for space: Towards an efficient IBE scheme with short(er) public parameters in the standard model. Dongho Won and Seungjoo Kim, editors. *Information Security and Cryptology—ICISC 2005, 8th International Conference*, Seoul, Korea, December 1–2, 2005, Revised Selected Papers, vol. 3935 of *Lecture Notes in Computer Science*, pp. 424–440, Springer, 2005.
52. Debrup Chakraborty and Palash Sarkar. A general construction of tweakable block ciphers and different modes of operations. In Helger Lipmaa, Moti Yung and Dongdai Lin, editors, *Inscrypt 2006*, vol. 4318 of *Lecture Notes in Computer Science*, pp. 88–102, Springer, 2006.

2 Introduction to Communication Security

Vesna Hassler

CONTENTS

2.1 INTRODUCTION

Although a number of studies on data security have been published in the last decade, many security breaches still occur because some newly introduced security protocols exhibit the vulnerabilities arising from the already known security problems. For example, the initial version of the IEEE 802.11 wired equivalent privacy (WEP) protocol allowed broadcasting of access point identifiers as cleartext so that a man-in-the-middle attack with a fake identifier was easily possible. Or, the encryption key length used in the WEP by default did not fulfill the requirements of the current standards for strong cryptography, that is, they used a 64-bit key instead of a 128-bit key. Therefore, it is of crucial importance to study and understand the fundamental security principles explained in this chapter.

2.2 SECURITY THREATS

This section is dedicated to the very source of our concerns about data security, namely security threats. Who would care about security if there were nothing to lose? Security threats are realized in the form of security attacks, which can be encountered in any communication network. The security threats to consider for a particular system should be determined within a process called *risk analysis*. Now we take a closer look at each of the general attacks. It must be borne in mind that today many different devices may be used for network communication and that there may be a network in place although you cannot see it.

Traffic analysis is probably the easiest way to carry out a security attack. The attacker only listens to the data exchanged between two communication partners and does not bother whether he can understand it or not, that is, whether the data are scrambled or encrypted. However, under certain circumstances, the fact that two partners start to communicate or intensify their communication may already be a valuable information. In addition, this attack may help you physically locate somebody or something in the network.

Eavesdropping is something that we can also encounter in the nonelectronic world. If you press your ear against a closed door behind which somebody is talking and you are not supposed to listen to the conversation, you may be accused of eavesdropping. In a similar way, if you intercept or in other ways collect electronic data exchanged between two communication partners in a computer network whereby the data are not meant to be read by you, you are an eavesdropper.

Masquerading can be fun if you disguise as someone else for a party, but in general it may be quite unpleasant if misused for cheating. In the networking world, you would be disguising by using a false electronic user or computer identification (ID) to obtain access to resources that you are not supposed to use.

Infiltration is a word known from the world of secret services where different agencies try to infiltrate each other. You can infiltrate a computer or a local area network by masquerading as a legitimate user or by misusing an error in a communication protocol.

Tampering with unprotected electronic messages is in general much easier than with messages written on paper because no changes can be seen. It may, however, cause significant damage to the sender. For example, imagine that you send an e-mail to your bank to transfer €10 to somebody and an interceptor changes it to transfer €10,000 instead. This type of attack is sometimes referred to as the man-in-the-middle attack because the adversary places himself between the communication partners. An alternative attack in this scenario is that the interceptor replays the same message 1000 times, with the same negative result for your bank account.

Privacy is the ability of an individual or group to control by whom and how their personal information is used. Browsing or shopping on the Internet often leaves tracks of the personal activities of a user, which others may use to spy on him. This phenomenon is referred to as *invasion of privacy* in the networking world.

Social engineering methods can be misused to carry out a security attack. For example, an attacker can phone or e-mail the employees of an enterprise pretending to be the system administrator, which is actually a form of masquerading. In this way, he can trick the users to tell him their passwords or other sensitive information. Phishing is a popular name for "password fishing" through social engineering.

Denial-of-service (DoS) attacks are relatively easy to carry out because in general no knowledge of complicated math is necessary. If you keep dialing a friend's phone number, he will not be able to call anybody, and nobody else will have a chance to reach him. You will effectively disable his phone service. In a network, good knowledge of communication protocols and the way they are implemented is necessary to carry out this type of attack, which can disable computers and whole networks. Distributed DoS attacks are especially unpleasant because they make it difficult to find where the attack originated from.

Denial-of-action can be considered a passive attack but nonetheless can cause damage. For example, you can send a message to an online shop to order 1000 DVDs, and later, after delivery, claim that you never ordered them.

The attacks we have described so far are general, but there are many other specialized attacks too. For example, certain attacks may misuse a short period of time before the computer time is switched to a daylight saving time period. Or, some specialized attacks on smart cards measure their power consumption to draw conclusions about the cryptographic computations carried out on the card. Those attacks are beyond the scope of this chapter, but be aware that clever attackers can misuse any vulnerability, no matter where it originates from.

2.3 SECURITY SERVICES

The basic attacks described in the previous section can be prevented by suitable security services that are described in this section.

Authentication ensures that a principal's identity or data origin is genuine. This service can help us prevent masquerading and infiltration attacks because we can be sure where a message comes from, or who we are communicating with.

Access control is a follow-up activity of authentication. As soon as the identity (ID) of a principal has been determined through authentication, a lookup table with IDs and access permissions tells us which rights the principal is allowed to gain. For example, he may be permitted to read a file, or to both read and write to it.

Message tampering can be prevented by a data integrity service. This service guarantees that no unauthorized principals have modified the data. Data confidentiality service is concerned with changing the electronic data in such a way that only authorized principals can read and understand it. It can help us to prevent eavesdropping, that is to hide the contents of a message or any other confidential information, such as the fact that two communication partners have started exchanging messages, which effectively prevents traffic analysis.

Nonrepudiation service, also defined in ISO/IEC 13888-1, is aimed for protection against denial-of-action attacks. The denied action can be to author a document, or send or receive a message. Protection of privacy service prevents intrusion of privacy attacks that can include all possible cases of misuse of personal data.

What can be done against replaying a message, such as in the bank account scenario from the previous section? This is a task to be accomplished by assuring message freshness, which can be achieved in several different ways as explained in Section 2.4.6.

DoS attacks are difficult to defend against because, basically, for each type of DoS attack a different defense strategy is needed. We can say that, in general, all those strategies belong to some sort of resource consumption control services.

Finally, organizational security services, such as employee education, can help defend against social engineering attacks and other similar soft security attacks. They are not further discussed in this chapter. For more information, see [1].

2.4 SECURITY MECHANISMS AND TECHNIQUES

To implement security services, we use security mechanisms, which are in turn realized by deploying cryptographic algorithms or other security techniques such as

- Encryption algorithms AES or RSA for encryption mechanisms
- Cryptographic hash function SHA-1 for data integrity mechanisms
- Message authentication code for data authentication mechanisms
- Authentication exchange protocols for peer entity authentication mechanisms
- Identity-based access control for access control mechanisms
- Public key algorithms RSA, DSA, or ECDSA for digital signature mechanisms
- Time stamps and nonces for message freshness mechanisms

- Random data for traffic padding mechanisms
- Trusted third parties for notarization mechanisms
- Anonymizers for anonymizing mechanisms

The following sections describe those mechanisms. For more detailed math, see [2].

2.4.1 Encryption Mechanisms

Encryption is a transformation that renders a message nonunderstandable for everyone who does not know the cryptographic key that is needed for decryption. Consequently, decryption is a transformation to bring the message back to its original form. A family of such transformations is referred to as the cryptosystem. Encryption is obviously perfectly suitable to ensure data confidentiality.

In a symmetric cryptosystem, encryption and decryption are identical or easily derived from each other. Note that the encryption key and the decryption key are the same. Practically, we deal with one key, which is also called the *secret* key, because it must remain secret to everybody except the sender and the recipient. This also means that if you send a symmetrically encrypted message over an insecure network, you must use another, secure, medium to communicate the key to the recipient. The usual notation for symmetric encryption, which transforms message M into ciphertext C by applying key K, is as follows:

$$\text{Encryption } E_K(M) = C,$$
$$\text{Decryption } D_K(C) = M.$$

The state-of-the-art symmetric encryption algorithm is the Advanced Encryption Standard (AES [3]), which replaced its predecessor, the Data Encryption Standard (DES [4], still in use as 3DES). AES is a block cipher since it encrypts data in 128-bit blocks, and its key length can vary among 128, 192, and 256 bits. For a high security level, 128-bit keys are not recommended.

In an asymmetric or public key cryptosystem, there are two cryptographic keys that cannot be derived from each other. For example, if you wish to encrypt a confidential message for a specific recipient, you can look up a public directory to find this recipient's public encryption key and carry out the encryption transformation. Even if an adversary intercepts the message and finds out who is the intended recipient, it is computationally infeasible for him to decrypt the message because only the recipient knows the corresponding private decryption key. The notation for public key encryption, which transforms message M into ciphertext C by applying public key PuK, and for decryption applying private key PrK, is as follows:

$$\text{Encryption } E_{PuK}(M) = C,$$
$$\text{Decryption } D_{PrK}(C) = M.$$

The state-of-the-art public key encryption algorithm is RSA [5] whose security arises from the computational difficulty of factoring large composite numbers. RSA computations are performed with a modulus, which is a product of two large primes. These two primes are in fact the private key, and the modulus can (and should) be known to everybody because it is the public key. The modulus must be long enough to be secure—at the time of this writing (April 2006) it is at least 1024 bits.

2.4.2 Data Integrity Mechanisms

Integrity mechanisms are used to ensure message integrity. The computationally fastest way to achieve this goal is to use a cryptographic hash function of message M, h(M). Such functions are applied to an input value of nearly any length yielding an output value of constant length, which is referred to as the cryptographic checksum, hashsum, or message digest. However, the hash function should be easy to compute only in one way. In other words, if you have a hashsum, it must be practically impossible to find the original input or any other message yielding the same hashsum. (Note that many different messages will have the same hashsum because the input can be of nearly any length.) Additionally, it must be extremely difficult to find two different messages with the same hashsum.

If a message and its hashum are sent over an insecure network, message integrity cannot be guaranteed because both the message and the hashsum can be tampered with. Consequently, the cryptographic hash functions are usually combined with additional mechanisms as explained in the following sections.

A popular cryptographic hash function is the Secure Hash Standard (SHA-1 [6]). It produces a 160-bit output, whereby the input message can be up to 2^{64} bits long. Shorter hashsums are not considered secure from the viewpoint of the today's computing technology.

2.4.3 Authentication Mechanisms

Data authentication can be implemented by using a cryptographic hash function. The so-called message authentication code (MAC) is based on a combination of the cryptographic hash functions and a secret key. A sender can send a message along with its MAC value to the recipient. If the recipient also has the corresponding secret key, he can check the authenticity of the message by performing the same MAC computation. Keyed hash is the mechanism used for many Internet security protocols such as IPSec and SSL/TLS [7].

Data authentication can also be implemented by encryption. In this way, the authenticity of the data is proven by applying a specific encryption key. Finally, there are some special authentication exchange mechanisms called *zero-knowledge protocols* in which a principal proves knowledge of a secret without revealing anything about the secret [8].

Peer entity authentication is usually carried out by applying an authentication exchange protocol based on mechanisms similar to data authentication described earlier (see, for example, [9]). Since the protocol messages are sent over an insecure network, they can be easily copied and resent by an intruder, even if they are encrypted (e.g., encrypted password). To prevent this replay attack, the protocol messages must always be fresh; this can be accomplished by using the techniques described in Section 2.4.6.

2.4.4 Access Control Mechanisms

As mentioned earlier, access control relies on the result of a successful authentication process. An authenticated principal, be it a user or a computer process, has been assigned an identification that is used as a basis to determine his access rights or permissions. This process is usually referred to as *authorization.*

Identity-based access control uses an access control matrix, where the principals (or subjects) are arranged in rows and the resources to be protected (or objects) are arranged in columns. For example, if you wish to know whether a user, Smith, can write to a file file.txt, you can find the row for Smith and the column for file.txt. If the intersection of the row and the column contains the access right "write," Smith is allowed to write to file.txt. If the intersection says only "read," write access must be denied. Since this type of authorization is performed at the discretion of the object owner, it is sometimes referred to as the discretionary access control.

If a system contains data with different security levels, such as, for example, protected, secret, and top secret, security cannot be enforced by an identity-based access control policy. This problem can be solved by a rule-based access control policy that defines some specific sensitivity classes. Each protected object in a system bears a security label, which defines its sensitivity class (e.g., protected, secret, and top secret). This type of policy is also called mandatory access control or information flow control [10].

2.4.5 Digital Signature Mechanisms

Digital signing has a similar purpose as handwritten signing, but there are some differences in their features. A digital signature can be easily copied because it is in electronic form, so there exist more than one original in contrast to the handwritten signature. For this reason, the digital signature must be document- and signer-dependent; otherwise, you could attach it to any document.

Digital signature mechanisms can be used to implement the nonrepudiation service against denial-of-action attacks. For example, if you digitally sign a document, you cannot later deny signing it. The legal action that is denied by denying a signature depends on the context in which the signature was created.

Public key cryptosystems are suitable as digital signature mechanisms. As mentioned in Section 2.4.1, a public key cryptosystem has two cryptographic keys that cannot be derived from each other. The private key is used to create a signature, and the public key is used to verify it. Signing is actually encrypting with the private key, and verifying is decrypting with the public key. However, since documents to be signed can be quite long, the signing operation is performed over the document hashsum, h(M). The notation for the digital signing that transforms message M into signature S by applying private key PrK is as follows:

$$\text{Signing } D_{PrK}(h(M)) = S.$$

The public key can be published, and the private key must be kept private by its owner. Ideally, the owner should also generate his public key pair to guarantee the private key's confidentiality. For example, many digital signature cards are equipped with key generators so that the private key never leaves the card.

The verifier receives the document and the signature. The public (verification) key and the information about which cryptographic hash function and which signature algorithm were used are available in a PKI directory (see also Section 2.6). To verify the signature, the verifier first computes the document hashsum and then applies the public key to the signature to obtain the hashsum that was actually signed. The two hashsums must be identical, otherwise the signature is not valid. The notation for verification of signature S of message M is as follows:

$$\text{Signature verification: compare } h(M) \text{ and } E_{PuK}(S) = h'(M).$$

The public key encryption algorithm RSA can be used to create digital signatures, but some organizations have a problem with the fact that RSA can be used for both encryption and signing. The Digital Signature Algorithm (DSA) can be used for signing only, and its security is based on the discrete logarithm problem. Another widely used DSA, Elliptic Curve Digital Signature Algorithm (ECDSA), is based on the difficulty of the elliptic curve discrete logarithm problem. Elliptic curve cryptography (ECC) has been adopted by several standardization organizations such as IEEE through the P1363 standard, ISO, and ANSI [11,12]. The ECC keys are much shorter than RSA keys for the same security level (160 bits for ECC as opposed to 1024 bits for RSA). This makes the ECC more suitable for signature devices with limited processing power such as smart cards. All three signature algorithms are also recommended by the U.S. Digital Signature Standard [13].

The date and time of signing usually play an important role too, so they should form part of the document to be signed. However, computer time can be easily manipulated. How can we be sure that the signature was really made at the alleged time? There are basically two approaches to solve this problem. One approach is to obtain a time stamp that is digitally signed by a trusted timeserver. Another approach is to use a tamper-proof security-certified time-stamping device connected to a computer on which the signature is created. The time-stamping device generates a digitally signed time stamp, which can be added to the document to be signed (e.g., TSS 400 by timeproof Time Signature Systems GmbH).

Another problem with digital signatures is to ensure that the signature is computed over the content shown to the signer on the computer screen. This approach is called "what you see is what you sign" and is not easy to implement. A solution can combine a so-called secure viewer program (e.g., trustview by IT Solution GmbH) in combination with a nonrewritable computer memory. It is also of crucial importance to secure the path between the secure viewer component and the signature creation device (e.g., smart card in a card reader) because otherwise the users cannot be sure that what they saw was really sent to the smart card. Finally, the computer on which the viewer is installed must be kept free of viruses and malicious programs.

2.4.6 MESSAGE FRESHNESS MECHANISMS

Message freshness mechanisms protect against replay attacks. Suppose you have sent a digitally signed message to your bank to transfer €1000 to person A's account. If A is malicious, he can intercept your message and resend it 10 more times. If the bank has no possibility to find out whether the message is fresh (i.e., unused), your account balance will show €10,000 less than you would expect. For this reason, it is of crucial importance to ensure that different messages with identical contents can be differentiated. This can be achieved by including a time-variant parameter before encrypting or signing:

- You can generate a random number (i.e., a nonce) and add it to the message.
- You can add a time stamp to the message.

If a nonce is used, the bank has to store the used nonces to recognize them. A combination of a nonce and a counter can be used as well, for example, by including a nonce in the first message, nonce +1 in the second message, and so on.

If a time stamp is used, your PC clock and the bank's computer clock should be synchronized because a tolerance interval introduces additional insecurity. Time critical applications may even require a time stamp from a trusted timeserver, as explained in Section 2.4.5.

2.4.7 Traffic Padding Mechanisms

An adversary can obtain valuable information even if he may only learn that two communication partners have exchanged data or that the amount of data in transfer has suddenly changed. We mentioned earlier that such attacks are called traffic analysis. Traffic padding mechanisms can offer some protection against them. They keep the data traffic rate approximately constant so that nobody can obtain information by purely observing it. For example, two communication partners can keep exchanging network packets of constant length whereby the packet payload contains encrypted random data.

2.4.8 Notarization Mechanisms

Notarization mechanisms can assure integrity, origin, time, or destination of data. They can be provided by a third party trusted by all participants and therefore called *trusted third party*. For example, several times throughout this chapter we mentioned time stamps and timeservers. Digitally signed documents should always bear a time stamp from a trusted source. The time stamp can be signed by the trusted time service and added to the message before signing. An alternative approach is to send the already signed message to the trusted timeserver, which then adds a time stamp and signs everything together, that is, the message, its signature, and the time stamp.

2.4.9 Anonymizing Mechanisms

Anonymizing mechanisms can protect our privacy. Traffic padding, as mentioned in Section 2.4.7, is also an anonymizing mechanism because it hides the information about whether the communication parties have really exchanged some meaningful messages.

A well-known example of such mechanisms are Web anonymizers. They are implemented as Web servers that receive a request from a Web client, remove all personal data from the request, forward it to the destination Web server, and forward the response to the client. Some anonymizers even assign anonymized identities so that the clients can fill out Web forms without giving away their personal data. In addition, the request (i.e., part of the URL) can be sent encrypted by the client to the anonymizer so that only the anonymizer can actually see which Web site the client is looking for.

However, real anonymity in the networking world requires network anonymity, which in turn requires a special infrastructure. There are models of how to do it, but they have no real implementations. For example, a network of interconnected anonymizing e-mail servers could be implemented in such a way that each server can see only the address of the next server to forward the message [14].

2.5 KEY MANAGEMENT

Security of cryptographic algorithms depends on both the difficulty of the mathematical problem they are based on, and the quality of the cryptographic keys and the key management methods. The following sections give some explanations about the relevant aspects.

2.5.1 Key Generation

The security of cryptographic algorithms depends on the computational complexity and not on the theoretical impossibility to find the cryptographic key by applying the brute force attack, that is, trying out all possible keys without any prior knowledge about the right key. Due to the constant development of the computing power, we are forced to use longer and longer cryptographic keys. The same applies to the hashsum length. The only exception to that rule is the encryption algorithm called the one-time pad, which is hardly used for practical reasons: a one-time pad key can be used only once and it must be at least as long as the message.

The main prerequisite for generation of good cryptographic keys is a high-quality random number generator. This requirement holds for all random data used in cryptographic computations. Otherwise, the number of possible keys would be seriously reduced, so that a brute force attacker would have to try out fewer keys than there are theoretically available for a certain cryptographic algorithm. However, it is difficult to provide a true source of randomness, and hence the typical generators used are pseudorandom sequence generators. Cryptographically strong sequences must be unpredictable so that they cannot be reliably reproduced [15].

As mentioned in Section 2.4.5, private signature keys must be kept secret, and the best way to achieve it is to never let them leave the tamper-proof device in which they are generated. For personal signatures, the key pair can be generated on a smart card (signature card). For server signatures, the signature module needs more computing power, so it is usually a bigger piece of hardware in which the key pairs are generated.

2.5.2 Key Exchange

Public key encryption is much slower than symmetric encryption and therefore almost never used to encrypt all data (i.e., for bulk data encryption). Symmetric encryption keys cannot be sent over an insecure network, and hence they are usually hidden in a message encrypted with the recipient's public key. If the symmetric encryption key (sometimes referred to as the *session key*) is generated by one participant and sent to another participant, then we have a key transport protocol. In some cases, however, both participants wish to participate in computing the session key to make sure that its randomness and security are satisfactory. For this purpose, key agreement protocols are used. One of the most widely used key agreement protocols is the Diffie–Hellman key exchange protocol [16]. Its security is based on the difficulty of computing the discrete logarithm in a finite field. Key exchange messages should also be protected against replay attacks by including time stamps, nonces, or counters.

2.6 PUBLIC KEY INFRASTRUCTURE AND DIGITAL CERTIFICATES

The public keys for encryption or for signature verification should be available to anybody wishing to send a confidential message or to verify a signature, respectively. In addition, it must be somehow guaranteed that a particular public key belongs to a particular principal (e.g., person, company, or server). In other words, there should be some public key infrastructure (PKI) in place to provide this functionality. A PKI is based on the following:

- Digital certificates carrying the information about the key owner, the relevant cryptographic algorithms, the public key, key validity, and other information [9]
- Certification authorities (CA) or certification service providers (CSPs), trusted third parties that issue, digitally sign, and manage digital certificates [17]
- Agreements between the CSPs about mutual recognition of digital certificates in the form of cross certificates [9]

These are only the basic concepts because a PKI can be quite complex. For example, if a private key has been compromised, there must be a possibility to revoke it. A revocation is published in the so-called certificate revocation list (CRL) by the CSP that issued the certificate.

Almost all member states of the European Union (EU) have their national PKIs including the accredited national CSPs. PKIs are based on the national digital signature laws, which are in compliance with the Electronic Signature Directive issued by the European Commission in 1999 [18].

2.7 SECURITY EVALUATION

Security evaluation is a discipline ensuring that secure computing and communication systems really do what they promise. However, it is practically impossible to prove that a system is secure because even for simple systems the proof is extremely computing intensive. Fortunately, it is possible to build verifiably correct secure systems if the verification is integrated into the system's specification, design, and implementation, as required by security evaluation criteria.

There have been several national collections of security evaluation criteria, from U.S. Department of Defense Trusted Computer System Evaluation Criteria (TCSEC) in the 1980s to Information Technology Security Evaluation Criteria (ITSEC) in 1991 [19]. All those collections became input to the state-of-the-art security evaluation criteria, the so-called Common Criteria (CC) for information technology security evaluation. The current version is 2.3 (2005), and version 2.2 was published as the 15408 ISO/IEC standards series.

For a particular IT product or system under evaluation (target of evaluation, TOE), the security requirements are described in the form of a security target (ST). An implementation-independent set of security requirements for a TOE family (e.g., operating systems, firewalls, and smart cards) is referred to as the protection profile (PP). One of the published and evaluated PPs can be used to write a particular ST [20]. The CC defines two types of security requirements:

- Security functional requirements define the desired security behavior.
- Security assurance requirements ensure that the alleged security measures are effective and implemented correctly.

Assurance is a measure of confidence that a system meets its security objectives. The CC defines seven evaluation assurance levels, from EAL1, which stands for functionally tested, up to EAL7, meaning formally verified, designed, and tested. EAL7 is military-level security; for commercial products the highest practical level is EAL4.

When the TOE contains security functions realized by a probabilistic or permutational mechanism, such as passwords or cryptographic hash functions, the function's minimum strength level strength of function (SOF) can be required. The SOF level (basic, medium, or high) corresponds to the minimum effort necessary to successfully attack the underlying security mechanism.

An internationally recognized CC evaluation may be carried out by an accredited evaluation laboratory, and the corresponding security certificate may be issued by an accredited certification body. More information can be found on the CC home page [20].

In Section 2.6, we mentioned the digital signature legislative in the EU. For the digital signatures that are a priori recognized by the national laws of the EU member states, only sufficiently evaluated hardware and software components may be used. For more information, see [18].

2.8 SECURITY AUDIT

Information security management systems can also be certified. The ISO/IEC 27001 standard [21], which recently replaced BS 7799, can be used within commercial or nonprofit organizations to formulate security requirements and objectives and as a framework for the implementation and management of controls ensuring that the specific security objectives are met. In addition, the standard can be applied by the internal and external auditors to determine the degree of compliance with the policies, directives, and standards adopted by an organization. Nationally accredited certification bodies can issue IT security management certificates. For example, certification bodies in Austria have to be accredited by the Federal Ministry of Economics and Labor.

REFERENCES

1. International Organization for Standardization, Information Technology—Security Techniques—Code of Practice for Information Security Management, ISO/IEC 17799 (see also BS7799-1), 2005.
2. Hassler, V., *Security Fundamentals for E-Commerce*, Artech House, Norwood, MA, 2000.
3. National Institute of Standards and Technology, Advanced Encryption Standard (AES), FIPS PUB 197, November 2001.
4. American National Standards Institute, American National Standard for Data Encryption Algorithm (DEA), ANSI X3.92, 1981.
5. Rivest, R.L., A. Shamir, and L.A. Adleman, A method for obtaining digital signatures and public-key cryptosystems, *Communications of the ACM*, 21(2), 1978, 120–126.
6. National Institute of Standards and Technology, Secure Hash Standard (SHS), FIPS PUB 180–2, August 2002.
7. Krawczyk, H., M. Bellare, and R. Canetti, HMAC: Keyed-Hashing for Message Authentication, The Internet Engineering Task Force, RFC 2104, February 1997.
8. International Organization for Standardization, Information Technology—Security Techniques—Entity Authentication—Part 5: Mechanisms Using Zero-Knowledge Techniques, ISO/IEC 9798-5, 2004.
9. International Organization for Standardization, Information Technology—Open Systems Interconnection—The Directory: Public-Key and Attribute Certificate Frameworks, ISO/IEC 9594-8, 2001.
10. Denning, D.E., *Cryptography and Data Security*, Addison-Wesley, Reading, MA, 1982.
11. International Organization for Standardization, Information Technology—Security Techniques—Cryptographic Techniques Based on Elliptic Curves, ISO/IEC 15946, Part 1–4, 2002–2004.
12. American National Standards Institute, Public Key Cryptography for the Financial Services Industry, The Elliptic Curve Digital Signature Algorithm (ECDSA), ANSI X9.62, 2005.
13. National Institute of Standards and Technology, Digital Signature Standard (DSS), FIPS PUB 186–2, January 2000.
14. Chaum, D., Untraceable electronic mail, return addresses and digital pseudonyms, *Communications of the ACM*, 2(24), 1981, 84–88.
15. Eastlake, D., J. Schiller, and S. Crocker, Randomness Requirements for Security, The Internet Engineering Task Force, RFC 4086, June 2005.

16. Rescorla, E., Diffie–Hellman Key Agreement Method, The Internet Engineering Task Force, RFC 2631, June 1999.
17. Housley, R., W. Polk, W. Ford, and D. Solo, Internet X.509 Public Key Infrastructure. Certificate and Certificate Revocation List (CRL) Profile, The Internet Engineering Task Force, RFC 3280, April 2002.
18. European Commission, Information Society: Electronic Signatures Directive 1999/93/CE, http://europa.eu.int/information_society/eeurope/2005/all_about/security/esignatures/index_en.htm, February 2006.
19. Hassler, V., M. Manninger, M. Gordeev, and C. Müller, *Java Card for E-Payment Applications*, Artech House, Norwood, MA, 2001.
20. Common Criteria Project, The Official Web site of the Common Criteria Project, http://www.commoncriteriaportal.org/, 2006.
21. International Organization for Standardization, Information Technology—Security Techniques—Information Security Management Systems—Requirements, ISO/IEC 27001, 2005.

3 Efficient VLSI Architectures for the Advanced Encryption Standard Algorithm

Xinmiao Zhang

CONTENTS

3.1 INTRODUCTION

The rapidly growing wireless communication industry faces an exploding need for security. With the ever-increasing computing speed brought by advanced technologies, higher and higher security level is required to counter various attacks. The data encryption standard (DES) has been the U.S. government standard since 1977. However, with the fast computing technology these days, it can be cracked quickly and inexpensively. In January 1997, the National Institute of Standards and Technology (NIST) invited proposals for new algorithms for the advanced encryption standard (AES). Fifteen preliminary algorithms were proposed in response. Among these preliminary candidates, MARS, RC6, Rijndael, Serpent, and Twofish were announced as the finalists on August 9, 1999. After further evaluating the security, as well as both software and hardware implementations of these finalists, NIST announced in October 2000 that Rijndael was selected as the AES algorithm [1].

The AES algorithm has broad applications in wireless communications, including cellular phones, smart cards, network servers, and surveillance systems. Compared with software implementations, hardware implementations of the AES algorithm not only can achieve higher speed and lower power consumption, but can also provide more physical security. This chapter addresses various optimization approaches for efficient hardware implementations of the AES algorithm. Generally, optimizations can be carried out in three levels: circuit, architectural, and algorithmic levels. Compared with circuit-level optimizations, algorithmic- and architectural-level optimizations can usually achieve much more significant improvements. Three architectural optimization techniques can be employed to speed up the hardware implementations of the AES algorithm. They are pipelining, subpipelining, and loop unrolling. The speedup factor and area consumption of each technique are provided in this chapter. Architectural- and algorithmic-level optimizations are inseparable and interactive. Successful applications of architectural optimization techniques depend on how the algorithm is transformed into hardware. Various algorithmic modifications can be employed to reduce the hardware complexity of the AES algorithm, such as substructure sharing and composite field arithmetic. These optimization methods are also presented in this chapter. In addition, resource-sharing issues between encryptors and decryptors are discussed. These issues become very important when both the encryptor and the decryptor need to be implemented in a small area.

The structure of this chapter is as follows. In Section 3.2, the AES algorithm is briefly introduced. Three architectural optimization approaches are investigated in Section 3.3. In Section 3.4, various algorithmic modifications for the AES algorithm are presented. Section 3.5 explores resource sharing between encryptors and decryptors and Section 3.6 concludes this chapter.

3.2 ADVANCED ENCRYPTION STANDARD ALGORITHM

The AES algorithm is a symmetric-key cipher, in which a single key is used in both encryption and decryption. The key length of the AES algorithm can be 128, 192, or 256 bits. The AES algorithm is also a block cipher. Messages are divided into blocks of 128 bits and the encryption or decryption is carried out on each block. The 128-bit block can be divided into sixteen 8-bit bytes in_0, in_1, $in_2, \ldots, in_{15}$. These bytes are mapped to a 4×4 array, called the State, as illustrated in Figure 3.1. The encryption or decryption is performed on the State, and at the end, the final value is mapped to the output bytes out_0, out_1, $out_2, \ldots, out_{15}$. Each byte in the State is denoted by $S_{i,j} (0 \leq i, j < 4)$ and

Input bytes

in_0	in_4	in_8	in_{12}
in_1	in_5	in_9	in_{13}
in_2	in_6	in_{10}	in_{14}
in_3	in_7	in_{11}	in_{15}

⇨

State array

$S_{0,0}$	$S_{0,1}$	$S_{0,2}$	$S_{0,3}$
$S_{1,0}$	$S_{1,1}$	$S_{1,2}$	$S_{1,3}$
$S_{2,0}$	$S_{2,1}$	$S_{2,2}$	$S_{2,3}$
$S_{3,0}$	$S_{3,1}$	$S_{3,2}$	$S_{3,3}$

⇨

Output bytes

out_0	out_4	out_8	out_{12}
out_1	out_5	out_9	out_{13}
out_2	out_6	out_{10}	out_{14}
out_3	out_7	out_{11}	out_{15}

FIGURE 3.1 Mapping of input bytes, State array, and output bytes. (From Zhang, X. and Parhi, K.K., *IEEE Circuits Syst. Mag.*, 2(4), 26, 2002. With permission.)

is considered as an element of finite field GF(2^8). Although all degree eight irreducible polynomials over GF(2) can be used to construct GF(2^8), the irreducible polynomial specified by the AES algorithm is $P(x) = x^8 + x^4 + x^3 + x + 1$. The key of the AES algorithm can be mapped to four rows of bytes in a similar way, except the number of bytes in each row, denoted by Nk, can be 4, 6, or 8 when the length of the key is 128, 192, or 256 bits, respectively. The AES algorithm is carried out in a number of rounds. The total round number, Nr, is 10 when Nk = 4, Nr = 12 when Nk = 6, and Nr = 14 when Nk = 8. Figure 3.2 illustrates the block diagram of the AES encryption and the straightforward decryption structures.

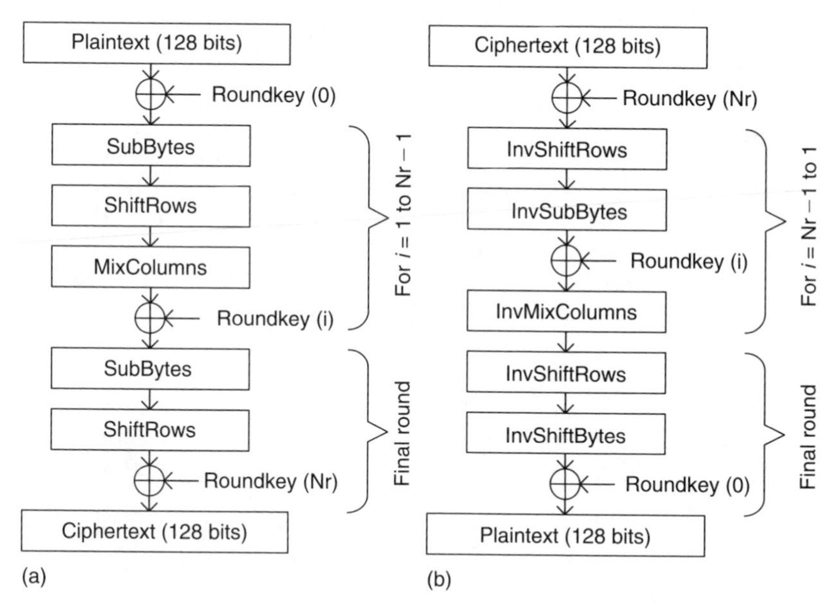

FIGURE 3.2 The AES algorithm. (a) Encryption structure. (b) Straightforward decryption structure. (From Zhang, X. and Parhi, K.K., *IEEE Circuits Syst. Mag.*, 2(4), 27, 2002. With permission.)

3.2.1 Encryption

The roundkeys used in encryption and decryption are generated from the key expansion process. The details of this process are introduced in the next subsection. In the encryption, after the initial roundkey addition, Nr rounds are carried out. The first Nr − 1 rounds are the same. As illustrated in Figure 3.2a, each of these rounds consists of four transformations: the SubBytes, the ShiftRows, the MixColumns, and the AddRoundKey. The only difference in the final round is that there is no MixColumns transformation.

The SubBytes is performed on each individual byte of the State. This transformation first computes the multiplicative inverse of each byte in GF(2^8), followed by an affine transformation. The SubBytes can be described by

$$S'_{i,j} = MS_{i,j}^{-1} + C, \tag{3.1}$$

where

$$M = \begin{bmatrix} 1 & 0 & 0 & 0 & 1 & 1 & 1 & 1 \\ 1 & 1 & 0 & 0 & 0 & 1 & 1 & 1 \\ 1 & 1 & 1 & 0 & 0 & 0 & 1 & 1 \\ 1 & 1 & 1 & 1 & 0 & 0 & 0 & 1 \\ 1 & 1 & 1 & 1 & 1 & 0 & 0 & 0 \\ 0 & 1 & 1 & 1 & 1 & 1 & 0 & 0 \\ 0 & 0 & 1 & 1 & 1 & 1 & 1 & 0 \\ 0 & 0 & 0 & 1 & 1 & 1 & 1 & 1 \end{bmatrix}, \quad C = \begin{bmatrix} 1 \\ 1 \\ 0 \\ 0 \\ 0 \\ 1 \\ 1 \\ 0 \end{bmatrix}.$$

The ShiftRows is a simple transformation. The bytes in the first row of the State do not change whereas those in the second, third, and fourth rows cyclically shift 1 byte, 2 bytes, and 3 bytes to the left, respectively. This transformation is illustrated in Figure 3.3.

The MixColumns is a columnwise transformation. The four bytes in each column of the State are considered as the coefficients of a degree three

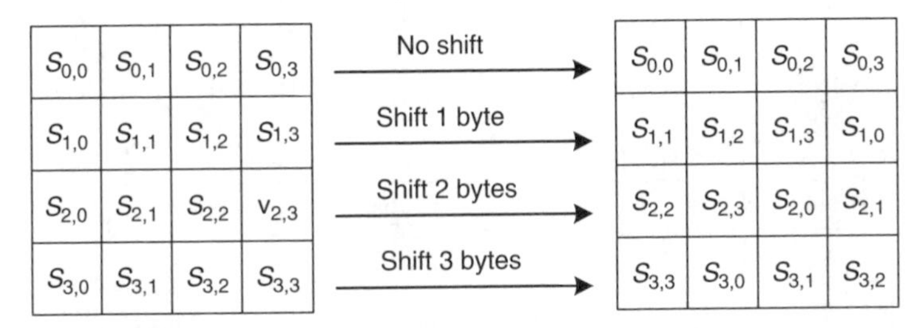

FIGURE 3.3 ShiftRows transformation. (From Zhang, X. and Parhi, K.K., *IEEE Circuits Syst. Mag.*, 2(4), 27, 2002. With permission.)

polynomial over GF(2^8). Then this polynomial is multiplied by $m(x)$ modulo x^4+1, where

$$m(x) = \{03\}_{16}x^3 + \{01\}_{16}x^2 + \{01\}_{16}x + \{02\}_{16}.$$

In the above equation, $\{y\}_{16}$ denotes the number y in hexadecimal form whereas $\{y\}_2$ used later in this chapter stands for y in binary form. In matrix form, the MixColumns can be expressed as

$$\begin{bmatrix} S'_{0,c} \\ S'_{1,c} \\ S'_{2,c} \\ S'_{3,c} \end{bmatrix} = \begin{bmatrix} \{02\}_{16} & \{03\}_{16} & \{01\}_{16} & \{01\}_{16} \\ \{01\}_{16} & \{02\}_{16} & \{03\}_{16} & \{01\}_{16} \\ \{01\}_{16} & \{01\}_{16} & \{02\}_{16} & \{03\}_{16} \\ \{03\}_{16} & \{01\}_{16} & \{01\}_{16} & \{02\}_{16} \end{bmatrix} \begin{bmatrix} S_{0,c} \\ S_{1,c} \\ S_{2,c} \\ S_{3,c} \end{bmatrix} 0 \leq c < 4. \quad (3.2)$$

Finally, in the AddRoundKey transformation, a 128-bit roundkey is added to the State by bitwise Exclusive-OR (XOR) operation.

3.2.2 Key Expansion

A total of $(\text{Nr}+1)$ roundkeys are needed for the encryption or decryption. Each 128-bit roundkey can be divided into four 4-byte words. The key is used as the initial set of Nk words, and the rest of the words are generated from the key iteratively through the key expansion process described by the pseudocode in Figure 3.4 [1]. The output of the key expansion is an array of 4-byte words denoted by $w(i)(0 \leq i < 4(\text{Nr}+1))$, and each roundkey can be formed by concatenating four words: roundkey $(i) = (w(4i), w(4i+1), w(4i+2), w(4i+3))$. In Figure 3.4, the function of the SubWord is to apply the SubBytes transformation to each byte in a word whereas RotWord cyclically rotates each byte in a word one byte to the left. For example, given the input to the RotWord as four bytes (a_0, a_1, a_2, a_3), RotWord would return (a_1, a_2, a_3, a_0). Rcon is the round constant word vector, and only the leftmost byte of each entry in Rcon is nonzero. The values of the leftmost bytes for Rcon(1) through Rcon(10) are $\{01\}_{16}, \{02\}_{16}, \{04\}_{16}, \{08\}_{16}, \{10\}_{16}, \{20\}_{16}, \{40\}_{16}, \{80\}_{16}, \{1b\}_{16}$, and $\{36\}_{16}$, respectively.

3.2.3 Decryption

As illustrated in Figure 3.2b, a straightforward decryption structure can be derived by inverting each transformation and the sequence of the transformations in the encryption structure. The inverse transformation of the SubBytes is the InvSubBytes, in which the following operation is performed on each byte of the State

$$S'_{i,j} = (M^{-1}(S_{i,j} + C))^{-1}. \quad (3.3)$$

```
KeyExpansion(byte key(4Nk), word w(4(Nr+1)), Nk)
begin
        word temp
        i = 0
        while (i < Nk)
            w(i) = word(key(4i), key(4i+1), key(4i+2), key(4i+3))
            i = i+1
        end while
        i = Nk
        while (i < 4(Nr+1))
            temp = w(i-1)
            if (i mod Nk=0)
                temp = SubWord(RotWord(temp)) XOR Rcon(i/Nk)
            else if (Nk > 6 and i mod Nk=4)
                temp = SubWord(temp)
            end if
            w(i) = w(i-Nk) XOR temp
            i=i+1
        end while
end
```

FIGURE 3.4 Pseudocode for key expansion. (From Zhang, X. and Parhi, K.K., *IEEE Circuits Syst. Mag.*, 2(4), 28, 2002. With permission.)

The inverse of the ShiftRows is the InvShiftRows. In this transformation, the first row of the State does not change whereas the rest of the rows are shifted cyclically to the right by the same offsets as those in the ShiftRows. The InvMixColumns perform the inverse function of the MixColumns. This transformation considers the four bytes in each column of the State as the coefficients of a polynomial and multiply this polynomial by $m^{-1}(x)$ modulo x^4+1, where

$$m^{-1}(x) = \{0b\}_{16}x^3 + \{0d\}_{16}x^2 + \{09\}_{16}x + \{0e\}_{16}.$$

In matrix form, the InvMixColumns can be written as

$$\begin{bmatrix} S'_{0,c} \\ S'_{1,c} \\ S'_{2,c} \\ S'_{3,c} \end{bmatrix} = \begin{bmatrix} \{0e\}_{16} & \{0b\}_{16} & \{0d\}_{16} & \{09\}_{16} \\ \{09\}_{16} & \{0e\}_{16} & \{0b\}_{16} & \{0d\}_{16} \\ \{0d\}_{16} & \{09\}_{16} & \{0e\}_{16} & \{0b\}_{16} \\ \{0b\}_{16} & \{0d\}_{16} & \{09\}_{16} & \{0e\}_{16} \end{bmatrix} \begin{bmatrix} S_{0,c} \\ S_{1,c} \\ S_{2,c} \\ S_{3,c} \end{bmatrix} 0 \le c < 4. \quad (3.4)$$

The inverse of the AddRoundKey is still bitwise XOR operations. Hence, the name is kept unchanged.

As can be observed from Figure 3.2a and Figure 3.2b, the straightforward decryption structure has a totally different sequence of transformations from that of the encryption structure. This difference puts an obstacle to resource

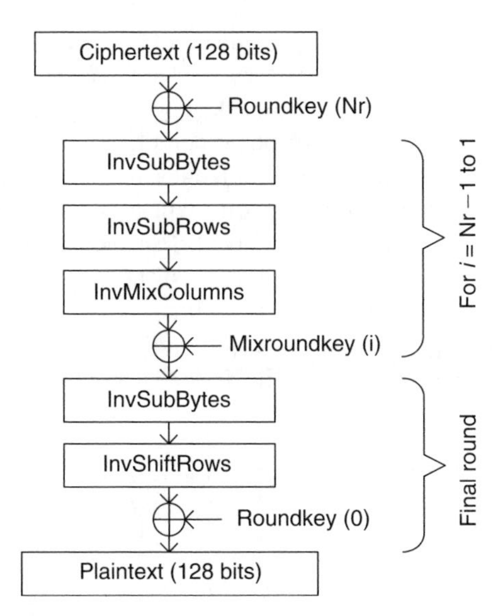

FIGURE 3.5 Equivalent decryption structure. (From Zhang, X. and Parhi, K.K., *IEEE Circuits Syst. Mag.*, 2(4), 29, 2002. With permission.)

sharing between the implementation of encryptors and decryptors. Fortunately, two features of the AES algorithm can be employed to change the sequence of the transformations in the decryption.

1. The positions of the InvShiftRows and InvSubBytes can be exchanged without affecting the decryption.
2. The addition of the roundkeys can be moved to after the InvMixColumns if the InvMixColumns transformation is applied to the roundkeys before they are added up.

Applying these two properties, the equivalent decryption structure as illustrated in Figure 3.5 can be derived. In Figure 3.5, mixroundkeys denote the modified roundkeys as a result of applying the InvMixColumns to the roundkeys. As can be observed, the sequence of the transformations in the equivalent decryption structure is exactly the same as that in the encryption structure. As a result, more efficient implementations of joint encryptors and decryptors are enabled.

3.3 ARCHITECTURAL OPTIMIZATIONS

The AES algorithm is a block cipher. The most commonly used modes of operation for block ciphers are electronic code book (ECB), counter (CTR), cipher block chaining (CBC), cipher feedback (CFB), and output feedback (OFB). The first two belong to nonfeedback modes, where the encryption or decryption of different blocks is independent of each other and can be carried out simultaneously. The other three modes are feedback modes. In these modes, due to the existence of feedback loops, the processing of the next block cannot start until the current block is completed. Architectural optimizations do not bring much improvement for feedback modes. This section focuses on the speedups that can be brought by different architectures for nonfeedback modes and the relative area requirements of these architectures.

For nonfeedback modes, speedups can be achieved by processing multiple data blocks simultaneously. Three types of architectures can be used for this purpose. They are pipelining, subpipelining, and loop unrolling. These architectures are illustrated in Figure 3.6 together with a

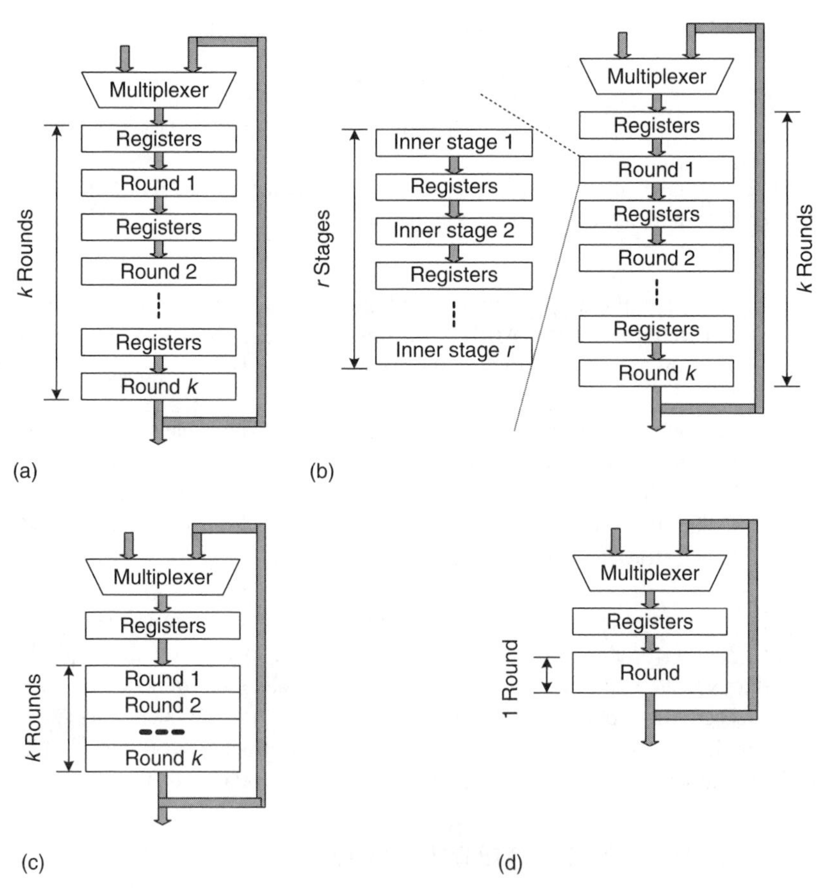

FIGURE 3.6 Three types of architecture for encryptor and decryptor with a basic reference architecture: (a) pipelined architecture, (b) subpipelined architecture, (c) loop-unrolled architecture, and (d) basic reference architecture. (From Zhang, X. and Parhi, K.K., *IEEE Circuits Syst. Mag.*, 2(4), 30, 2002. With permission.)

basic reference architecture. The performance of these architectures is analyzed in [2–4].

The speed of a digital system is usually measured by throughput, which is defined as the average number of bits processed per second. For the AES algorithm, throughput can be also computed as

$$\text{Throughput} = \frac{128}{\text{Average number of clock cycles to process one block} \times \text{clock period}}.$$

The basic architecture illustrated in Figure 3.6d can only process one block of data at a time, and one round of encryption or decryption is carried out in each clock cycle. Hence, this architecture needs Nr clock cycles to process one block of data. In addition, the minimum achievable clock period is decided by the path with the longest computational time, which is also called the critical path, between each pair of adjacent registers. Therefore, the maximum throughput that can be achieved by the basic architecture can be computed as

$$\text{Throughput}_{\text{basic}} = \frac{128}{\text{Nr} \times t_{\text{basic}}},$$

where $t_{\text{basic}} = t_{\text{round}} + t_{\text{mux}} + t_{\text{setup}} + t_{\text{prop}}$. t_{round} stands for the delay of the combinational logic in each round unit and t_{mux} denotes the delay of a multiplexer whereas t_{setup} and t_{prop} are the setup time and propagation delay of a register, respectively. In the following, the speedups that can be achieved by pipelining, subpipelining, and loop unrolling over the basic architecture are provided. In addition, the area consumptions of these architectures are discussed.

1. *Pipelining*. Pipelining inserts rows of registers between each round unit. The combinational logic between adjacent registers is called pipelining stages. In this architecture, the number of the copies of the round unit, k, is usually chosen to be a divisor of Nr. During each clock cycle, the partially processed data block moves to the next pipelining stage and its place is taken by the subsequent block. Hence, after an initial delay of k clock cycles, k blocks of data are processed simultaneously. When a partially processed block reaches the kth round, it is fed back to the first round until all the Nr rounds are performed on this block. Therefore, after the initial delay, k blocks of data are processed for every $k \times (\text{Nr}/k) = \text{Nr}$ clock cycles. Accordingly, the average number of clock cycles to process one data block is Nr/k. As shown in Figure 3.6a and Figure 3.6d, the minimum achievable clock period of the pipelined architecture is the same as that of the basic architecture. As a result, the k-round pipelined architecture can achieve k times

speedup over the basic architecture whereas the area requirement is approximately k times of the basic reference architecture.

2. *Subpipelining*. Subpipelining inserts registers not only between each round unit but also inside each round unit. Assume that the critical path of each round unit is further broken into r segments by registers. Then $r \times k$ blocks of data can be processed simultaneously after the initial delay of $r \times k$ clock cycles. In addition, all the Nr rounds for these $r \times k$ blocks can be carried out in $r \times k \times \mathrm{Nr}/k = r \times \mathrm{Nr}$ clock cycles. Hence, the average number of clock cycles to process one block of data is $(r \times \mathrm{Nr})/(r \times k) = \mathrm{Nr}/k$. Moreover, if the r substages in each round unit have equal delay, then the minimum achievable clock period of the r-substage subpipelining is

$$t_{\text{subpipelining}} = t_{\text{round}}/r + t_{\text{mux}} + t_{\text{setup}} + t_{\text{prop}}.$$

Let

$$\tau = \frac{t_{\text{setup}} + t_{\text{prop}} + t_{\text{mux}}}{t_{\text{round}}}.$$

Then the speedup of k-round r-substage subpipelining over the basic architecture is

$$\frac{\text{Throughput}_{\text{subpipelining}}}{\text{Throughput}_{\text{basic}}} = \frac{kr(1+\tau)}{1+r\tau}.$$

Usually τ is small. Hence, if each round unit can be divided into r substages with equal delay, the k-round subpipelining can achieve almost $k \times r$ times speedup over the basic architecture. Compared with pipelining, subpipelining can achieve almost additional r times speedup at the expense of slightly increased area caused by extra registers and control logic. However, the speedup that can be achieved by subpipelining is limited by the indivisible combinational component with the longest delay in the round unit. Breaking the critical path of the rest of the round unit into shorter segments does not reduce the minimum achievable clock period. Although more blocks of data are processed simultaneously, the average number of clock cycles to process one block of data is increased by the same proportion. In this case, the overall speed does not improve despite the increased area caused by additional registers.

3. *Loop Unrolling*. In a loop-unrolled architecture as illustrated in Figure 3.6c, only one block of data is processed at a time. However, multiple rounds are performed in each clock cycle. The unrolling factor, k, is

usually chosen as a divisor of Nr also. It takes Nr/k clock cycles to process one data block. In addition, it can be observed from Figure 3.6c that the minimum achievable clock period of the loop-unrolled architecture is

$$t_{\text{loop unrolling}} = kt_{\text{round}} + t_{\text{mux}} + t_{\text{setup}} + t_{\text{prop}}.$$

Therefore, it can be derived that the speedup can be achieved by a k-round loop-unrolled architecture over the basic architecture is

$$\frac{\text{Throughput}_{\text{loop-unrolled}}}{\text{Throughput}_{\text{basic}}} = \frac{1+\tau}{1+\tau/k}. \tag{3.5}$$

As can be observed from Equation 3.5, since τ is usually small, the loop-unrolled architecture does not achieve much speedup over the basic architecture, despite the almost k times area requirement.

In summary, the speed and area consumption of pipelining, subpipelining, and loop unrolling are listed in Table 3.1. The numbers in this table are normalized with respect to the speed and area of the basic architecture. The speed of the subpipelined architecture is computed based on the assumption that each round unit can be divided into r substages with equal delay. In addition, ρ and σ are the fractions of the area of a 128-bit register and a 128-bit 2-to-1 multiplexer over the total area of a basic architecture, respectively. Usually, ρ and σ are small. It can be observed from Table 3.1 that the subpipelined architecture can achieve the maximum speedup and optimum speed over area ratio in nonfeedback modes. Employing subpipelining with 10 copies of round unit and seven substages in each round unit, a throughput of 21.56 Gbps has been achieved on Xilinx FPGA devices [5]. In small area applications, subpipelined architecture with only one

TABLE 3.1
Speed and Area of Pipelining, Subpipelining, and Loop Unrolling

Architecture	Speed	Area
Basic	1	1
k-Round pipelining	k	$k-\sigma(k-1)$
r-Substage k-round subpipelining	$\dfrac{kr(1+\tau)}{1+r\tau}$	$k-\sigma(k-1)+\rho k(r-1)$
k-Round loop unrolling	$\dfrac{1+\tau}{1+\tau/k}$	$k-(k-1)(\rho+\sigma)$

round unit can be employed. If the subpipelined architecture can achieve a speed that is higher than the application requirement, lower power supply voltage can be employed to reduce power consumption.

3.4 ALGORITHMIC OPTIMIZATIONS

This section introduces the algorithmic strength on the optimization of individual transformations of the AES algorithm. Since no logic operations are involved in the ShiftRows or InvShiftRows, and the AddRoundKey only costs a bitwise XOR operation, no optimization needs to be done on these transformations.

3.4.1 IMPLEMENTATIONS OF SUBBYTES AND INVSUBBYTES

The SubBytes and InvSubBytes can be implemented by two approaches. One of them is based on lookup tables (LUTs) [2,3,6–10]. The inverse value of every GF(2^8) element can be precomputed and stored in an LUT of $2^8 \times 8 = 2$ K bits. Then the inverse of a given element can be read out from the LUT by using proper addresse. Each SubBytes or InvSubBytes needs 16 such tables. Hence, the memory requirement of this approach becomes very large when multiple round units need to be implemented. In addition, the delay of the memory access is unbreakable. This feature prohibits each round unit from being divided into multiple substages with equal delay. As a result, utilizing LUTs in SubBytes or InvSubBytes implementations prohibits taking further advantages of subpipelining to achieve higher speed.

Another approach is to employ combinational logic only in the implementation of the multiplicative inversion. In this approach, the elements with unbreakable delay are individual gates. Hence, each round unit can be divided into multiple substages with equal delay. However, the computation of multiplicative inverse in GF(2^8) is hardware demanding. In order to reduce complexity, composite field arithmetic can be employed [11]. The idea of applying composite field arithmetic to the AES algorithm is first proposed in [12] and is explored in detail in [5,13–19]. Applying composite field arithmetic, the elements of large-order fields are mapped to those of small-order fields in which the field operations can be carried out in a simpler way.

3.4.1.1 Composite Field Implementations of Multiplicative Inversion

Given an irreducible polynomial $R(x)$ over GF(p) with degree q, the set $\{1, x, x^2, \ldots, x^{q-1}\}$ forms a standard basis of GF(p^q), where x is a root of $R(x)$. Using standard basis, an element $a \in$ GF(p^q) can be represented in the form of $a_0 + a_1x + \cdots + a_{q-1}x^{q-1}$, where $a_0, a_1, \ldots, a_{q-1} \in$ GF(p). Finite fields have two associate operations: the additive operation and the multiplicative operation. With the field elements represented in polynomial form, the additive operation can be defined as polynomial addition whereas the multiplicative operation can be defined as polynomial multiplication modulo $R(x)$. In this case, we say GF(p^q) is constructed from GF(p) by $R(x)$, and $R(x)$ is called the field polynomial of GF(p^q).

The pair $\{GF(2^n), Q(y) = y^n + \sum_{i=0}^{n-1} q_i y^i, q_i \in GF(2)\}$ and $\{GF((2^n)^m), P(x) = x^m + \sum_{i=0}^{m-1} p_i x^i, p_i \in GF(2^n)\}$ is called a composite field [11] if

- GF(2^n) is constructed from GF(2) by $Q(y)$
- GF($(2^n)^m$) is constructed from GF(2^n) by $P(x)$

Composite fields are denoted by GF($(2^n)^m$), and a composite field GF($(2^n)^m$) is isomorphic to the field GF(2^q) for $q = nm$. Composite fields can also be built iteratively from lower-order fields. For example, the composite field of GF(2^8) can be built iteratively from GF(2) by the following irreducible polynomials:

$$\begin{aligned} &GF(2) \Rightarrow GF(2^2) : P_0(x) = x^2 + x + 1, \\ &GF(2^2) \Rightarrow GF((2^2)^2) : P_1(x) = x^2 + x + \phi, \\ &GF((2^2)^2) \Rightarrow GF(((2^2)^2)^2) : P_2(x) = x^2 + x + \lambda, \end{aligned} \tag{3.6}$$

where $\phi \in GF(2^2)$ and $\lambda \in GF((2^2)^2)$. In addition, to maintain additive and multiplicative homomorphisms, an isomorphic mapping function $f(a) = \delta \times \underline{a}$ and its inverse need to be applied to map the representation of an element in GF(2^n) to its composite field and vice versa. Here, $\underline{a} = [a_{n-1}, a_{n-2}, \ldots, a_0]^T$ is the n-bit column vector formed by the coefficients in the standard basis representation of a, and δ is an $n \times n$ binary matrix. The entries in δ are decided by both the irreducible polynomial used for the construction of GF(2^n) from GF(2) and those for the composite field. For example, assume $\phi = \{10\}_2$ and $\lambda = \{1100\}_2$, the δ matrix corresponding to $P(x) = x^8 + x^4 + x^3 + x + 1$ and the field polynomials in Equation 3.6 can be found as follows [13]:

$$\delta = \begin{bmatrix} 1 & 0 & 1 & 0 & 0 & 0 & 0 & 0 \\ 1 & 1 & 0 & 1 & 1 & 1 & 1 & 0 \\ 1 & 0 & 1 & 0 & 1 & 1 & 0 & 0 \\ 1 & 0 & 1 & 0 & 1 & 1 & 1 & 0 \\ 1 & 1 & 0 & 0 & 0 & 1 & 1 & 0 \\ 1 & 0 & 0 & 1 & 1 & 1 & 1 & 0 \\ 0 & 1 & 0 & 1 & 0 & 0 & 1 & 0 \\ 0 & 1 & 0 & 0 & 0 & 0 & 1 & 1 \end{bmatrix}. \tag{3.7}$$

In the composite field GF($(2^4)^2$), an element can be written as $S(x) = s_h x + s_l$, where $s_h, s_l \in GF(2^4)$ and x is a root of $P_2(x)$. Computing the multiplicative inverse of $S(x)$ modulo $P_2(x)$ is equivalent to finding polynomials $A(x)$ and $B(x)$ satisfying the following equation:

$$A(x)P_2(x) + B(x)S(x) = 1. \tag{3.8}$$

Then $S^{-1}(x) = B(x)$. The extended Euclidean algorithm can be applied to solve this problem. First, we need to write $P_2(x)$ in terms of the quotient and remainder of the polynomial division by $S(x)$. By long division, it can be derived that

$$P_2(x) = \left(s_h^{-1}x + (1 + s_h^{-1}s_l)s_h^{-1}\right)S(x) + \left(\lambda + (1 + s_h^{-1}s_l)s_h^{-1}s_l\right). \tag{3.9}$$

Multiply s_h^2 and $\Theta = \left(s_h^2\lambda + s_h s_l + s_l^2\right)^{-1}$ to both sides of Equation 3.9; it follows that

$$\Theta s_h^2 P_2(x) = \Theta(s_h x + (s_h + s_l))S(x) + 1.$$

Comparing the above equation with Equation 3.8, it can be observed that

$$S^{-1}(x) = s_h\Theta x + (s_h + s_l)\Theta. \tag{3.10}$$

According to Equation 3.10, the multiplicative inversion involved in the SubBytes and InvSubBytes can be implemented by the architecture illustrated in Figure 3.7. For a given set of irreducible polynomials used for composite field construction, the δ matrix is fixed. Hence, we can precompute the product of δ^{-1} and the M matrix such that the inverse isomorphic mapping and the affine transformation can be combined.

The multiplication in GF(2^4) can be further decomposed into GF($(2^2)^2$) to reduce hardware complexity. Assume two elements $a,b \in$ GF($(2^2)^2$) can be expressed as $a_hx + a_l$ and $b_hx + b_l$, respectively, where $a_h,a_l,b_h,b_l \in$ GF(2^2) and x is a root of $P_1(x)$. Then the product of a and b can be computed as

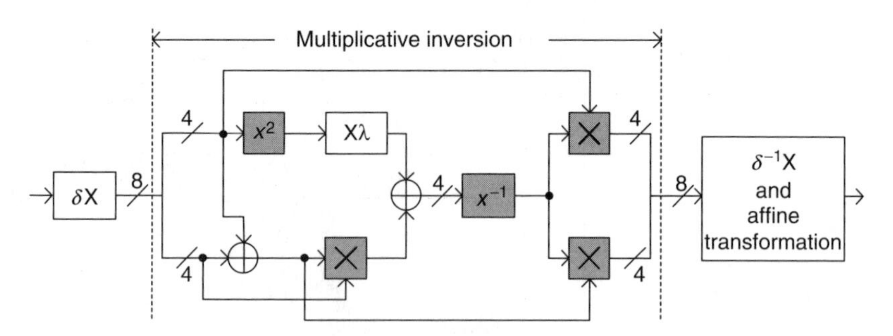

FIGURE 3.7 Implementation of the SubBytes transformation. (From Zhang, X. and Parhi, K.K., *IEEE Trans. VLSI Syst.*, 12(9), 957, 2004. With permission.)

$$
\begin{aligned}
&(a_hx + a_l)(b_hx + b_l) \bmod P_1(x) \\
&\quad = a_hb_hx^2 + (a_hb_l + a_lb_h)x + a_lb_l \bmod P_1(x) \\
&\quad = a_hb_h(x + \phi) + (a_hb_l + a_lb_h)x + a_lb_l \\
&\quad = ((a_h + b_h)(a_l + b_l) + a_lb_l)x + (a_hb_h\phi + a_lb_l).
\end{aligned} \tag{3.11}
$$

Accordingly, the GF(2^4) multiplier can be implemented by the architecture illustrated in Figure 3.8a. By using an equation similar to Equation 3.11, the multiplication in GF(2^2) can also be decomposed. In GF(2), multiplications are simply AND operations. Hence, the GF(2^2) multiplier can be implemented by the architecture shown in Figure 3.8b.

A square operation can be considered as a multiplication with two equal operands. Compared to a general multiplier, the implementation of a squarer is much more simple. In GF(2^2), a_h and a_l can be expressed as $a_3y + a_2$ and $a_1y + a_0$, respectively, where y is a root of $P_0(x)$. Replace b_h and b_l with a_h and a_l, respectively in Equation 3.11, and cancel out common terms, simple equations can be derived for the squarer. For example, in the case of $\phi = \{10\}_2$, assume the four bits associated with a are $\{a_3, a_2, a_1, a_0\}$, the bits in $a^2 = \{a'_3, a'_2, a'_1, a'_0\}$ can be computed as

$$
\begin{aligned}
a'_3 &= a_3, \\
a'_2 &= a_3 + a_2, \\
a'_1 &= a_2 + a_1, \\
a'_0 &= a_3 + (a_1 + a_0).
\end{aligned} \tag{3.12}
$$

Therefore, a squarer in GF(2^4) can be implemented by only four XOR gates with two XOR gates in the critical path when $\phi = \{10\}_2$. Similarly, for given

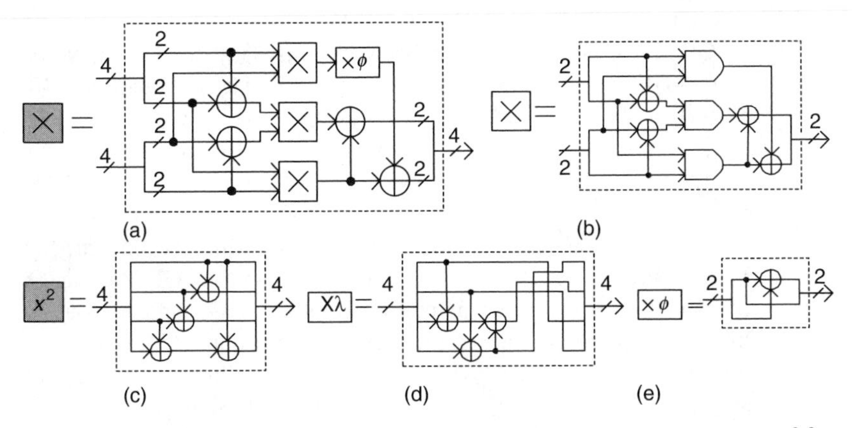

FIGURE 3.8 Implementations of individual blocks: (a) multiplier in GF($(2^2)^2$), (b) multiplier in GF(2^2), (c) squarer in GF(2^4), (d) constant multiplication by $\lambda = \{1100\}_2$, and (e) constant multiplication by $\phi = \{10\}_2$. (From Zhang, X. and Parhi, K.K., *IEEE Trans. VLSI Syst.*, 12(9), 961, 2004. With permission.)

values of ϕ and λ, the constant multiplications by ϕ and λ can also be simplified. For example, in the case of $\phi = \{10\}_2$, ϕ can be expressed as y, where y is a root of $P_0(x)$. Assume that $c \in \mathrm{GF}(2^2)$ can be expressed as $c_1y + c_0$ ($c_1, c_0 \in \mathrm{GF}(2)$). Then multiplying c by ϕ can be computed as

$$(c_1y + c_0)y = c_1y^2 + c_0y = (c_1 + c_0)y + c_1. \tag{3.13}$$

Hence, the constant multiplication by $\phi = \{10\}_2$ can be implemented by one XOR gate. In addition, when $\lambda = \{1100\}_2$, it can be derived that the product of λ and $b = \{b_3, b_2, b_1, b_0\}$ can be computed as

$$\begin{aligned} b'_3 &= (b_2 + b_0), \\ b'_2 &= (b_3 + b_1) + (b_2 + b_0), \\ b'_1 &= b_3, \\ b'_0 &= b_2. \end{aligned} \tag{3.14}$$

Sharing the term $b_2 + b_0$ in Equation 3.14, the constant multiplication by $\lambda = \{1100\}_2$ can be implemented by three XOR gates with two XOR gates in the critical path. In summary, the architectures for squarer in $\mathrm{GF}(2^4)$, the constant multiplier by $\phi = \{10\}_2$, and $\lambda = \{1100\}_2$ are illustrated in Figure 3.8c through Figure 3.8e, respectively.

The inversion in $\mathrm{GF}(2^4)$ can be implemented by different approaches:

1. Since for $s \in \mathrm{GF}(2^q)$, $s^{2^q-1} = 1$, then $s \times s^{2^q-2} = 1$. Hence, s^{-1} can be computed as s^{2^q-2}. Therefore, the inverse of $s \in \mathrm{GF}(2^4)$ can be computed as $s^{-1} = s^{14} = ((s^2)^2)^2\,(s^2)^2\,s^2$. Accordingly, the inversion in $\mathrm{GF}(2^4)$ can be implemented by repeat squaring and multiplying as illustrated in Figure 3.9a.
2. In $\mathrm{GF}((2^2)^2)$, an element can be written as $S'(x) = s'_hx + s'_l$, where $s'_h, s'_l \in \mathrm{GF}(2^2)$ and x is a root of $P_1(x)$. Similar to Equation 3.10, $S'^{-1}(x)$ can be computed as

$$S'^{-1}(x) = s'_h\Theta'x + (s'_h + s'_l)\phi, \tag{3.15}$$

where $\Theta' = (s'^2_h\phi + s'_hs'_l + s'^2_l)^{-1}$. This decomposition is illustrated in Figure 3.9b. It can be derived that the squarer in $\mathrm{GF}(2^2)$ can be combined with the constant multiplier block ($\times\phi$), and the output bits of the combined block are the same as the two input bits with their bit positions switched when $\phi = \{10\}_2$. In addition, the inverse of $s' = \{s'_1, s'_0\} \in \mathrm{GF}(2^2)$ is $\{s'_1, s'_1 + s'_0\}$. This inversion can be implemented by one XOR gate.
3. Based on Figure 3.9b, the expressions can be derived for the output bits in terms of the input bits, and Boolean algebra can be applied to

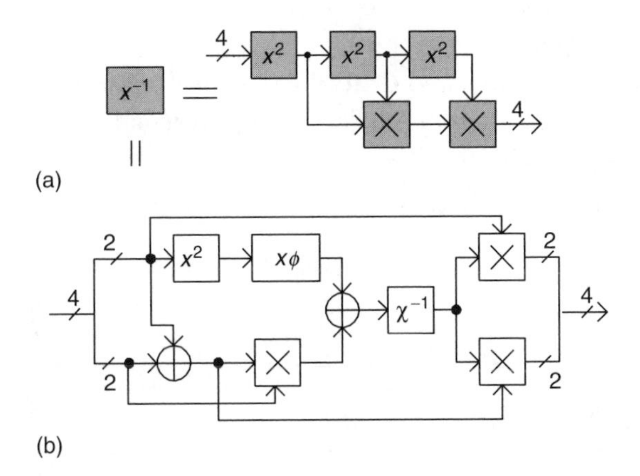

FIGURE 3.9 Implementations of inversion in GF(2^4): (a) square-multiply approach and (b) multiple decomposition approach. (From Zhang, X. and Parhi, K.K., *IEEE Trans. VLSI Syst.*, 12(9), 961, 2004. With permission.)

simplify these expressions. For example, in the case of $\phi = \{10\}_2$, taking the four bits of $s \in \mathrm{GF}(2^4)$ as $\{s_3, s_2, s_1, s_0\}$, it can be derived that the bits in $s^{-1} = \{s_3^{-1}, s_2^{-1}, s_1^{-1}, s_0^{-1}\}$ can be computed by the following equations:

$$\begin{aligned}
s_3^{-1} &= s_3 + s_3s_2s_1 + s_3s_0 + s_2, \\
s_2^{-1} &= s_3s_2s_1 + s_3s_2s_0 + s_3s_0 + s_2 + s_2s_1, \\
s_1^{-1} &= s_3 + s_3s_2s_1 + s_3s_1s_0 + s_2 + s_2s_0 + s_1, \\
s_0^{-1} &= s_3s_2s_1 + s_3s_2s_0 + s_3s_1 + s_3s_1s_0 + s_3s_0 + s_2 \\
&\quad + s_2s_1 + s_2s_1s_0 + s_1 + s_0.
\end{aligned} \tag{3.16}$$

Applying substructure sharing, the number of gates needed for the implementation of Equation 3.16 can be further reduced.

The gate counts and the critical paths for the three approaches to implement the multiplicative inversion in GF(2^4) are summarized in Table 3.2. As it can be observed from the table, the third approach based on direct implementation of the derived equation requires the least number of gates and has the shortest critical path. Composite field decomposition may be applied to reduce the hardware complexity when the order of the involved field is large. However, it may not be the optimum approach when the field order is small.

The complexity of the multiplicative inversion architecture in Figure 3.7 can be further reduced by combining the GF(2^4) squarer block and the

TABLE 3.2
Gate Counts and Critical Paths for the Three Implementation Approaches of Inversion in GF(2^4)

Approach	Total Gate Number	Critical Path
1	54 XOR + 18 AND	12 XOR + 2 AND
2	17 XOR + 9 AND	7 XOR + 2 AND
3	14 XOR + 8 AND	3 XOR + 2 AND

Source: From Zhang, X. and Parhi, K.K., *IEEE Trans. VLSI Syst.*, 12(9), 963, 2004. With permission.

constant multiplier ($\times\lambda$) block, as well as sharing common terms among the GF($(2^2)^2$) multipliers [19]. Replacing b_0, b_1, b_2, b_3 in Equation 3.14 by a'_0, a'_1, a'_2, a'_3 in Equation 3.12, it can be derived that

$$\begin{aligned} b'_0 &= a_2 + a_1 + a_0, \\ b'_1 &= a_3 + a_0, \\ b'_2 &= a_3, \\ b'_3 &= a_3 + a_2. \end{aligned} \tag{3.17}$$

Hence, the combined squarer and constant multiplier ($\times\lambda$) can be implemented by four XOR gates with two XOR gates in the critical path. In addition, from Figure 3.7, it can be observed that each of the GF($(2^2)^2$) multiplier pairs on the right and bottom share a common operand. When two GF($(2^2)^2$) multipliers have a common operand, the result of the bitwise addition carried out in one of the adders on the left of Figure 3.8a can be shared, and each of the three pairs of GF(2^2) multiplier inside have a common input. Furthermore, when two GF(2^2) multipliers have a common operand, the result of the single-bit addition carried out in one of the XOR gates on the left of Figure 3.8b can be shared. Therefore, for each pair of GF($(2^2)^2$) multipliers with a common operand, an area reduction of five XOR gates can be achieved by sharing the common terms.

3.4.1.2 Constructions of Optimum Composite Fields for the Advanced Encryption Standard Algorithm

Employing composite field arithmetic in the computation of the multiplicative inversion in the SubBytes and InvSubBytes transformations of the AES algorithm not only reduces hardware complexity, but enables deep subpipelining such that higher speed can be achieved.

Different irreducible polynomials can be used to construct the composite field of the same order. References [13] and [16] proposed one way to construct the composite field for the AES algorithm. However, there exist other construction schemes with smaller gate count and shorter critical path. In [14], different constructions of the composite field are compared to find the one with minimum gate count. Nevertheless, this design implements all the transformations of the AES algorithm in composite fields. Taking the isomorphic mapping into consideration, the complexity of the AddRoundKey does not change if the isomorphic mapping of the key is precomputed, while the complexity of MixColumns and InvMixColumns can be much higher. For example, using the isomorphic mapping matrix in Equation 3.7, the constant multiplications by $\{02\}_{16}$ and $\{03\}_{16}$ in the MixColumns transformation are mapped to the constant multiplications by $\{5f\}_{16}$ and $\{5e\}_{16}$, respectively. Compared with $\{02\}_{16}$ and $\{03\}_{16}$, there are more nonzero bits in $\{5f\}_{16}$ and $\{5e\}_{16}$. In addition, the nonzero bits in $\{5f\}_{16}$ and $\{5e\}_{16}$ have higher weights. Therefore, the constant multiplications by $\{5f\}_{16}$ and $\{5e\}_{16}$ require larger area and have longer critical path. It is the same case for the constant multiplications involved in the InvMixColumns transformation. Although the isomorphic mapping matrix changes with the irreducible polynomials used for composite field construction, generally the constant multiplications in the MixColumns and InvMixColumns are mapped to more complicated multiplications in the composite field. Therefore, it is more efficient to carry out only the multiplicative inversion in the SubBytes and InvSubBytes in the composite field. In this case, the construction scheme selected by Rudra et al. [14] is no longer optimum.

Optimum composite field construction schemes have been discussed in [17–19]. The approach in [18] only considers the cases when the value of ϕ in Equation 3.6 is $\{10\}_2$, and the optimum construction is selected based on the number of nonzero entries in the isomorphic mapping matrices. However, applying substructure sharing, the matrix with the least number of nonzero entries does not always lead to minimum gate count. The approach in [19] optimizes for overall area requirement. In addition, this work proposed to use normal basis representation for finite field elements such that the sharing of one extra common operand between $GF((2^2)^2)$ multipliers is enabled. Nevertheless, the critical path issue is not considered. Zhang and Parhi [17] introduced possible schemes to construct the composite fields for the AES algorithm by using irreducible polynomials in the form of Equation 3.6. The complexities of field operations depend on the coefficients of the field polynomials. Results are provided in [17] on how these coefficients affect the complexity of each subfield operation involved in the composite field implementation of the multiplicative inversion. In addition, for each construction scheme, there exist multiple isomorphic mappings with various complexities. Based on the complexities of both the subfield operations and the isomorphic mapping, the optimum constructions of the composite field for the AES algorithm can be selected to minimize gate count and critical path.

The composite field for $GF(2^8)$ can be constructed iteratively from GF(2) by using irreducible polynomials other than those in the form of Equation 3.6. However, in case the coefficients for x in $P_1(x)$ and $P_2(x)$ are not identity, higher hardware complexity is required. Therefore, we only consider the irreducible polynomials in the form of Equation 3.6 in the construction of the composite field. $P_0(x)$ is the only degree two irreducible polynomial over GF(2). Hence, it is the only choice for constructing $GF(2^2)$ from GF(2). The values of $\phi \in GF(2^2)$ and $\lambda \in GF((2^2)^2)$ need to satisfy that $P_1(x)$ is irreducible over $GF(2^2)$ and $P_2(x)$ is irreducible over $GF((2^2)^2)$.

According to the definition, a polynomial is irreducible if it cannot be factored into nontrivial polynomials over the same field. For a degree two polynomial $F(x)$ over $GF(2^q)$, if it can be factored into nontrivial polynomials, it must be factored into the form of $F(x) = (x + \mu)(x + \nu)$, where $\mu, \nu \in GF(2^q)$. Hence, for a given value ϕ, the testing of whether $P_1(x)$ is irreducible can be done by examining if any elements of $GF(2^2)$ are roots of $P_1(x)$. In addition, if an element $\tau \in GF(2^2)$ is a root of $P_1(x)$, then $\tau^2 + \tau + \phi = 0$. Hence, alternatively, the testing of irreducibility can be done by evaluating $x^2 + x$ over all elements of $GF(2^2)$. A list of the evaluation results can be derived. The values of ϕ, which make $P_1(x)$ irreducible over $GF(2^2)$, consist of all the elements of $GF(2^2)$ not equaling any of the evaluation results. Using this scheme, it can be derived that the only values of ϕ that make $x^2 + x + \phi$ irreducible over $GF(2^2)$ are $\phi = \{10\}_2$ and $\phi = \{11\}_2$. Depending on the value of ϕ, the elements of $GF((2^2)^2)$ can be represented differently. Using the same irreducibility testing scheme, it can be derived that there are eight possible values of λ that make $P_2(x)$ irreducible over $GF((2^2)^2)$ when $GF(2^2)$ is constructed by using either $\phi = \{10\}_2$ or $\phi = \{11\}_2$. These values of λ are

$$\begin{aligned} \lambda &= \{1000\}_2, & \lambda &= \{1100\}_2, \\ \lambda &= \{1001\}_2, & \lambda &= \{1101\}_2, \\ \lambda &= \{1010\}_2, & \lambda &= \{1110\}_2, \\ \lambda &= \{1011\}_2, & \lambda &= \{1111\}_2. \end{aligned} \tag{3.18}$$

Altogether, there are $2 \times 8 = 16$ ways to construct $GF(((2^2)^2)^2)$ by using irreducible polynomials in the form of Equation 3.6.

The values of ϕ and λ may affect the complexity of the composite field implementation of the multiplicative inversion. Next, we analyze how the complexity of each involved subfield operation changes with ϕ and λ.

1. $(\times\phi)$ block
   Using standard basis representation, $c \in GF(2^2)$ can be written as $c_1 y + c_0$, where $c_1, c_0 \in GF(2)$ and y is a root of $P_0(x)$. Similarly, $\phi = \{11\}_2$ can be written as $y + 1$. Hence,

$$c \times \phi = (c_1 y + c_0)(y + 1) = c_1 y^2 + (c_0 + c_1)y + c_0 = c_0 y + (c_0 + c_1).$$

Accordingly, the constant multiplier $(\times\phi)$ can be implemented by one XOR gate when $\phi=\{11\}_2$. From Equation 3.13, the constant multiplication by $\phi=\{10\}_2$ also takes one XOR gate. Therefore, the complexity of the $(\times\phi)$ block is the same for the two possible values of ϕ.

2. Multiplier in GF(2^2) and GF($(2^2)^2$)
   From Figure 3.8a and Figure 3.8b, it can be observed that the complexity of the multiplier in GF(2^2) is independent of the values of ϕ and λ. In addition, the only block in the GF($(2^2)^2$) multiplier that might be affected by the values of ϕ and λ is the $(\times\phi)$ block. From the previous discussion, the complexity of the $(\times\phi)$ block is the same for the two possible values of ϕ. Therefore, the complexity of the multiplier in GF($(2^2)^2$) does not change with the construction of the composite field.
3. Squarer in GF(2^4) and the $(\times\lambda)$ blocks
   The squarer in GF(2^4) and the $(\times\lambda)$ block can be combined to reduce hardware complexity. Hence, we consider the effects of ϕ and λ on the complexity of the combined block. Assume that the input to the squarer is a_hx+a_h and λ can be expressed by $\lambda_hx+\lambda_l$ ($a_h, a_l, \lambda_h, \lambda_l \in$ GF(2^2)). Then the output of the combined block, b_hx+b_l, can be computed as

$$\begin{aligned} b_hx+b_l &= (a_hx+a_l)^2(\lambda_hx+\lambda_l) \\ &= (a_h^2x^2+a_l^2)(\lambda_hx+\lambda_l) \\ &= (a_h^2x+(a_l^2+a_h^2\phi))(\lambda_hx+\lambda_l) \\ &= a_h^2\lambda_hx^2+(a_h^2\lambda_l+(a_l^2+a_h^2\phi)\lambda_h)x+(a_l^2+a_h^2\phi)\lambda_l \\ &= (a_h^2(\lambda_l+\lambda_h)+(a_l^2+a_h^2\phi)\lambda_h)x+(a_l^2\lambda_l+a_h^2\phi(\lambda_l+\lambda_h)). \end{aligned} \tag{3.19}$$

   Hence, two values need to be computed in the combined squarer and $(\times\lambda)$ block

$$\begin{aligned} b_h &= a_h^2(\lambda_l+\lambda_h)+(a_l^2+a_h^2\phi)\lambda_h, \\ b_l &= a_l^2\lambda_l+a_h^2\phi(\lambda_l+\lambda_h). \end{aligned} \tag{3.20}$$

   Based on Equation 3.20, expressions can be derived for each bit in b_h and b_l. Canceling common terms and applying substructure sharing, the gate number needed for the implementation of the combined squarer and $(\times\lambda)$ block are listed in Table 3.3 for each possible value of ϕ and λ. In addition, the critical path for each implementation is two XOR gates.
4. Multiplicative inversion in GF(2^4)
   The equations for directly computing the multiplicative inverse in GF(2^4) can be derived from Figure 3.9b. As it can be observed from

TABLE 3.3
Gate Count for Combined Squarer and $(\times\lambda)$ Implementation

ϕ	λ	Gate Number	ϕ	λ	Gate Number
	$\{1000\}_2$	3 XOR		$\{1000\}_2$	3 XOR
	$\{1001\}_2$	3 XOR		$\{1001\}_2$	4 XOR
	$\{1010\}_2$	3 XOR		$\{1010\}_2$	3 XOR
$\{10\}_2$	$\{1011\}_2$	4 XOR	$\{11\}_2$	$\{1011\}_2$	3 XOR
	$\{1100\}_2$	4 XOR		$\{1100\}_2$	3 XOR
	$\{1101\}_2$	5 XOR		$\{1101\}_2$	4 XOR
	$\{1110\}_2$	3 XOR		$\{1110\}_2$	5 XOR
	$\{1111\}_2$	4 XOR		$\{1111\}_2$	3 XOR

this figure, the complexity of the inversion in GF(2^4) is only dependent on ϕ. In the case of $\phi = \{11\}_2$, the bits in $s^{-1} = \{s_3^{-1}, s_2^{-1}, s_1^{-1}, s_0^{-1}\}$ can be computed by the following equation:

$$
\begin{aligned}
s_3^{-1} &= s_2 + s_0s_3 + s_1s_2s_3, \\
s_2^{-1} &= s_3 + s_0s_3 + s_1s_2 + s_0s_2s_3 + s_1s_2s_3, \\
s_1^{-1} &= s_1 + s_2 + s_0s_2 + s_0s_3 + s_1s_2 + s_1s_3 + s_1s_2s_3 + s_0s_1s_3, \\
s_0^{-1} &= s_0 + s_1 + s_3 + s_0s_2 + s_0s_3 + s_1s_2 + s_0s_1s_2 \\
&\quad + s_0s_1s_3 + s_0s_2s_3 + s_1s_2s_3.
\end{aligned}
\tag{3.21}
$$

Applying substructure sharing, Equation 3.21 can be implemented by 14 XOR gates and 8 AND gates with 3 XOR gates and 2 AND gates in the critical path. Compared to the complexity in the case of $\phi = \{10\}_2$, which is listed in Table 3.2, the complexity of the inversion in GF(2^4) is the same when $\phi = \{11\}_2$.

From the previous discussion, the only subfield operation whose complexity is affected by the values of ϕ and λ is the combined squarer and the $(\times\lambda)$ block. In addition, the complexity of the isomorphic mapping may also change with ϕ and λ. Isomorphic mappings are needed to map the elements in the original field to its composite field, such that both multiplicative and additive homomorphisms are preserved. For a fixed construction of the composite field, there exist multiple isomorphic mappings, and the complexities of these mappings vary. The entries of the isomorphic mapping matrices are decided by the irreducible polynomials used in the construction of the original fields and the composite fields, as well as to which elements of the composite field the base elements in the original field are mapped.

Assume the set $\{1, \alpha, \alpha^2, \ldots, \alpha^7\}$ forms a standard basis for GF(2^8), where α is a root of $P(x) = x^8 + x^4 + x^3 + x + 1$. The basic idea of finding an isomorphic mapping between GF(2^8) and GF($((2^2)^2)^2$) is to find eight elements $1, \beta, \beta^2, \ldots, \beta^7$ of GF($((2^2)^2)^2$), to which the base elements $1, \alpha, \alpha^2, \ldots, \alpha^7$ are mapped. Then the jth column of the isomorphic mapping matrix is formed by the binary vector representation of $\beta^{(8-j)}$. Additive homomorphism always holds for arbitrary mapping matrices. Assume $a,b,c \in$ GF(2^8) and δ is an 8×8 binary matrix. If $a = b + c$, then $\delta \underline{a} = \delta(\underline{b} + \underline{c}) = \delta \underline{b} + \delta \underline{c}$, where a, b, and c are the column vectors formed by the bits in the standard basis representation of a, b, and c, respectively. Hence, additive homomorphism does not add any constraints to the isomorphic mapping matrices. However, in order for the multiplicative homomorphism to hold, α cannot be mapped to any β. Instead $P(\beta) = 0$ needs to be satisfied [11]. Such a β can be found by exhaustive search. Nevertheless, this approach has very high complexity. One property of finite field elements is that if β is not a root of $P(x)$, then none of the conjugates of β are roots of $P(x)$. This property can be employed to reduce the number of trials in the searching for β. Accordingly, the values of $\beta \in$ GF(2^q) satisfying $P(\beta) = 0$ can be found by the algorithm described by the pseudocodes listed in Algorithm 1 [17].

Algorithm 1

Initialization: $t = 1$, stop $= 0$, flag(i) $= 0$ for $i = 1, 2, \ldots, 2^q - 1$

while stop $==$ 0

{

$\omega =$ dectobin(t,q)

compute $P(\omega)$

if $P(\omega) ==$ 0

    mapping found, output $\beta = \omega$

    stop $= 1$

else

    index (j) $=$ bintodec(ω^{2^j}), for $j = 0, 1, 2, \ldots, q - 1$

    flag(index(j)) $= 1$, for $j = 0, 1, 2, \ldots, q - 1$

find the minimum integer $l > t$, such that flag(l) $= 0$

$t = l$

}

Algorithm 1 is based on an algorithm proposed in [11], which only considers the cases when α is a primitive element. However, since the irreducible polynomial $P(x) = x^8 + x^4 + x^3 + x + 1$ specified by the AES algorithm is not primitive, its root, α, is not primitive. Algorithm 1 includes the testing for nonprimitive elements. In Algorithm 1, $\omega =$ dectobin(t,q) converts the integer t to a q-bit binary vector and takes this vector as the standard basis representation for ω in the composite field. Similarly, $t =$ bintodec(ω)

implements the reverse function. In addition, the evaluation of $P(x)$ on ω is carried out based on the operations specified in the composite field. If ω is not a root of $P(x)$, then none of the conjugates of ω, which are ω^{2^j} for $j = 1, 2, \ldots, q-1$, are roots of $P(x)$. By setting the flags for these elements, they are excluded from the next checking. Similarly, if an element is a root of $P(x)$, then all its conjugates are roots of $P(x)$, and α can be mapped to any of them. The number of elements in a conjugacy class must be a divisor of q. It turns out for each combination of ϕ and λ there exist eight isomorphic mappings for GF(2^8).

Example 1 Find the isomorphic mapping matrices between GF(2^8) constructed by $P(x) = x^8 + x^4 + x^3 + x + 1$ and GF($((2^2)^2)^2$) constructed by Equation 3.6 with $\phi = \{11\}_2$ and $\lambda = \{1010\}_2$.

We start with $t = 1$. In this case, $\omega = \{00000001\}_2$ and it can be computed that $P(\omega) = 1 \neq 0$. Hence, $\{00000001\}_2$ is not an element α can be mapped to. $\{00000001\}_2$ does not have any other conjugates. Therefore, no other element can be excluded from the checking as a result of this iteration.

Next, we consider $t = 2$. The corresponding ω is $\{00000010\}_2$. Following the operations of the composite field, it can be computed that

$$\omega^3 = \{00000001\}_2,$$
$$\omega^4 = \{00000010\}_2,$$
$$\omega^8 = \{00000011\}_2.$$

Therefore, $P(\omega) = \{00000011\}_2 \neq 0$ for $t = 2$. The only other element in the same conjugacy class as $\{00000010\}_2$ is $\{00000011\}_2$. Hence, $t = 3$ can be excluded from the checking.

Since $t = 3$ is excluded from checking, the next value that needs to be checked is $t = 4$ with the corresponding ω equals $\{00000100\}_2$. In this case, it can be computed that $P(\{00000100\}_2) = \{00001101\}_2$. The other elements in the same conjugacy class as $\{00000100\}_2$ are $\{00000111\}_2$, $\{00000101\}_2$, and $\{00000110\}_2$. Therefore, $t = 5, 6, 7$ can be excluded from checking.

The process is carried on until $t = 72$, whose corresponding ω is $\{01001000\}_2$. In this case,

$$\omega^3 = \{01110101\}_2,$$
$$\omega^4 = \{01010110\}_2,$$
$$\omega^8 = \{01101010\}_2.$$

It can be computed that $P(\{01001000\}_2) = 0$. Hence, α can be mapped to $\beta = \{01001000\}_2$, and the corresponding isomorphic mapping matrix is

$$\delta = \begin{bmatrix} 1 & 0 & 1 & 0 & 0 & 0 & 0 & 0 \\ 0 & 1 & 1 & 1 & 1 & 1 & 1 & 0 \\ 0 & 0 & 0 & 0 & 1 & 1 & 0 & 0 \\ 0 & 1 & 1 & 1 & 1 & 1 & 0 & 0 \\ 1 & 1 & 0 & 0 & 0 & 1 & 1 & 0 \\ 0 & 1 & 0 & 1 & 1 & 0 & 0 & 0 \\ 1 & 0 & 0 & 1 & 0 & 1 & 0 & 0 \\ 0 & 1 & 0 & 0 & 1 & 0 & 0 & 1 \end{bmatrix}. \quad (3.22)$$

α can be also mapped to the conjugates of $\{01001000\}_2$, which are $\{01101010\}_2$, $\{01001100\}_2$, $\{01111101\}_2$, $\{01010011\}_2$, $\{01111010\}_2$, $\{01101100\}_2$, and $\{01010110\}_2$. A different isomorphic mapping matrix can be derived for each conjugate.

For each combination of ϕ and λ, there are eight isomorphic mapping matrices, and the optimum one can be selected based on minimum gate number and shortest critical path. The complexities for the optimum isomorphic mappings, as well as combined inverse mapping and affine are listed in Table 3.4 for each combination of ϕ and λ. These numbers are derived after applying substructure sharing.

TABLE 3.4
Complexity of Optimum Isomorphic Mapping and Inverse

		Isomorphic Mapping		Inverse Mapping + Affine	
ϕ	λ	Gate Count	Critical Path	Gate Count	Critical Path
$\{10\}_2$	$\{1000\}_2$	11 XOR	3 XOR	16 XOR	3 XOR
	$\{1001\}_2$	10 XOR	3 XOR	19 XOR	3 XOR
	$\{1010\}_2$	13 XOR	3 XOR	16 XOR	5 XOR
	$\{1011\}_2$	15 XOR	3 XOR	16 XOR	3 XOR
	$\{1100\}_2$	11 XOR	3 XOR	18 XOR	5 XOR
	$\{1101\}_2$	13 XOR	4 XOR	16 XOR	3 XOR
	$\{1110\}_2$	12 XOR	4 XOR	17 XOR	3 XOR
	$\{1111\}_2$	11 XOR	5 XOR	19 XOR	3 XOR
$\{11\}_2$	$\{1000\}_2$	11 XOR	5 XOR	17 XOR	4 XOR
	$\{1001\}_2$	12 XOR	3 XOR	17 XOR	3 XOR
	$\{1010\}_2$	11 XOR	3 XOR	17 XOR	3 XOR
	$\{1011\}_2$	11 XOR	3 XOR	17 XOR	3 XOR
	$\{1100\}_2$	11 XOR	3 XOR	18 XOR	3 XOR
	$\{1101\}_2$	13 XOR	3 XOR	16 XOR	3 XOR
	$\{1110\}_2$	11 XOR	3 XOR	18 XOR	4 XOR
	$\{1111\}_2$	12 XOR	3 XOR	17 XOR	3 XOR

Source: From Zhang, X. and Parhi, K.K., *IEEE Trans. Circuits Syst. II*, submitted, 53(10), 1157, October 2006. With permission.

In the selection of the optimum constructions of the composite fields for the AES algorithm, both the complexity of the involved subfield operations and that of the isomorphic mapping need to be considered. The only subfield operation whose complexity changes with λ or ϕ is the combined squarer and $(\times\lambda)$ computation. Adding up the corresponding gate counts in Table 3.3 and Table 3.4, it can be derived that the construction using $\phi = \{10\}_2$ and $\lambda = \{1000\}_2$ is optimum. For this construction, the lowest complexity of isomorphic mapping and inverse can be achieved when the root of $P(x)$ is mapped to $\beta = \{01111010\}_2$. In addition, this construction also leads to the shortest critical path.

3.4.2 Implementations of MixColumns and InvMixColumns

Architectures for the implementations of the MixColumns and InvMixColumns have been proposed in [4,5,7,13,20]. In the architecture presented in [7], MixColumns and InvMixColumns are implemented according to the bit-level expressions derived for the involved constant multiplications. The complexity of this approach can be reduced by applying substructure sharing to the bit-level expressions [4]. However, in this approach, it is very hard to find the terms that can be shared among the computations of different bytes in a column of the State. Alternatively, substructure sharing can be applied in byte level [5,13,20]. Any constant multiplication can be decomposed into multiplications by integer powers of two. Hence, the multiplications by $\{02\}_{16}, \{04\}_{16}, \{08\}_{16}$ can be first computed and shared among the constant multiplications in the MixColumns and InvMixColumns. The architecture in [5,13] can achieve the lowest gate count and shortest critical path. The efficiency of this architecture comes from applying substructure sharing to both the computation of a byte and among the computations of the four bytes in a column of the State.

In MixColumns, the constant multiplications by $\{02\}_{16}$ and $\{03\}_{16}$ need to be implemented. An element of GF(2^8) can be represented in standard basis as $S(x) = s_7x^7 + s_6x^6 + s_5x^5 + s_4x^4 + s_3x^3 + s_2x^2 + s_1x + s_0$, where $s_0, s_1, \ldots, s_7 \in$ GF(2), and x is a root of the field polynomial $P(x)$. Hence, $\{02\}_{16}$ can be expressed as x and

$$\begin{aligned}\{02\}_{16}S = xS &= s_7x^8 + s_6x^7 + s_5x^6 + s_4x^5 + s_3x^4 + s_2x^3 + s_1x^2 + s_0x \bmod P(x)\\ &= s_6x^7 + s_5x^6 + s_4x^5 + (s_3 + s_7)x^4 + (s_2 + s_7)x^3 + s_1x^2 + (s_0 + s_7)x + s_7.\end{aligned}$$

Therefore, the constant multiplication by $\{02\}_{16}$ can be implemented by three XOR gates with one XOR gate in the critical path. Once $\{02\}_{16}S$ has been computed, $\{03\}_{16}S$ can be computed as $\{02)_{16}S + S$. To apply substructure sharing, Equation 3.2 can be rewritten as

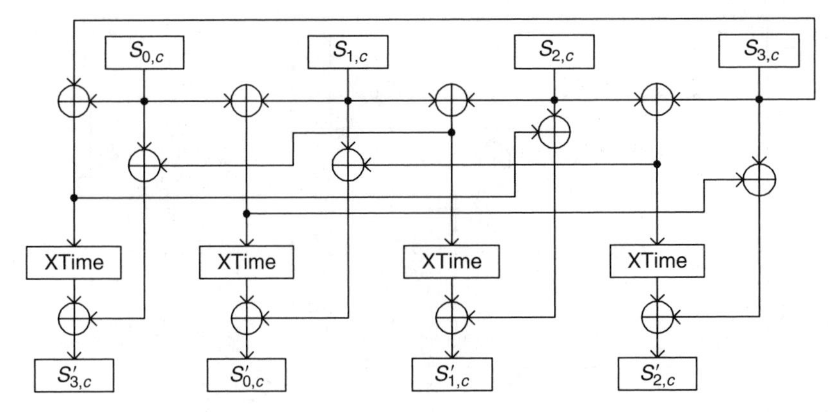

FIGURE 3.10 An efficient implementation of the MixColumns transformation. (From Zhang, X. and Parhi, K.K., *IEEE Trans. VLSI Syst.*, 12(9), 963, 2004. With permission.)

$$\begin{aligned} S'_{0,c} &= \{02\}_{16}(S_{0,c} + S_{1,c}) + (S_{2,c} + S_{3,c}) + S_{1,c}, \\ S'_{1,c} &= \{02\}_{16}(S_{1,c} + S_{2,c}) + (S_{3,c} + S_{0,c}) + S_{2,c}, \\ S'_{2,c} &= \{02\}_{16}(S_{2,c} + S_{3,c}) + (S_{0,c} + S_{1,c}) + S_{3,c}, \\ S'_{3,c} &= \{02\}_{16}(S_{3,c} + S_{0,c}) + (S_{1,c} + S_{2,c}) + S_{0,c}. \end{aligned} \tag{3.23}$$

According to Equation 3.23, the Mixcolumns transformation can be implemented by the architecture illustrated in Figure 3.10 [5]. In this figure, the function of the XTime is to implement the constant multiplication by $\{02\}_{16}$. It follows that the MixColumns can be implemented by 108 XOR gates with 3 XOR gates in the critical path.

The computations in the InvMixColumns are more complicated. Equation 3.4 can be rewritten as the equations listed below to facilitate substructure sharing.

$$\begin{aligned} S'_{0,c} =& (\{02\}_{16}(S_{0,c}+S_{1,c})+(S_{2,c}+S_{3,c})+S_{1,c}) \\ &+(\{02\}_{16}(\{04\}_{16}(S_{0,c}+S_{2,c})+\{04\}_{16}(S_{1,c}+S_{3,c}))+\{04\}_{16}(S_{0,c}+S_{2,c})), \\ S'_{1,c} =& (\{02\}_{16}(S_{1,c}+S_{2,c})+(S_{3,c}+S_{0,c})+S_{2,c}) \\ &+(\{02\}_{16}(\{04\}_{16}(S_{0,c}+S_{2,c})+\{04\}_{16}(S_{1,c}+S_{3,c}))+\{04\}_{16}(S_{1,c}+S_{3,c})), \\ S'_{2,c} =& (\{02\}_{16}(S_{2,c}+S_{3,c})+(S_{0,c}+S_{1,c})+S_{3,c}) \\ &+(\{02\}_{16}(\{04\}_{16}(S_{0,c}+S_{2,c})+\{04\}_{16}(S_{1,c}+S_{3,c}))+\{04\}_{16}(S_{0,c}+S_{2,c})), \\ S'_{3,c} =& (\{02\}_{16}(S_{3,c}+S_{0,c})+(S_{1,c}+S_{2,c})+S_{0,c}) \\ &+(\{02\}_{16}(\{04\}_{16}(S_{0,c}+S_{2,c})+\{04\}_{16}(S_{1,c}+S_{3,c}))+\{04\}_{16}(S_{1,c}+S_{3,c})). \end{aligned} \tag{3.24}$$

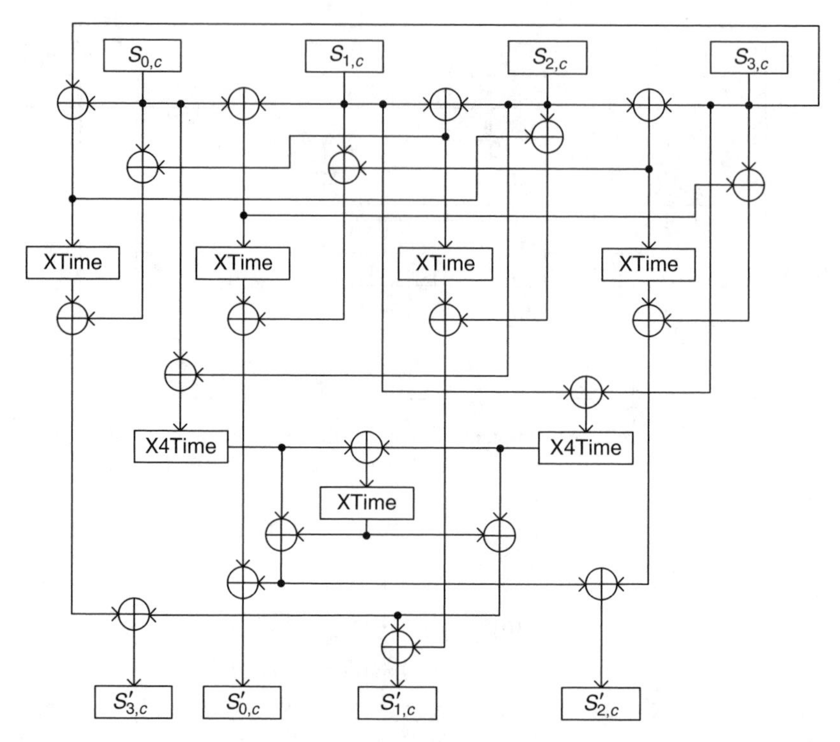

FIGURE 3.11 An efficient implementation of the InvMixColumns transformation. (From Zhang, X. and Parhi, K.K., *IEEE Trans. VLSI Syst.*, 12(9), 963, 2004. With permission.)

According to Equation 3.24, the InvMixColumns can be implemented by the architecture illustrated in Figure 3.11 [5]. The function of the X4Time block in this figure is to compute the constant multiplication by $\{04\}_{16}$. This block can be implemented by two serially concatenated XTime blocks, which consist of six XOR gates. Alternatively, bit-level expression can be directly derived for this multiplication. In polynomial form, $\{04\}_{16}$ can be written as x^2. Hence,

$$
\begin{aligned}
\{04\}_{16}S = x^2S &= s_7x^9 + s_6x^8 + s_5x^7 + s_4x^6 + s_3s^5 + s_2x^4 + s_1x^3 + s_0x^2 \bmod P(x) \\
&= s_5x^7 + s_4x^6 + (s_3 + s_7)x^5 + (s_2 + (s_6 + s_7))x^4 + (s_1 + s_6)x^3 \\
&\quad + (s_0 + s_7)x^2 + (s_6 + s_7)x + s_6.
\end{aligned}
$$

In the above equation, the term $s_6 + s_7$ can be shared. Hence, the constant multiplication by $\{04\}_{16}$ can be implemented by five XOR gates with two XOR gates in the critical path. Therefore, it can be derived from Figure 3.11 that the InvMixColumns can be implemented by 193 XOR gates with 7 XOR gates in the critical path.

3.4.3 Implementations of Key Expansion

Roundkeys can be either generated beforehand and stored in memory or generated on the fly. In the former approach, a memory of Nr $\times$ 128 bits is required to store all the roundkeys. The roundkeys can be read out from the memory by using proper addresses when they are needed. In addition, there is no extra latency associated with the roundkey generation in the decryption process. However, this approach is not suitable for the applications where the key changes from time to time. Furthermore, the unbreakable delay of reading the roundkeys out of memory may offset the speedup achieved by the subpipelined round units.

Figure 3.12 illustrates a key expansion architecture suitable for subpipelined AES algorithm with 128-bit key. At the "start" signal, the initial key is loaded into the registers in the first column with the least significant bit in the

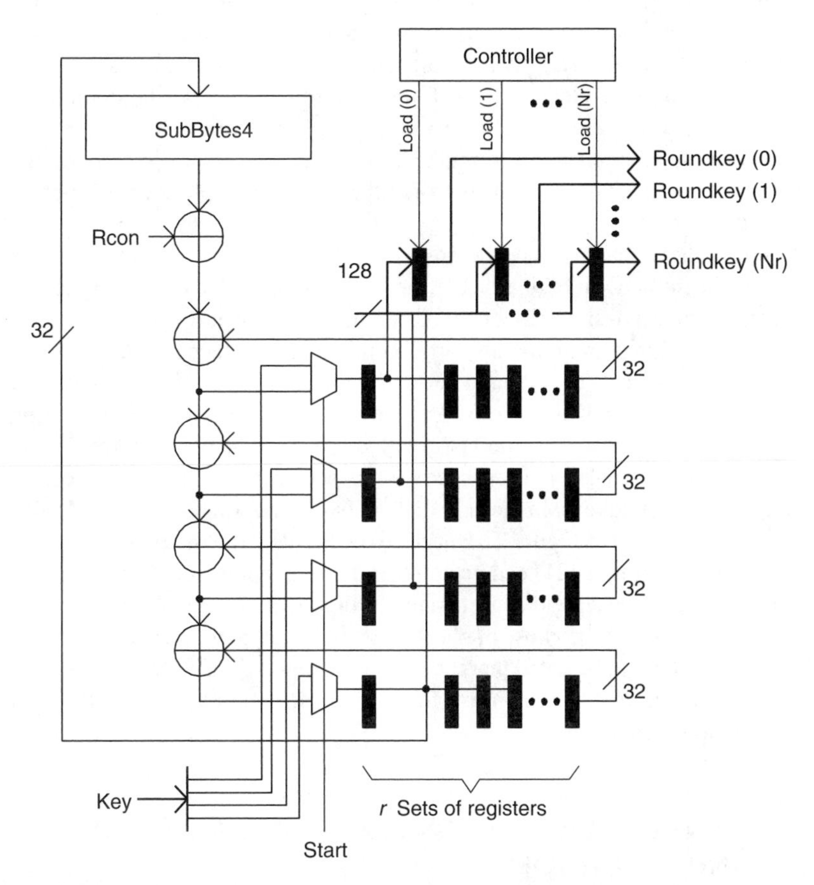

FIGURE 3.12 The key expansion architecture for r-substage subpipelined AES algorithm with 128-bit key. (From Zhang, X. and Parhi, K.K., *IEEE Trans. VLSI Syst.*, 12(9), 964, 2004. With permission.)

top register, and the key expansion process begins. From the key expansion algorithm in Figure 3.4, the computation of every fourth word needs to go through the Rotword and SubWord function. Hence, the output from the bottom 32-bit register is fed into the SubBytes4 block, which consists of four copies of the SubBytes architecture as illustrated in Figure 3.7. Since the RotWord can be implemented by switching the wire connections, it is not explicitly shown in Figure 3.12. Assume that the critical path of this key expansion architecture, which consists of the SubBytes4 block, five XOR gates, and one multiplexer, is divided into r substages. Then the computation of the words "$w(i)$" in the key expansion algorithm needs to wait for r clock cycles until the value of "temp" is available. Hence, r sets of registers are inserted after each multiplexer. In this case, at clock cycle $r \times i$, the output of the registers in the first column is the corresponding "roundkey (i)." The controller in Figure 3.12 generates "load (i)" signals, which go to 1 in clock cycle $r \times i$, and stays at "1" afterward. Such a controller can be easily implemented by two serially concatenated Johnson counters. Using these load signals as the clock input of the top row registers, the roundkeys are loaded into the top row registers when they are generated. After $r \times \mathrm{Nr}$ clock cycles, all the $\mathrm{Nr} + 1$ roundkeys are available at the output of the top row registers and are held there for the entire encryption or decryption process.

When the roundkeys are generated on-the-fly by architectures such as the one illustrated in Figure 3.12, the encryption and the key expansion processes can start simultaneously. In addition, we need to divide the key expansion architecture into the same number of substages with the same maximum delay as in the round unit to avoid extra buffers and delay. In decryption, the roundkeys are used in reverse order. Hence, the decryption process can start only after the last roundkey is generated. Furthermore, the InvMixColumns transformation needs to be performed on the roundkeys to derive the mixroundkeys. In the case that the path consisting of five XOR gates and one multiplexer in Figure 3.12 needs to be divided into multiple substages, the retiming technique [21] can be employed. For example, the key expansion architecture can be retimed as illustrated in Figure 3.13 to break the path into two substages. For the purpose of clarity, the irrelevant parts are excluded from Figure 3.13. It might be noted that the number of registers in each row, r, equals the total number of substages in the SubBytes4 block and the five XOR gates and one multiplexer path.

3.5 JOINT IMPLEMENTATION ISSUES OF ENCRYPTORS AND DECRYPTORS

In the applications where both the encryptor and the decryptor need to be implemented in a small area, resource sharing between encryptors and

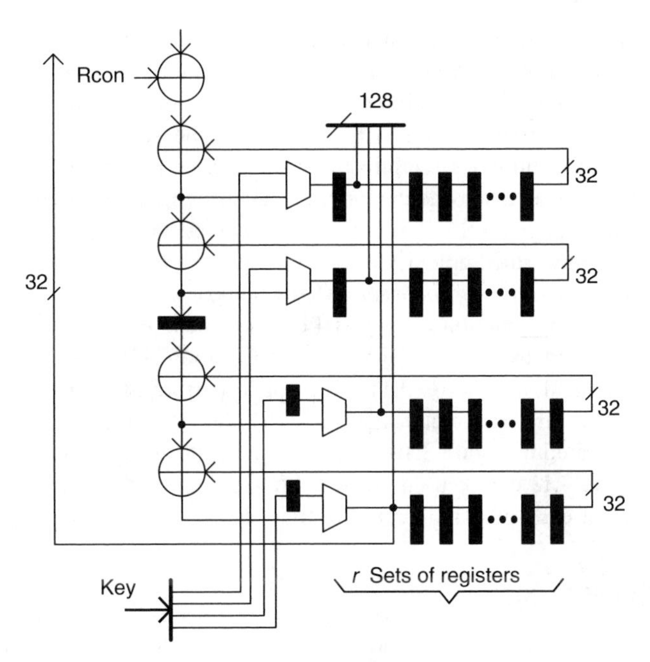

FIGURE 3.13 Retimed key expansion architecture.

decryptors becomes important. While algorithmic modifications described in the previous section can be employed to reduce area, more significant area savings can be achieved by sharing resources between encryptors and decryptors. Employing the equivalent decryption structure as illustrated in Figure 3.5, resource sharing between each of the corresponding transformations is enabled.

1. *Resource sharing between SubBytes and InvSubBytes.* Comparing Equation 3.1 and Equation 3.3, it can be derived that the SubBytes and InvSubBytes can share the multiplicative inverse computation. Accordingly, a joint SubBytes and InvSubBytes transformation can be implemented by the architecture illustrated in Figure 3.14. During encryption, the multiplexers select the top branches: the multiplicative inverse is computed for the input, then the affine transformation is carried out on the inverse value. The computations in the bottom branches are selected during decryption. In this case, the inverse affine

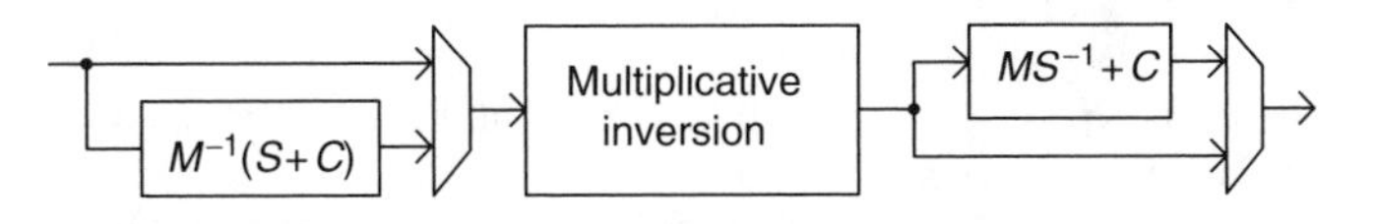

FIGURE 3.14 Joint implementation of SubBytes and InvSubBytes.

transformation is first carried out on the input, then the result is input to the multiplicative inversion block. The multiplicative inversion can be implemented by either LUTs or the composite field arithmetic approach as illustrated in Figure 3.7. In the latter approach, as discussed in previous sections, $M \times \delta^{-1}$ can be precomputed, such that the inverse isomorphic mapping can be combined with the affine transformation. Similarly, $\delta \times M^{-1}$ can be precomputed to reduce the hardware complexity of decryption.

2. *Resource sharing between MixColumns and InvMixColumns.* As shown in Equation 3.23 and Equation 3.24, Equation 3.23 can be computed as the first part of the Equation 3.24. Accordingly, as illustrated in Figure 3.11, the upper part of the InvMixColumns architecture is exactly the same as the MixColumns architecture illustrated in Figure 3.10. Therefore, a single architecture as shown in Figure 3.11 can be used for both MixColumns and InvMixColumns in a joint encryptor and decryptor.

3.6 CONCLUSION

Architectural and algorithmic optimization approaches for efficient hardware implementations of the AES algorithm have been addressed in this chapter. Among the three architectural-level optimization techniques, subpipelining can achieve the highest speed and optimum speed over area ratio. In addition, in a subpipelined architecture, speed and area trade-offs can be easily achieved by changing the number of round units and the number of substages in each round unit. In order to reduce hardware complexity and enable deep subpipelining, composite field arithmetic can be employed to implement the multiplicative inversion. Furthermore, this chapter analyzed how the complexities of the involved subfield operations and the isomorphic mapping change with the coefficients of the irreducible polynomials used for field construction. Another algorithmic-level optimization technique that can be employed is substructure sharing. This technique is applied whenever possible to further reduce the area requirement. Joint implementations of encryptors and decryptors were also discussed in this chapter. Employing the equivalent decryption structure, the SubBytes and InvSubBytes can share a multiplicative inversion block, and a single InvMixColumns architecture can be used to implement both the MixColumns and InvMixColumns.

REFERENCES

1. Advanced Encryption Standard (AES), Federal Information Processing Standards Publication 197, November 26, 2001.
2. Elbirt, A.J. et al., An FPGA implementation and performance evaluation of the AES block cipher candidate algorithm finalist, *Proceedings of the Third Advanced*

Encryption Standard (AES) Candidate Conference, New York, April 2000, pp. 13–27.
3. Gaj, K. and Chodowiec, P., Comparison of the hardware performance of the AES candidates using reconfigurable hardware, *Proceedings of the Third Advanced Encryption Standard (AES) Candidate Conference*, New York, April 2000, pp. 40–56.
4. Zhang, X. and Parhi, K.K., Implementation approaches for the advanced encryption standard algorithm, *IEEE Circuits and Systems Magazine*, 2(4), 24–46, 2002.
5. Zhang, X. and Parhi, K.K., High-speed VLSI architecture for the AES algorithm, *IEEE Transactions on VLSI Systems*, 12(9), 957–967, 2004.
6. McLoone, M. and McCanny, J.V., High performance single-chip FPGA Rijndael algorithm implementation, *Proceedings of Cryptographic Hardware and Embedded Systems 2001*, Paris, France, May 2001, pp. 65–76.
7. Fischer, V. and Drutarovsky, M., Two methods of Rijndael implementation in reconfigurable hardware, *Proceedings of CHES 2001*, Paris, France, May 2001, pp. 77–92.
8. Kuo, H. and Verbauwhede, I., Architectural optimization for a 1.82 Gbits/sec VLSI implementation of the AES Rijndael algorithm, *Proceedings of Cryptographic Hardware and Embedded Systems 2001*, Paris, France, May 2001, pp. 51–64.
9. Standaert, F. et al., Efficient implementation of Rijndael encryption in reconfigurable hardware: improvements and design tradeoffs, *Proceedings of Cryptographic Hardware and Embedded Systems 2003*, Cologne, Germany, September 2003, pp. 334–350.
10. Saggese, G.P. et al., An FPGA based performance analysis of the unrolling, tiling and pipelining of the AES algorithm, *Proceedings of FPL 2003*, Portugal, September 2003.
11. Paar, C., Efficient VLSI Architecture for Bit-Parallel Computations in Galois Field, Ph.D. thesis, Institute for Experimental Mathematics, University of Essen, Germany, 1994.
12. Rijmen, V., Efficient implementation of the Rijndael S-box, Available at http://homes.esat.kuleuven.be/rijmen/rijndael/sbox.pdf
13. Satoh, A. et al., A compact Rijndael hardware architecture with S-box optimization, *Proceedings of ASIACRYPT 2001*, Gold Coast, Australia, December 2000, pp. 239–254.
14. Rudra, A. et al., Efficient implementation of Rijndael encryption with composite field arithmetic, *Proceedings of Cryptographic Hardware and Embedded Systems 2001*, Paris, France, May 2001, pp. 171–184.
15. Jarvinen, K.U., Tommiska, M.T., and Skytta, J.O., A fully pipelined memoryless 17.8 Gbps AES-128 encryptor, *Proceedings of International Symposium on Field-Programmable Gate Arrays (FPGA 2003)*, Monterey, CA, February 2003, pp. 207–215.
16. Wolkerstorfer, J., Oswald, E., and Lamberger, M., An ASIC implementation of the AES S-boxes, *Proceedings of the RSA Conference*, San Jose, CA, February 2002, pp. 67–78.
17. Zhang, X. and Parhi, K.K., On the optimum construction of the composite field for the AES algorithm, Submitted to *IEEE Transactions on Circuits and Systems II*, 53(10), pp. 1153–1157, October 2006.

18. Mentens, N. et al., A systematic evaluation of compact hardware implementations for the Rijndael S-box, *Proceedings of Topics in Cryptology—CT-RSA 2005*, San Francisco, CA, February 2005, pp. 323–333.
19. Canright, D., A very compact S-box for AES, *Proceedings of Cryptographic Hardware and Embedded Systems 2005*, Edinburgh, UK, September 2005, pp. 441–455.
20. Lu, C.C. and Tseng, S.Y., Integrated design of AES (advanced encryption standard) encrypter and decrypter, *Proceedings of the IEEE International Conference on Application-Specific Systems, Architectures and Processors*, 2002, pp. 277–285.
21. Parhi, K.K., *VLSI Digital Signal Processing Systems, Design and Implementations*, Wiley & Sons, New York, 1999.

4 Hardware Design Issues in Elliptic Curve Cryptography for Wireless Systems

Apostolos P. Fournaris and O. Koufopavlou

CONTENTS

4.1 INTRODUCTION

The immense growth of wireless system applications in today's communication environment cannot be ignored. Wireless systems are gradually replacing many traditional communication systems because of the increasing need for mobility in high-end technology applications. However, wireless systems remain a relatively new trend in the communication world. A great deal of improvement can still be made to better the functionality of those systems. One still open issue in wireless mobile systems is security. Wireless security is an essential factor for every wireless system. However, the very constrained resource technological environment of a wireless system poses strict limits on the security that such a system can support. Because of this, wireless security in many cases is not considered adequate for enterprise needs. Attempts have already been made to construct a more secure wireless environment by using more recent and efficient cryptographic algorithms. Moreover, this requires an increase in computational performance, power, and memory, factors that are restrictive in wireless systems.

A cryptographic system should be able to provide the following for the involved entities:

- Confidentiality of two entities' transactions and data exchanges
- Authentication of each entity's identity and its data transferred through a communication channel
- Data integrity so that no unauthorized user can alter those data
- Nonrepudiation of an entity's identity so that its transactions are legally binding

In the communication world and especially wireless systems, cryptographic demands are satisfied by providing a personal certificate for each communicating entity, encrypting the transmitted message, and generating an appropriate key for initialization of this encrypted transaction and certificate generation. For message encryption–decryption, a fast cryptographic algorithm, usually a symmetric key stream algorithm, is required. For the other two operations, digital signature schemes are employed along with a corresponding key exchange suite. The certificates must be digitally signed by a trusted third-party certificate authority on receiving the public key of an entity. Then, using a public key infrastructure (PKI) implementation, the certificate authority can perform a set of operations including registration, certification, key generation, key pair recovery—update, and revocation [1,2]. For most of those operations, key pairs have to be generated and a digital signature algorithm has to be chosen.

Wireless LAN networks, such as the 802.11x series, and tools that allow access to the Internet through mobile wireless devices, such as wireless application protocol (WAP), require a serious security status. However, the existing proposed and used solutions on the security of those protocols are still inadequate. For example, with relatively small CPU computation capacity and network traffic analysis, the secret encoded wired equivalent privacy (WEP) keys of 802.11g protocol can be determined. WAP in the wireless transport layer security (WTLS) uses public key cryptography. Following the PKCS #1 standard, in WTLS, the 512-bit RSA public key exchange and Diffie–Hellman (DH) key pairs are employed for key establishment and certification. Additionally, in WTLS, for the first time in wireless systems, 113-bit elliptic curve digital signature algorithm (ECDSA) and elliptic curve DH key pairs are proposed [3]. As technology in cryptanalysis evolves, the key lengths of those algorithms will eventually increase. In the future, a secure RSA cryptosystem would require 1024- to 2048-bit keys, meaning that 1024- to 2048-bit numbers would have to be used for computationally demanding mathematical operations like modular multiplication and exponentiation inside the cryptographic calculations. The resulting computational cost of such a cryptosystem would be very high thus making this public key cryptosystem solution impractical.

The above remark highlights the major problem of public key cryptographic algorithms and especially RSA. The key's length in public key cryptography is big and the required mathematical calculations complex. A considerable amount of research is done on simplifying the mathematical algorithms for achieving better performance of public key cryptographic operations, with very promising results. However, the key-length problem has no solution when traditional public key cryptographic algorithms, such as RSA or El Gamal, are employed.

Recently, Koblitz [4] and Miller [5] have proposed a different solution to the above public key problem, elliptic curve cryptography (ECC). When using elliptic curves for representing a plain text message (not encrypted), the required encryption key has a small length to achieve the same security level as that of other public key cryptographic algorithms. This major decrease in key length, shown in Table 4.1, is extremely useful in the wireless systems environment where the computation and memory resources are limited.

Table 4.1 shows such key-length comparisons where the security strength is evaluated by the required breaking time using the fastest known cryptanalytic methods (Pollard's rho method). For example, to achieve the security level of a 1024-bit RSA cryptosystem, ECC requires only 160-bit key length.

TABLE 4.1
Key Sizes (in Bits) for Various Public Key Cryptographic Systems

	Key Size				
RSA systems	1,024	2,048	3,072	8,192	15,360
Discrete logarithm systems	1,024	2,048	3,072	8,192	15,360
Elliptic curve systems	160	224	256	384	512

In this chapter, after a brief description of the basic principles of public key cryptography, we focus our analysis on the basic aspects of ECC. We analyze how this cryptographic approach can be used in the design of efficient cryptosystems that can be introduced to the future wireless security needs. A brief mathematical analysis of elliptic curves is given along with several required number and group theory principles. After solidifying a mathematical framework, we focus our analysis on how this mathematical framework can be employed in designing and implementing an elliptic curve cryptosystem. Design problems of elliptic curve cryptosystems are presented along with algorithms and methods of solving such problems.

4.2 BASIC PRINCIPLES OF PUBLIC KEY CRYPTOGRAPHY

Public key cryptography was first introduced to solve two major problems of the conventional symmetric key cryptography (secret key cryptography), key distribution, and key management. Key distribution in symmetric key cryptography is a problem because the channel needed for key transmission has to be secure in order to maintain the secrecy of one or many transmitted keys. Key management in symmetric key cryptography is another major problem. Secure communication between many entities requires management of a considerable amount of keys especially if each entity has to communicate with a considerable amount of other entities (requiring a different secret key for each such communication). Instead of relying on the secrecy of one or more keys that need to be dealt between several entities, public key protocols suggest using a pair of keys for each involved entity. The first key, called public key, is not secret and characterizes the involved entity, whereas the second key, called private key, is known only to this entity and no one else. If users want to send an encrypted message to the involved entity, they use this entity's public key. The decryption is performed by applying the involved entity's private key to the encrypted message. Therefore, no secrecy of any shared key is involved in the whole process.

The security strength of the public key cryptography lies in the computational infeasibility of finding each entity's private key from information about the public key or the public key itself. The problem of deriving the private key from the public key is equivalent to solving a computational problem that is considered intractable. Three such problems are used in public key cryptography:

1. Integer factorization problem (IFP): If n is a positive integer, find its prime factorization: meaning $n = p_1^{e_1} p_2^{e_2} \ldots p_k^{e_k}$, where the p_i are pairwise distinct prime integer numbers and each $e_i \geq 1$.
2. Discrete logarithm problem (DLP): If p is a prime number, a is a generator element of Z_p^* $(0 < a < p)$, and β is an element of Z_p^*, find integer x, $0 \leq x \leq p-2$, such that $a^x \equiv \beta \pmod{p}$.
3. Elliptic curve discrete logarithm problem (ECDLP): It is a generalization of the DLP. If there is an elliptic curve E defined over a field F and if there is a point $P \in E(F)$ of order n, and a point $Q \in E(F)$, find an integer s, $0 \leq s \leq n-1$, such that $Q = sP$, provided that such an integer exists.

Public key algorithms use a complex mathematical background and require a considerable amount of modular operations (addition, multiplication, and inversion). Due to the fact that such operations are performed over very big numbers (1024-bit length at least, in the case of RSA), a resulting public key cryptosystem is considered slow and with considerable hardware resource needs. This is the reason why public key cryptography is not used for message encryption–decryption but rather in coherence with symmetric key cryptography or for digital signature schemes. In the first case, a public key cryptosystem is used for encryption–decryption of the secret key of a symmetric key algorithm. Therefore, public key cryptography is employed only once per session for encryption–decryption of a small value (the secret key is usually 128 to 256-bits long in symmetric key cryptography). Therefore, the time demanding message encryption–decryption handling is appointed to the symmetric key algorithms that require less hardware resources and are usually fast. In the case of digital signature schemes, public key cryptography is employed for certifying the authenticity of a message and its owner.

Digital signature is a digital string for providing authentication. Commonly, in public key cryptography, it is a digital string that binds a public key to a message in the following way: only the person knowing the message and the corresponding private key can produce the string, and anyone knowing the message and the public key can verify that the string was properly produced. A digital signature may or may not contain the information necessary to recover the message itself. More on public key cryptography can be found in [6,7].

4.3 BASIC PRINCIPLES OF GROUP THEORY

We can define the additive group $(G, +)$ as a set of elements G and the arithmetic operation $+$. This means $(G, +)$ has the following properties:

- Associativity, meaning that $(a+b)+c=a+(b+c)$ for all $a, b, c \in G$.
- Identity, meaning that there is an element $0 \in G$ such that $a+0=0+a=a$ for all $a \in G$.
- Inverse, meaning that for every $a \in G$ there exists an element $-a \in G$ such that

$$a+(-a) = -a+a = 0.$$

Accordingly, we can define the multiplicative group $(G, \times)$ as a set of elements G and the arithmetic operation $\times$. Such a group has the following properties:

- Associativity, meaning that $a \times (b \times c) = (a \times b) \times c$ for all $a, b, c \in G$.
- Identity, meaning that there is an element $1 \in G$ such that $a \times 1 = 1 \times a = a$ for all $a \in G$.
- Inverse, meaning that for every $a \in G$ there exists an element $a^{-1} \in G$ such that

$$a \times a^{-1} = a^{-1} \times a = 1 \quad \text{for all } a \in G.$$

A group is called abelian (commutative) if $a+b=b+a$ or $a \times b = b \times a$ for all $a, b \in G$, according to the arithmetic operation that defines that group.

A group $(F, +, \times)$ is called a field and has a set of elements F with the arithmetic operations $+$ and $\times$. A field has the following properties:

- $(F, +)$ is an abelian group with identity 0.
- $\times$ operation is associative, meaning $(a \times b) \times c = a \times (b \times c)$ for all $a, b, c \in F$.
- There exists an identity element $1 \in F$ with $1 \neq 0$ such that $1 \times a = a \times 1 = a$ for all $a \in F$.
- Operation $\times$ is distributive over $+$, meaning that $a \times (b+c) = (a \times b) + (a \times c)$ and $(b+c) \times a = (b \times a) + (c \times a)$ for all $a, b, c \in F$.
- $(F, \times)$ is abelian, meaning that $a \times b = b \times a$, with identity 1.
- For every $a \neq 0$, $a \in F$, there exists an element $a^{-1} \in F$ such that $a^{-1} \times a = a \times a^{-1} = 1$.

There are two types of fields, infinite and finite fields. Infinite fields use an infinite underlined set of elements. Infinite fields are real numbers, rational numbers, or complex numbers. Finite fields have a finite set of elements. We also call such fields Galois fields, GF(q), in honor of the mathematician who first mentioned them. Finite fields are extremely useful in a variety of different computer applications including error code detection and ECC [8]. They have the following form GF(q) = $\{0,1,\ldots, q-1\}$.

A finite field exists only when $q = p^k$, where p is a prime number, called the characteristic *Char*(GF(q)) of the finite field, and k is a positive integer. If $k = 1$ then the finite field is called prime field GF(p) and when $k > 1$ the finite field is called extension finite field. When $p = 2$, the finite field, denoted as GF(2^k), is called binary extension finite field and is extremely useful in computer applications. The arithmetic operations defined on a GF(2^k) binary extension field are significantly simpler than those defined over GF(p) prime fields. We also define the order of a finite field, *Order*(GF(q)), as the number of elements of a finite field.

In ECC, the elliptic curve E is defined over a GF(q) or GF(2^k). Therefore, E has a finite set of rational points that form the group E(GF(q)), respectively, as is analyzed later in this chapter.

4.4 BASIC PRINCIPLES OF ELLIPTIC CURVES

While the theory of elliptic curves is very extensive, we focus our analysis only on those elliptic curves that are useful in cryptography and are defined over finite (GF(p) or GF(2^k)) fields.

An elliptic curve E defined over a field F is the set of solutions (x, y) where $x, y \in F$, of the long Weierstrass equation E: $y^2 + a_1xy + a_3y = x^3 + a_2x^2 + a_4x + a_5$ along with the point at infinity denoted as ∞. The variables a_1, a_2, a_3, a_4, $a_5 \in F$ and $\Delta \neq 0$, where Δ is the discriminant of E. More about the long Weierstrass equation, its properties, and more general information on elliptic curves can be found in [9].

We can say that two elliptic curves E_1 and E_2 defined over F are isomorphic if there exists a transformation between each x and y of E_1 and each x', y' of E_2 of the form $x = u^2x' + r$, $y = u^3y' + su^2x' + t$, where u, s, r, $t \in F$. This transformation, referred to as admissible change in variables, leads to a different equation defining the elliptic curve. This equation, denoted as the short Weierstrass equation, can have several different forms according to the field F defining the elliptic curve E.

Before we present the short Weierstrass equation of an elliptic curve defined over finite fields, supersingular and nonsupersingular curves must be defined. An elliptic curve E defined over a field F is supersingular if the characteristic of the field F divides t, where t is the trace. If the characteristic of a field F does not divide t, then E is nonsupersingular. There are evidences that supersingular elliptic curves are weak for cryptography [10]; therefore, we will not refer to them anymore.

For nonsupersingular elliptic curves defined over the GF(q) field, where q is a prime and has the characteristic $Char(\mathrm{GF}(q)) \neq 2$ or 3, the short Weierstrass equation has the form E: $y^2 = x^3 + ax + b$, where a, $b \in \mathrm{GF}(q)$ and $\Delta = -16(4a^3 + 27b^2) \neq 0$.

For nonsupersingular elliptic curves defined over the GF(2^k) field, the characteristic $Char(\mathrm{GF}(2^k)) = 2$ and the short Weierstrass equation has the form E: $y^2 + xy = x^3 + ax^2 + b$, where a, $b \in \mathrm{GF}(2^k)$ and $\Delta = b \neq 0$.

There is also one special type of elliptic curve, called Koblitz curves, that is defined over GF(2^k) fields. Such elliptic curves have a short Weierstrass equation of the form E: $y^2 + xy = x^3 + ax^2 + b$, where a, $b \in \mathrm{GF}(2)$. Koblitz elliptic curves are anomalous binary curves that have one very interesting property. Using Koblitz elliptic curves, point multiplication can be performed without any point doubling operation.

The number of rational points on an elliptic curve defined over a finite field is finite and is denoted as $\#E(\mathrm{GF}(q))$ or $\#E(\mathrm{GF}(2^k))$ accordingly.

4.5 GROUP LAW

If E is an elliptic curve (EC) defined over a field F, the point addition operation is performed by adding two points of the elliptic curve to get a third point. The set of EC points, together with the addition rule, forms an abelian group $E(F)$ of type $(G, +)$ with the identity element the point at infinity. Point doubling is the addition of one elliptic curve point with itself and can be considered a special case of point addition.

There is a geometric rule for finding the sum of two EC points, called chord and tangent rule. Suppose that we want to add EC point $P_1 = (x_1, y_1)$ to EC point $P_2 = (x_2, y_2)$ to get a third EC point $P_3 = (x_3, y_3) = P_1 + P_2$ of the elliptic curve E. We can find EC point P_3 by drawing a line through EC points P_1 and P_2, then mark the EC point on the curve that this line intersects. The reflection of the marked point over the x-axis is the EC point P_3. Point doubling follows a similar rule. Suppose that we want to find the double of an EC point $P_1 = (x_1, y_1)$, which is $P_3 = P_1 + P_1 = 2P_1 = (x_3, y_3)$ on an elliptic curve E. We draw the tangent line of P_1 and mark the EC point where this line intersects the elliptic curve. The reflection of this EC point over the x-axis is EC point $P_3 = 2P_1$. The above general geometric rules are shown in Figure 4.1 for an elliptic curve defined over real numbers.

The above geometric description can be translated into algebraic equations. Such equations describe point addition and point doubling following analytic geometry principles. They depend on the form of the elliptic curve equation and the field defining the curve. Since we are only interested in ECC, our analysis is focused only on nonsupersingular elliptic curves defined over finite fields (GF(p) or GF(2^k)).

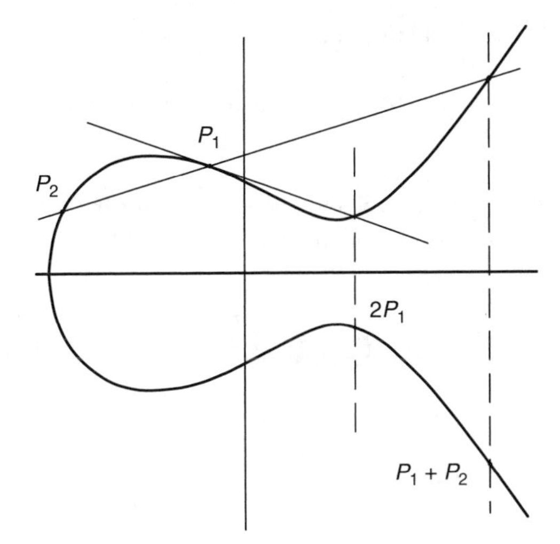

FIGURE 4.1 EC geometrical point addition and doubling for elliptic curve $y^3 = x^2 - 3x + 3$ defined over real numbers.

For nonsupersingular elliptic curves defined over GF(p) fields, the characteristic of the field is $Char(\mathrm{GF}(p)) > 3$. Therefore, the short Weierstrass equation of the elliptic curve would be E: $y^2 = x^3 + ax + b$, as analyzed in the previous section. In that case, we can define point addition ($P_3 = P_1 + P_2$) and point doubling ($P_3 = 2P_1$) as follows.

When $P_1 \neq P_2$ (point addition) the slope λ of the line between P_1 and P_2 would be $\lambda = (y_2 - y_1)/(x_2 - x_1)$ for $x_2 \neq x_1$ and the point $P_3 = P_1 + P_2 = (x_3, y_3)$ would be

$$x_3 = \lambda^2 - x_1 - x_2,$$
$$y_3 = \lambda(x_1 - x_3) - y_1 = \lambda(2x_1 + x_2 - \lambda^2) - y_1.$$

When $P_1 \neq P_2$ but $x_2 = x_1$ the slope is ∞, meaning that the line between P_1 and P_2 is vertical to the x-axis and, therefore, intersects the elliptic curve in point at infinity. In this case, $P_3 = P_1 + P_2 = \infty$.

When $P_1 = P_2$ (point doubling) and $y_1 \neq 0$, the slope of the tangent line in P_1 would be $\lambda = 3x_1^2 + a/2y_1$ and the point $P_3 = P_1 + P_2 = 2P_1 = (x_3, y_3)$ would be

$$x_3 = \lambda^2 - x_1 - x_2 = \lambda^2 - 2x_1,$$
$$y_3 = \lambda(x_1 - x_3) + y_1 = \lambda(3x_1 - \lambda^2) - y_1.$$

When $P_1 = P_2$ (point doubling) and $y_1 = 0$, the tangent line is vertical to the x-axis and therefore, $P_3 = 2P_1 = \infty$.

Since the point at infinity is the identity element of the $E(F)$ group, $P_3 = P_1 + \infty = P_1$. Point subtraction can be performed using the point $-P_2$ instead of P_2, where $-P_2 = (x_2, -y_2)$.

For nonsupersingular elliptic curves defined over binary extension fields GF(2^k), the characteristic of the field is $Char(\text{GF}(2^k)) = 2$. Therefore, the short Weierstrass equation of the elliptic curve would be E: $y^2 + xy = x^3 + ax^2 + b$, as analyzed in the previous section. In that case, we can define point addition ($P_3 = P_1 + P_2$) and point doubling ($P_3 = 2P_1$) as follows:

When $P_1 \neq P_2$ (point addition) the slope λ of the line between P_1 and P_2 would be $\lambda = y_2 + y_1/x_2 + x_1$ for $x_2 \neq x_1$ and the point $P_3 = P_1 + P_2 = (x_3, y_3)$ would be

$$
\begin{aligned}
x_3 &= \lambda^2 + \lambda + a + x_1 + x_2, \\
y_3 &= \lambda(x_1 + x_3) + x_3 + y_1.
\end{aligned}
$$

When $P_1 \neq P_2$ but $x_2 = x_1$ the slope is ∞, meaning that the line between P_1 and P_2 is vertical to the x-axis and, therefore, intersects the elliptic curve in point at infinity. In this case, $P_3 = P_1 + P_2 = \infty$.

When $P_1 = P_2$ (point doubling) and $y_1 \neq 0$, the slope of the tangent line in P_1 would be $\lambda = x_1^2 + y_1/x_1$ and the point $P_3 = P_1 + P_2 = 2P_1 = (x_3, y_3)$ would be

$$
\begin{aligned}
x_3 &= \lambda^2 + \lambda + a = x_1^2 + \frac{b}{x_1^2}, \\
y_3 &= \lambda(x_1 + x_3) + x_3 + y_1 = x_1^2 + \lambda x_3 + x_3.
\end{aligned}
$$

When $P_1 = P_2$ (point doubling) and $y_1 = 0$, the tangent line is vertical to the x-axis and therefore, $P_3 = 2P_1 = \infty$.

Since the point at infinity is the identity element of the $E(F)$ group, $P_3 = P_1 + \infty = P_1$. Point subtraction can be performed using the point $-P_2$ instead of P_2, where $-P_2 = (x_2, x_2 + y_2)$.

4.6 POINT MULTIPLICATION

If E is an elliptic curve defined over a field F, multiplication between an integer s and an EC point P results in a new EC point $Q = sP$. This operation is called point multiplication or scalar multiplication. Point multiplication is a repeated process that can be analyzed in a series of point additions and point doublings using Algorithm 1.

4.7 GENERAL ECC DESIGN METHODOLOGY

In this section, a general plan for designing an elliptic curve cryptosystem is described. First of all, an appropriate ECC protocol and algorithm for encryption–decryption, digital signature, or authentication scheme must be chosen. This algorithm is based on ECDLP and as a result requires the use of point multiplication operation once or multiple times.

Point multiplication is the major design bottleneck in ECC. It requires a series of other mathematical operations that have increased mathematical complexity. As presented in the previous section, point multiplication uses other EC point operations. Those operations are point addition ($P_1 + P_2$) and point doubling ($P_2 = 2P_1$).

Point addition and doubling follow the Group Law and use the mathematical framework of the finite field on which the EC is defined (GF(q) or GF(2^k)). Therefore, all mathematical operations between the coordinates (x, y) of EC points P_1, P_2, as dictated by the Group Law, are performed using GF(q) or GF(2^k) field arithmetic.

In finite field arithmetic four mathematical operations can be identified. Those operations are addition–subtraction, multiplication, squaring, and inversion–division. Each such operation has a different computational and hardware resources cost (measured in throughput, critical path delay, gate—storage element number, and power dissipation). Such cost is higher for inversion–division, whereas addition–subtraction has the lowest. The notable cost of multiplication in finite fields is of great importance since this mathematical operation can also be used under certain circumstances for inversion–division.

Following the above remarks, a design plan for an elliptic curve cryptosystem is presented in the pyramid of Figure 4.2. There are four different design levels, each one depending on the lower level's mathematic framework. The base of the pyramid is formed by the finite field mathematic framework that includes operations between elements of a finite field. On top of the finite field mathematic framework, the point addition–doubling layer is located, using the finite field mathematic framework for EC point operation following the Group Law. This layer is used for the calculation of the point multiplication product that forms the homonymous design layer. The point multiplication layer is employed in the highest design level of the ECC algorithm and protocols. The design methodology begins from the construction of the lowest design layer and proceeds gradually toward the top of the pyramid.

The pyramidal scheme of Figure 4.2 also symbolizes the frequency of the used operations in each design layer. For one EC protocol (the highest design layer), few EC point multiplications would be required. For each point multiplication, many point additions and doublings (depending on the integer s of $Q = sP$) would have to be performed and for each such operation a series

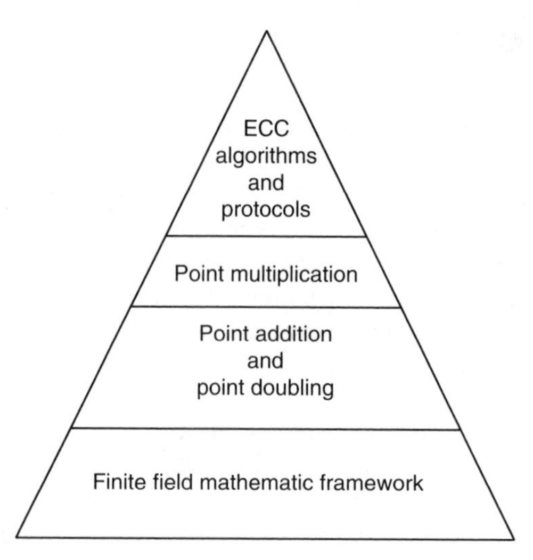

FIGURE 4.2 The ECC design methodology pyramid.

of finite field calculations are needed as described by the Group Law. Therefore, the number of required arithmetic operations of each design layer increases dramatically as we move toward the base of the pyramid. Efficient designing in terms of power consumption, speed, and hardware component number has an increasing effect on the overall system as we move toward the base of the pyramid. The finite field mathematical framework design layer, forming the base of the pyramid, plays a crucial role in the design of the overall ECC system.

In the rest of this chapter, we address problems and solutions in efficient designing of the design layers of the pyramid in Figure 4.2, with special interest in the lowest design layers. Those layers are usually designed in hardware, leaving the more abstract higher layers (ECC algorithms and protocols) to software.

4.8 FINITE FIELDS

Finite fields fit into two major categories, as described earlier, prime fields, denoted as GF(p) fields and extension fields, denoted as GF(p^k) fields. Each GF(p^k) field can be described as a vector space of k dimension with each vector element belonging to GF(p) field. When $k=1$, each element is a one-dimensional vector of GF(p). A specific type of GF(p^k) fields that has $p=2$ stands out among all the extension fields. These finite fields, called binary extension fields or GF(2^k) fields, are described as k-dimensional vectors over GF(2). They have some very interesting properties that fit well into the binary logic of modern computer applications. For that reason, GF(2^k) fields are the dominant kind of extension fields used in ECC. In this section, we focus our analysis on GF(p) and GF(2^k) fields. The mathematic operations of each finite field category are analyzed and methods for optimizing those operations are presented.

4.8.1 GF(p) Fields

An element a of a GF(p) field, also called prime field, is $a \in \{0,1,2,\ldots, p-1\}$. Therefore, each mathematical operation defined in GF(p) fields must follow the following principle. If the outcome of any such operation exceeds the range of the GF(p) field, this outcome is returned into the field by applying a modular reduction operation (mod p). Each element a of the GF(p) field is considered a binary vector number $a = \{a_{n-1}, a_{n-2}, \ldots, a_1, a_0\}$, where $a_i \in$ GF(2) and $0 \leq a \leq p-1$.

4.8.1.1 GF(p) Field Addition–Subtraction

If a, b are elements of a GF(p) field, $a, b \in \{0, 1, 2, \ldots, p-1\}$ and $a > b$, then addition and subtraction between those two elements have the following form (modular addition–subtraction):

$$(a+b) \bmod p = \begin{cases} a+b & \text{if } a+b < p \\ a+b-p & \text{if } a+b \geq p \end{cases} \qquad (a-b) \bmod p = a-b.$$

The two operations involved in modular addition–subtraction are integer addition–subtraction. Modular subtraction is identical to integer subtraction while modular addition requires at most one integer addition and one integer subtraction. Since subtraction between integers in binary form can be performed by additions using two's complement numbers [11], integer addition is the key operation in modular addition–subtraction.

Many hardware designs exist for integer addition, like ripple carry, carry lookahead, carry-save, carry-select, or carry-skip adders. Additional information on the topic can be found in well-known books for computer arithmetic or hardware design [11,12].

4.8.1.2 GF(p) Field Multiplication

Multiplication in GF(p) fields is always performed on modulus p, $((a \cdot b) \bmod p)$. There are two different approaches to modular multiplication design. The first approach consists of two steps:

Step 1. Perform integer multiplication.
Step 2. Perform mod p reduction of the integer multiplication product of step 1.

In step 1, integer multiplication algorithms can be used for obtaining the multiplication product, which is not in the GF(p) field. Such algorithms can be taken from general computer arithmetic theory and resulting multiplier architectures can be designed. One such algorithm, used extensively in ECC applications, is the Karatsuba–Ofman multiplication method.

In step 2, where reduction is performed, the classic approach of test division [11] cannot be applied in efficient modular multiplication architectures. The dominant algorithms for modular reduction are Barret's reduction and Montgomery modular reduction method. However, by applying rules of special prime numbers p, such as those proposed by NIST [13] and IEEE 1362 [14] (NIST Primes), reduction can be simplified.

The second approach consists of only one step where both multiplication and modular reduction are performed. Montgomery modular multiplication algorithm is a well-known method that is employed in efficient modular multiplication architectures and therefore is widely used in GF(p) elliptic curve applications.

All the modular multipliers in GF(p) fields are of bit serial or digit serial nature, meaning that they perform multiplication by processing bits or digits of data per round at a given number of multiplication rounds greater than 1.

4.8.1.2.1 Karatsuba–Ofman Multiplication

The Karatsuba–Ofman multiplication algorithm employs a divide and conquer technique for performing multiplication [15]. It is especially useful when multiplying very large numbers; this makes the method extremely beneficial in cryptography where big numbers are involved.

Suppose that A, B are integer numbers in n-bit binary form, where $A = \{a_{n-1}, a_{n-2}, \ldots, a_1, a_0\}, B = \{b_{n-1}, b_{n-2}, \ldots, b_1, b_0\}$, and $n = 2m$. We can rewrite A, B as $A = A_1 2^m + A_2$ and $B = B_1 2^m + B_2$, where A_1, A_2, B_1, and B_2 are m-bit numbers of the form $A_1 = \{a_{n-1}, a_{n-2}, \ldots, a_{m+1}, a_m\}$, $A_2 = \{a_{m-1}, a_{m-2}, \ldots, a_1, a_0\}$ and $B_1 = \{b_{n-1}, b_{n-2}, \ldots, b_{m+1}, b_m\}$, $B_2 = \{b_{m-1}, \ldots, b_1, b_0\}$. In that case, multiplying $A \cdot B$ becomes

$$\begin{aligned} A \cdot B &= (A_1 2^m + A_2) \cdot (B_1 2^m + B_2) \\ &= (A_1 \cdot B_1) 2^{2m} + (A_1 \cdot B_2 + A_2 \cdot B_1) 2^m + (A_2 \cdot B_2). \end{aligned}$$

Instead of performing one multiplication between n-bit numbers, using Karatsuba–Ofman method three parallel multiplications, two parallel additions, and two subtractions of m-bit numbers are required, where $m = n/2$. We compute $C = A_1 \cdot B_1$, $D = A_2 \cdot B_2$ and then

$$\begin{aligned} (A_1 + A_2) \cdot (B_1 + B_2) = A_1 \cdot B_2 + A_2 \cdot B_1 + A_1 \cdot B_1 + A_2 \cdot B_2 \Leftrightarrow A_1 \cdot B_2 + A_2 \cdot B_1 \\ = (A_1 + A_2) \cdot (B_1 + B_2) - C - D. \end{aligned}$$

Multiplication between the m-bit numbers can not only be performed using the long multiplication method (the well-known school method) [11], but can also be performed using again the Karatsuba–Ofman method. In the second case, the Karatsuba–Ofman method is used recursively until a multiplication bit length is reached, beyond which this method is not affordable. This method works even if n is not even. By padding with zeros, the multiplicand or multiplier can fit to the appropriate bit length.

4.8.1.2.2 Barrett's Reduction

Barrett's reduction method is a fast way for modular reduction when many reductions are required with a single modulus. That makes this method highly applicable in GF(p) fields where all modular calculations are performed with the modulus p (mod p). A special value has to be precomputed in Barrett's reduction algorithm denoted as $\mu = \lfloor b^{2n}/p \rfloor$, where b is the chosen radix (base) of the involved numbers (Algorithm 2).

It has to be noted that Algorithm 2 is highly efficient for high radix values, $b > 3$. This does not mean that it cannot be used for radix 2 hardware architectures. Barrett's reduction algorithm does not employ any division but more simple operations (addition and subtraction) and many shiftings that can be performed with minimal computational cost.

4.8.1.2.3 Montgomery Reduction and Montgomery Multiplication

In 1985, Peter Montgomery introduced a new method for modular reduction and multiplication [16]. Montgomery's approach avoids the time-consuming trial division, which is the common bottleneck of other algorithms. His method has been proven to be efficient in terms of computational speed and hardware resources. Thus, it has been used in many implementations of modular multiplication in hardware as well as software.

The Montgomery modular reduction [16] is used for calculation of the value $MontR(x, p) = c = x \cdot r^{-1}$ mod p, where r is a constant number (usually $r = b^n$) and b is the base (radix) of the involved numbers. The n-bit value p has to be an integer filling the condition $gcd(r, p) = 1$. Since p is a prime, the above constraint is always true. There is a one-to-one correspondence between each element $x \in$ GF(p) and its representation $c = x \cdot r^{-1}$ mod p (Algorithm 3). The Montgomery modular reduction is usually not used independently but as a part of the Montgomery modular multiplication method (Algorithm 4).

As shown earlier, the resulting Montgomery multiplication product $MontM(x, y, p)$ includes the r^{-1} number in the multiplication product. In order to get a result free of the r^{-1} factor, a precomputated procedure has to be followed. To compute $x \cdot y$ mod p, the value $x' = x \cdot r$ mod p has to be calculated by performing $MontM(x, r^2, p)$. We say then that x' is in the Montgomery or p-residue domain. Using x' in the Montgomery modular multiplication method, the correct result $x \cdot y$ mod p is calculated by $MontM(x', y)$.

One of the problems in the Montgomery multiplication algorithm is the calculation of p'. Depending on the base b, on which the involved numbers are defined, the p' value might require the use of inversion operation. To simplify this operation or avoid it completely, there are several bases that can be used. The most commonly used base is binary representation ($b = 2$) or similar higher-order radix representations ($b = 2^2$, $b = 2^4$). In the case of $b = 2, p' = 1$ and $MontM(x, y, p)$ is greatly simplified (Algorithm 5).

4.8.1.2.4 NIST Special Primes for Multiplication–Reduction

NIST proposes in FIPS 186-2 standard [13] the use of some special GF(p) fields for simplifying the reduction process in multiplication. This happens because the proposed prime p can be extended to a sum or difference of powers of 2. This special ability leads to fast reduction (faster than Montgomery's method) and is especially applicable to machines with word size of 32-bits. Those fields are

$$
\begin{aligned}
\mathrm{GF}(p_{192})\colon p_{192} &= 2^{192} = 2^{64} - 1,\\
\mathrm{GF}(p_{224})\colon p_{224} &= 2^{224} - 2^{96} + 1,\\
\mathrm{GF}(p_{256})\colon p_{256} &= 2^{256} - 2^{224} + 2^{192} + 2^{96} - 1,\\
\mathrm{GF}(p_{384})\colon p_{384} &= 2^{384} - 2^{128} - 2^{96} + 2^{32} - 1,\\
\mathrm{GF}(p_{521})\colon p_{521} &= 2^{521} - 1.
\end{aligned}
$$

More on this topic can be found in [13,14,17].

4.8.1.2.5 Hardware Design of Modular Multipliers in GF(p) Fields

In the area of modular multiplication for GF(p) fields, Montgomery modular multiplication dominates among the other offered alternatives. In this algorithm, through correct choice of r ($r = 2^n$) the multiplication process is fast and unmatched by any other method when multiplying many times with the same modulus. Among the first proposed hardware architectures on Montgomery modular multiplication method is the work of Eldridge and Walter [18], in which they prove the advantage of the Montgomery algorithm when compared with other techniques in speed because of its small critical path. From this point on, Montgomery modular multiplication became one of the most widely analyzed and researched computer algebra algorithms because of the advantages presented here. Therefore, in this subsection we focus our analysis on how hardware architectures are designed based on the Montgomery modular multiplication algorithm.

Two methods are employed for designing Montgomery multipliers, systolic arrays using various different encoding methods (redundant schemes [19], Booth encoding [20]) and residue number system (RNS) arithmetic.

A systolic array can be defined as a set of processing elements (PEs) arranged in an n-dimensional array formation. In every clock cycle, each such PE receives data from its neighboring PEs, uses those data for a specific function, and pumps the results to its neighboring PEs that perform similar or the same function. Through this procedure throughput can be increased dramatically.

Systolic arrays fit well with the Montgomery modular multiplication algorithm. The *MontMb*(x, y, p) algorithm is a repeated loop of step 2.1 and step 2.2, which can be easily represented by a series of 1-bit input–output PEs calculating those steps. A two-dimensional systolic array can be constructed in that fashion, as shown in Figure 4.3. This architecture is a typical example of carry-save redundancy used for Montgomery multipliers. Every processing element has to perform 5 to 3 additions using the current x and q value. However, the processing element of bit 0 has an additional role to perform (the calculation of q value) and requires two more gates. Every different column of the systolic array represents the value of the same bit through the algorithmic calculations, whereas each different row of the systolic array represents the values on another round of the *MontMcs*(x, y, p) algorithm. The carry$_2$ ($C^{(2)}$) signal of a PE is connected with the next PE of the next row. Carry$_1$ ($C^{(1)}$) signal is connected with the same PE of the next row (the same column), whereas the sum (S) signal is connected with the previous PE of the next row starting from bit 0. The outcome is in carry-save format through the signals carry$_2$, carry$_1$, and sum. This outcome is checked for the $c > p$ condition to subtract p, and a final addition is performed using an adder so as to return from the carry-save to normal number format.

There exist many designs, similar to that in Figure 4.3, achieving optimization using carry-save logic and relevant adders [22–24] because carry-save adders reduce the critical path delay significantly when employed in systolic Montgomery multipliers.

In addition, to decrease the critical path of the PE and increase the speed, many researchers propose a precomputation phase where certain values are calculated once to be used in all the rounds of the *MontMb*(x, y, p) algorithm, like in [22] where instead of y, the $y + p$ value is used. The combination of a different choice of r along with precomputation has also been employed for optimizing *MontMb*(x, y, p) algorithm. In [25], combining $r = 2^{n+8}$ and precomputation of the value $T = (((8 - p_2p_1p_0)^{-1} \bmod 8)P + 1)/8$, a fully pipelined systolic multiplier is designed with increased parallelism. However, more multiplication rounds in such a design are needed.

High radix Montgomery multipliers ($b = 2^k$, $k > 1$) have also been proposed [19,26,27], using *MontM*(x, y, p) algorithm with a change of the $p' = -p^{-1} \bmod b$ into $\tilde{p} = p'p = (-p^{-1} \bmod b)p, r = b^n$ and a replacement of y with $b \cdot y = 2^k \cdot y$. The problem with the above designs lies in the fact that digit adders have to be constructed for calculating each high radix sum. To solve this problem, in [19] a different approach is undertaken.

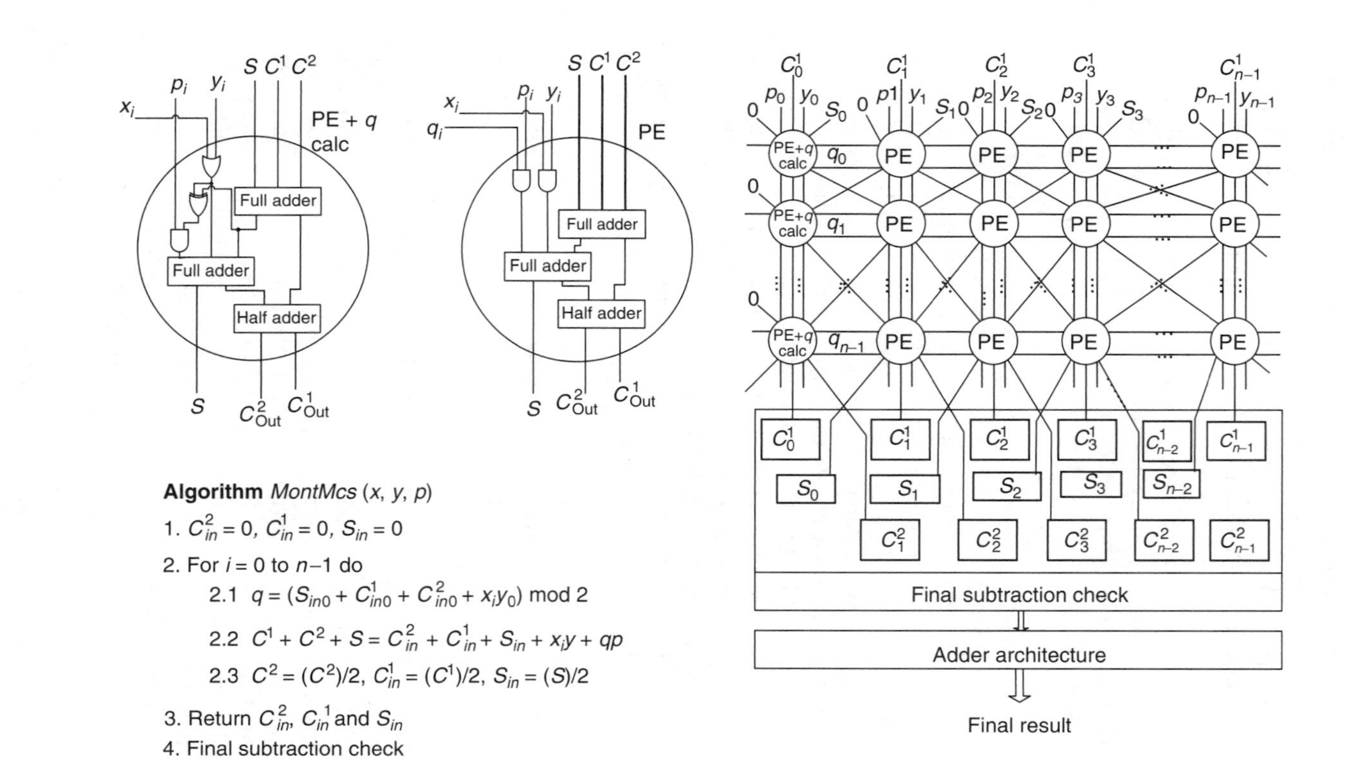

FIGURE 4.3 Systolic array of a Montgomery multiplier using carry-save logic, along with two types of PE (PE and PE with calculation of q). (Modified from Fournaris A.P. and Koufopavlou O., *Proceedings of 46th IEEE Midwest Symposium on Circuits and Systems '03 (MWSCAS 2003)*, Egypt, 2003.)

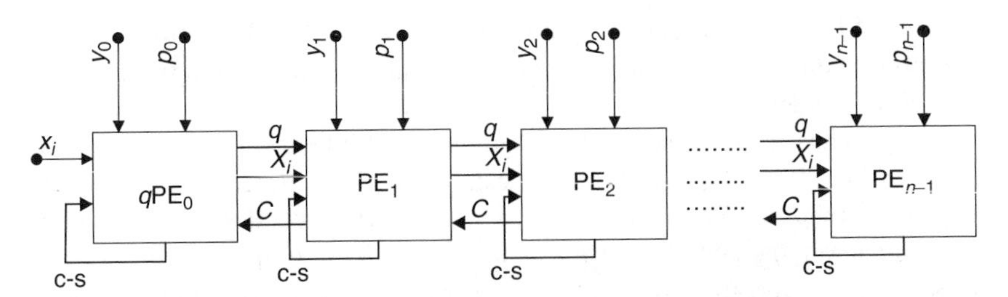

FIGURE 4.4 One-dimensional systolic array for Montgomery multiplier.

The multiplier uses the $MontMb(x, y, p)$ with $b = 2$ but processes values of word length.

Concerning systolic arrays, it should be remarked that the achieved high throughput comes as a result of a considerable hardware resource increase. That tradeoff is not always affordable especially in wireless design where system resources are limited. To delegate the problem in the above case, one-dimensional instead of two-dimensional systolic arrays could be used, like the one presented in Figure 4.4.

Another approach to modular multiplication is the use of the RNS system deriving from the Chinese Remainder Theorem (CRT). In RNS an integer x is described by a series of positive integers x_i as $x_{\mathrm{RNS}} = \{x_0, x_1, x_2, \ldots, x_k\}$ defined over an RNS base of relatively prime m_i numbers $B = \{m_0, m_1, m_2, \ldots, m_k\}$. Each x_i is defined as $x_i = x \bmod m_i$ for $i = 0,1,2, \ldots, k$. To reconstruct x from its RNS equivalent, CRT is used in the equation

$$x = \sum\nolimits_{i=0}^{k} x_i \cdot M_i \cdot \left|M_i^{-1}\right|_{m_i} \bmod M,$$

where $M = \prod_{i=0}^{k} m_i$, $M_i = M/m_i$, and M_i^{-1} are the invert of M_i mod m_i. The arithmetic operations of addition–subtraction and multiplication can be defined in RNS representation as follows:

$$x_{\mathrm{RNS}} \pm y_{\mathrm{RNS}} = \{|x_0 \pm y_0|_{m_0}, |x_1 \pm y_1|_{m_1}, \ldots, |x_k \pm y_k|_{m_k}\},$$
$$x_{\mathrm{RNS}} \times y_{\mathrm{RNS}} = \{|x_0 \times y_0|_{m_0}, |x_1 \times y_1|_{m_1}, \ldots, |x_k \times y_k|_{m_k}\}.$$

The benefit of the RNS system lies in the inherent parallelism of this system. Instead of calculating one modular multiplication with modulus p, RNS uses k parallel multiplications with modulus m_i, where each m_i is a smaller number than p. The problem of this approach lies in the conversion from RNS to non-RNS format, which requires modulus M reduction. Since $M > p$, modulus M reduction can be time and resource consuming. However, in case many modular multiplications are needed to be performed (such as GF(p) ECC) this drawback is counterbalanced by the overall gain in the RNS multiplication process.

In [28], an elliptic curve cryptosystem is proposed taking advantage of the above remarks with considerable performance gain in comparison with non-RNS designs.

Bajard has proposed several Montgomery multipliers in RNS [29,30]. The choice of $r = M$ is proposed in those designs and two different RNS bases, B_1 and B_2, are used for calculating q value. The above RNS Montgomery multiplication method, however, requires two base transformations, from base B_1 to B_2 and from B_2 to B_1, after certain calculations are done. Those transformations can be time consuming.

4.8.1.3 GF(p) Field Squaring

Squaring in GF(p) fields follows the general principles of modular multiplication. The methods described in Section 4.8.1.2 apply for a squaring operation with appropriate simplification due to the fact that in $c = x \cdot y \bmod p$ both x, y are the same ($x = y$).

4.8.1.4 GF(p) Field Inversion

Consider two elements x, a of GF(p) such that $a \cdot x \equiv 1 \pmod p$. If such an x exists, then it is unique, and a is said to be invertible, or a unit. The inverse of a is denoted by a^{-1} and is called multiplicative inverse of a. The process of finding a multiplicative inverse is called inversion. The multiplicative inverse a^{-1} exists as long as p and a are coprime, and since p is prime the above constraint is always valid.

Finding multiplicative inverses involves a considerable number of computations and ways of bypassing that operation are always examined, especially in ECC design. However, inversion cannot always be avoided, so several methods have been proposed for performing this operation with reduced computational and resource cost.

The dominant technique of calculating a multiplicative inverse of a number is the extended Euclidean algorithm (EEA) for greater common divisor (GCD). Kaliski in [31] proposes another approach by using the mathematical background of the Montgomery modular multiplication method. This approach results in the Montgomery inversion algorithm for calculating the multiplicative inverse in the Montgomery domain.

4.8.1.4.1 Extended Euclidean Algorithm

Equation $a \cdot x \equiv 1 \pmod p$ can be rewritten as $a \cdot a^{-1} + \hat{p} \cdot p = 1$. This equation is a direct representation of the outcome of the EEA for GCD, with inputs a, p. The EEA for GCD calculates $d = gcd(a, p)$, x and y values of $a \cdot x + y \cdot p = d$. However, when we work in GF(p) fields there is $d = 1$ since p is prime and $a \cdot x + y \cdot p = d \Leftrightarrow a \cdot x + y \cdot p = 1$, which is the equation of inversion.

The EEA is a repeated process of divisions and subtractions. Suppose that r_i, q_i, $u_i \in$ GF(p), where i is an integer ($i = \{0,1,2,\ldots\}$) representing the current round number of the algorithm. Initially, $r_{-1} = p$, $r_{-2} = a$, $u_{-2} = 1$, and $u_{-1} = 0$. In each round of the algorithm, r_{i-1} divides r_{i-2} giving a quotient q_i and a remainder r_i, $r_{i-1} = q_i \cdot r_{i-2} + r_i$. Those values are used in the calculation of $u_i = q_i \cdot u_{i-1} + u_{i-2}$. The process is repeated until the remainder r_i is zero (Algorithm 6).

At first glance, the main design problem in the *ExEucl*(a,p) algorithm is the extended number of required divisions. However, several researchers have described the algorithm in binary form where the division operation has been replaced by a number of shift operations (division or multiplication by 2). The binary EEA uses the theory of long division for integers in binary form (Algorithm 7).

4.8.1.4.2 Montgomery Inversion Algorithm

A number a converted in the Montgomery domain or p-residue domain becomes $a \cdot r$ mod p. If $r = 2^n$ then a number in the Montgomery domain becomes $a2^n$ mod p. To convert a number in the Montgomery domain, one Montgomery multiplication of number a with r^2 mod p is required (*MontMb*(a, r^2 mod p, p)). Following the above definitions, a multiplicative inverse of a number a in the Montgomery domain would be $a^{-1} \cdot r$ mod p. In the Montgomery inverse algorithm, proposed by Kaliski [31], a procedure similar to the EEA is employed for calculating the multiplicative inverse of a number a in the Montgomery domain. The algorithm consists of two phases (Algorithm 8 and Algorithm 9). In phase I, usually denoted as Montgomery almost inverse phase, the value $a^{-1}2^k$ mod p, where k is an integer and $n \leq k \leq 2k$ is calculated. However, this outcome is not a valid value in the Montgomery domain; therefore, a correction phase is required. This is phase II of the Montgomery inversion algorithm (Algorithm 9).

While the number of rounds in *MontAI*(a,p) is not constant, it is well constrained between $(n+1)$ and $(2n+2)$ rounds. Phase II of the Montgomery inverse algorithm is completed after k–n rounds.

If the input a of *MontAI*(a,p) is in the Montgomery domain, then the outcome of phase II is not in the Montgomery domain but rather a^{-1} mod p. If the input a of *MontAI*(a,p) is not in the Montgomery domain, then the final result after phase II would have to be multiplied using Montgomery multiplication with 1. However, by taking a result in phase II after $k-1$ rounds instead of $k-m$ rounds, no final multiplication is required for obtaining a result not in the Montgomery domain (Algorithm 10).

4.8.1.4.3 GF(p) Field Division

Division in GF(p) fields can be considered a combination of inversion and multiplication following the form $d = a/b = a \cdot (1/b)$, where a, b, $d \in$ GF(p). Therefore, performing one GF(p) inversion operation and using its output for

GF(p) modular multiplication, the division outcome can be calculated. These two operations can be either in normal domain or in the Montgomery domain following a combination of Montgomery inversion algorithm and Montgomery modular multiplication algorithm as presented in [32].

There are, however, modifications of the EEA that can be employed to support modular division. Takagi [33] proposes such a modified extended Euclidean algorithm (MEEA) for modular division (Algorithm 11).

Modular inversion or division is difficult to design in hardware as well as software since they require the use of all the other GF(p) operations. The fact that the outcome is not calculated after a constant number of iterations prohibits the use of systolic arrays and limits the use of pipelining. Few hardware architectures exist for modular inversion and most of them are direct designing of the corresponding algorithms using architectures for modular reduction multiplication, addition, and subtraction, as those presented earlier. Redundant, signed digit and two's complement representation, high radix bases (radix 4/2), and carry-save or carry-select addition–subtraction architectures are employed in such designs [34–36] to achieve high computational speed and reduced hardware resources.

4.8.2 GF(2^k) Fields

An element $a \in \text{GF}(2^k)$ field is defined over a base B of the form $B = \{b_{k-1}, b_{k-2}, \ldots, b_1, b_0\}$. Therefore, the element a can be written as a linear combination of the b_i of the base B as

$$a = a_{k-1}b_{k-1} + a_{k-2}b_{k-2} + a_{k-3}b_{k-3} + \cdots + a_1b_1 + a_0b_0.$$

Since the characteristic of the GF(2^k) field is $Char(\text{GF}(2^k)) = 2$, the coefficients a_i of a represented in the base B are defined in the GF(2) field. The element a can also be described in vector notation as $a = \{a_{k-1}, a_{k-2}, a_{k-3}, \ldots, a_1, a_0\}$.

The base B element representation of the GF(2^k) field can have many different forms. Each base form specifies the interaction of a field element with the other field elements. Therefore, the choice of the GF(2^k) field base drastically affects the mathematic operations in the GF(2^k) field. For this reason, the element representation choice in GF(2^k) fields plays an important role in the design of efficient architectures for the various GF(2^k) field mathematic operations.

The most prominent bases for representing GF(2^k) field elements are polynomial basis (standard basis) representation, normal basis representation, or double basis representation. Among those bases, widely used in modern applications are polynomial basis representation and a special case of normal basis representation called optimal normal basis (ONB). In the following subsections we analyze those two element representations of GF(2^k) fields and describe the corresponding mathematic operations for each such representation.

4.8.2.1 Polynomial Basis Representation

The GF(2^k) field is isomorphic to GF(2)[x]/($f(x)$), where $f(x)$ is a degree k monic irreducible polynomial of the form

$$f(x) = x^k + \sum_{i=0}^{k-1} f_i x^i$$

with coefficients $f_i \in$ GF(2).

According to the polynomial basis representation, an element a of a GF(2^k) field is a polynomial of degree at most $k-1$ defined over a basis $\{x^{k-1}, \ldots, x^3, x^2, x, 1\}$ with coefficients $\alpha_i \in$ GF(2), where x is a root of the irreducible polynomial $f(x)$. This can be written as

$$a(x) = \sum_{i=0}^{k-1} a_i x^i = a_{k-1}x^{k-1} + a_{k-2}x^{k-2} + \cdots + a_1 x + a_0.$$

4.8.2.1.1 GF(2^k) Field Addition–Subtraction in Polynomial Basis Representation

Suppose that $a(x)$, $b(x)$ polynomials are elements of a GF(2^k) field defined over the irreducible polynomial $f(x)$ in polynomial basis representation. Then, we define addition as $s(x) = a(x) + b(x) = \sum_{i=0}^{k-1} s_i x^i$, where $s(x) \in$ GF(2^k) and $s_i \in$ GF(2). Since each element in GF(2^k) fields can be described as a k-dimensional vector over GF(2), each s_i would be $s_i = (a_i + b_i)$ mod 2. Similarly, subtraction can be defined as $r(x) = a(x) - b(x) = \sum_{i=0}^{k-1} r_i x^i$, where $r(x) \in$ GF(2^k), $r_i \in$ GF(2), and $r_i = (a_i - b_i)$ mod 2. It can be noted that $(a_i - b_i)$ mod 2 $= (a_i + b_i)$ mod 2, so subtraction is identical to addition while both operations can be interpreted as an XOR operation between $a(x)$ and $b(x)$. As a result, no carry value exists for addition–subtraction in GF(2^k) fields. The above fact makes GF(2^k) fields highly advantageous when compared with other types of finite fields and affects all mathematic operations of this field regardless of the element representation base.

4.8.2.1.2 GF(2^k) Field Multiplication in Polynomial Basis Representation

Suppose that $a(x)$, $b(x)$ polynomials are elements of a GF(2^k) field defined over the irreducible polynomial $f(x)$ in polynomial basis representation. Then, we define multiplication as $c(x) = \sum_{i=0}^{k-1} c_i x^i = a(x) \cdot b(x) = a(x) \cdot b(x)$ mod f(x), where $c(x) \in$ GF(2^k) and $c_i \in$ GF(2). GF(2^k) field multiplication is a modular operation.

Following the principles of modular multiplication, there are two approaches in performing GF(2^k) field multiplication. The first approach consists of two steps:

Step 1. Perform polynomial multiplication resulting in a $2k-1$ degree polynomial.
Step 2. Perform mod $f(x)$ reduction of the polynomial multiplication product of step 1.

In step 1, similarly to GF(p) multiplication, polynomial multiplication algorithms can be used for obtaining the product that is not in the GF(2^k) field. Such algorithms can be taken from general computer arithmetic theory and resulting multiplier architectures can be designed. As in the case of GF(p) fields, the Karatsuba–Ofman multiplication method has been used extensively in finding the product of step 1, adapted to the carry-free GF(2^k) field arithmetic.

In step 2, where reduction is performed, the classic approach of test division [11] is excluded as a possible design solution. Reduction is performed using the definition of polynomial basis GF(2^k) fields. We know that the GF(2^k) field is defined over an irreducible polynomial $f(x)$ and that x is a root of $f(x)$. In that case, $f(x) = x^k + \sum_{i=0}^{k-1} f_i x^i = 0 \Rightarrow x^k = \sum_{i=0}^{k-1} f_i x^i$, since addition is identical to subtraction. Then, $x \cdot x^k = x^{k+1} = x \cdot \sum_{i=0}^{k-1} f_i x^i = f_{k-1} \cdot x^k + \sum_{i=1}^{k-1} f_{i-1} x^i$, but since $x^k = \sum_{i=0}^{k-1} f_i x^i$ we can find x^{k+1} as $x^{k+1} = f_{k-1} \cdot \sum_{i=0}^{k-1} f_i x^i + \sum_{i=1}^{k-1} f_{i-1} x^i = \sum_{i=1}^{k-1} (f_{k-1} f_i + f_{i-1}) x^i + f_{k-1} f_0$.

Following the above procedure, we can gradually replace all x^i, where $k \leq i \leq 2k-1$, with combinations of x^i and coefficients f_i of the irreducible polynomial $f(x)$, where $0 \leq i \leq k-1$, thus reducing the product of step 1 into a polynomial with $k-1$ degree of the GF(2^k) field.

To represent the reduction process, a $k \times k$ reduction matrix R can be constructed, as proposed by Mastrovito in [8]. Each row of R can be constructed recursively from the irreducible polynomial $f(x)$ following the form:

$$r_{i,j} = \begin{cases} f_j & \text{for } i = 0, j = 0, 1, \ldots, k-1, \\ r_{i-1,k-1} & \text{for } i = 1, 2, \ldots, k-2, j = 0, \\ r_{i-1,j-1} + r_{i-1,k-1} \cdot r_{0,j} & \text{for } i = 1, 2, \ldots, k-2, j = 1, 2, \ldots, k-1. \end{cases}$$

The resulting R matrix can then be used for mapping all the x^i, where $k \leq i \leq 2k-1$, to x^i, where $0 \leq i \leq k-1$ as follows:

$$\begin{bmatrix} x^k \\ x^{k+1} \\ \cdots \\ x^{2k-1} \end{bmatrix} \equiv \begin{bmatrix} r_{0,0} & r_{0,1} & \cdots\cdots & r_{0,k-1} \\ r_{1,0} & r_{1,1} & \cdots\cdots & r_{1,k-1} \\ \cdots & \cdots & \cdots\cdots & \cdots\cdots \\ r_{k-1,0} & r_{k-1,0} & \cdots\cdots & r_{k-1,k-1} \end{bmatrix} \cdot \begin{bmatrix} x^0 \\ x^1 \\ \cdots \\ x^{k-1} \end{bmatrix} \bmod f(x).$$

The second approach to polynomial GF(2^k) field multiplication consists of only one step where both polynomial multiplication and reduction are performed.

Multipliers in GF(2^k) fields can be categorized in bit serial, digit serial, and bit parallel designs. Bit serial multipliers require k clock cycles to come up with a multiplication product and process data bit by bit. Digit serial multipliers require $D < k$ clock cycles to come up with a multiplication product and process data in d-bit digits. Bit parallel multipliers require one clock cycle to come up with a multiplication product and process data in k-bit values. Several different GF(2^k) field multiplication algorithms exist for each type of multipliers. Such algorithms are summarized in Figure 4.5 and analyzed independently in the following subsections.

Before we proceed with the analysis, some important aspects of GF(2^k) field multiplication should be highlighted. The form of the irreducible polynomial plays a very important role in the efficiency of the multiplication

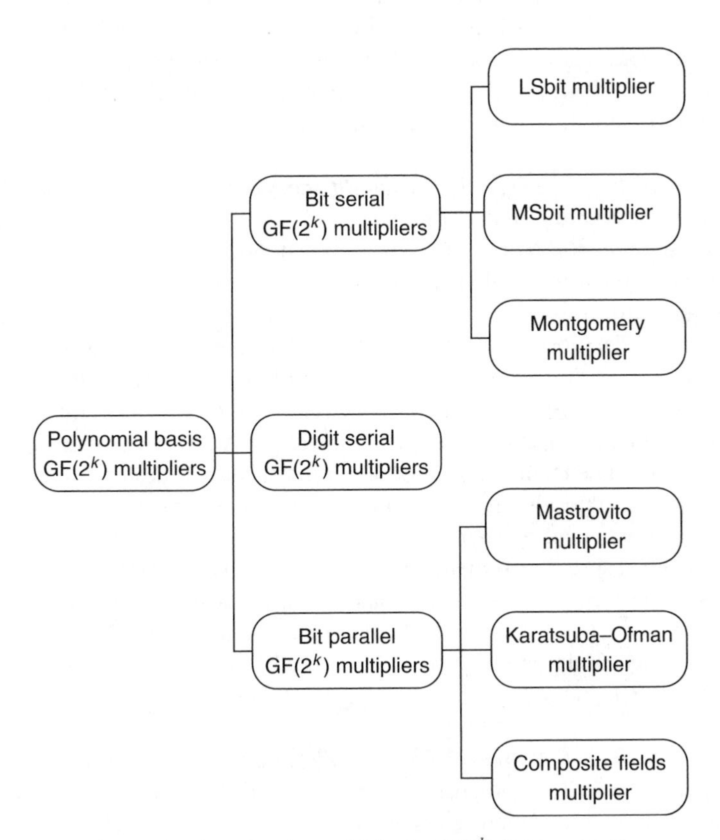

FIGURE 4.5 Categorization of polynomial basis GF(2^k) field multipliers.

process. A great deal of research has been undertaken in finding an appropriate irreducible polynomial that can optimize the multiplication process. Irreducible trinomials, equally spaced polynomials (ESP), or all one polynomials (AOP) have special properties that result in important optimizations in hardware as well as software multiplier designs [8].

Suppose that we have an irreducible polynomial $f(x) = x^{n_m} + x^{n_{m-1}} + x^{n_{m-2}} + \cdots + x^{n_1} + x^{n_0}$ then $n_m = k$ and $n_i - n_{i-1} = t_i$. If $n_i - n_{i-1} = t$ for all $i = \{0,1,2,\ldots,m\}$, then the polynomial $f(x)$ is called ESP. If all $t_i = 1$ the polynomial $f(x)$ is called AOP. Trinomials are the polynomials that have only three nonzero terms. NIST proposes some specific special type irreducible polynomials for cryptographic use in GF(2^k) fields with prime k. More can be found in [13,14].

However, the use of special irreducible polynomials restricts the reusability of a resulting multiplier. To avoid this problem, the notion of reconfigurability has been introduced in the GF(2^k) field multiplication process. Assume that a GF(2^k) field multiplier is able to handle generic type of irreducible polynomials and be able to perform multiplication not only for the underlined GF(2^k) field but for all GF(2^m) fields, where $0 < m < k$. In that case, the multiplier is called reconfigurable, versatile, or that it can handle arbitrary GF(2^k) fields and generic irreducible polynomials [37]. Not all the GF(2^k) field multipliers shown in Figure 4.5 can easily be made versatile.

4.8.2.1.2.1 Bit Serial Least Significant Bit and Most Significant Bit Multipliers

Using the polynomial reduction method in GF(2^k) fields along with the bit serial multiplication process, two well-known modular GF(2^k) field multiplication algorithms are obtained [38]. These algorithms follow the shift and add principle but process the multiplier $b(x)$ beginning from the least significant bit (LSB) or the most significant bit (MSB). Thus, those multipliers are called LSB or MSB multiplier, respectively (Algorithm 12 and Algorithm 13).

Algorithm 12 and Algorithm 13 consist only of shift operations ($x \cdot a$), XOR operations (+), and AND operations ($b_i \cdot a$). A bit serial architecture can easily be designed following these algorithms. Such a design for the MSB multiplication algorithm is shown in Figure 4.6, consisting of two input AND, XOR gates and 1-bit registers.

The use of special type irreducible polynomial can simplify the multiplication process. However, reconfigurability cannot be easily achieved in MSB and LSB multiplication algorithms since both algorithms use the MSB of a or c. Special circuitry is required for finding the MSB if we are to use a GF(2^k) field multiplier for GF(2^m) field multiplication with polynomials of degree m less than $k-1$.

4.8.2.1.2.2 Bit Serial Montgomery Multiplication for GF(2^k) Fields

Koç and Acar in [39] proposed a bit serial and digit serial version of the Montgomery multiplication algorithm for GF(2^k) fields. This algorithm is similar to the well-known algorithm for GF(p) fields. Instead of $a(x)b(x) \bmod f(x)$

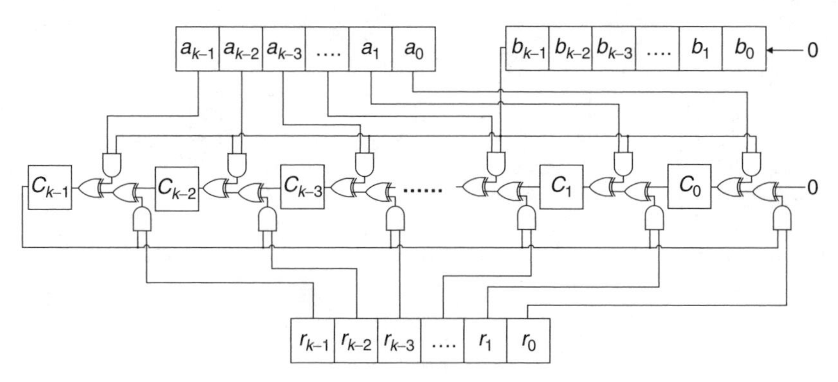

FIGURE 4.6 Bit serial MSB GF(2^k) field multiplier architecture.

the algorithm calculates $a(x)b(x)r^{-1}(x)$ mod $f(x)$, where $r(x)$ is a constant value. It is required that $gcd(r(x), f(x)) = 1$. Since $f(x)$ is irreducible the above condition is always valid.

Through the correct choosing of the value $r(x)$, the algorithm becomes less complex and can give efficient hardware architectures. The more appropriate choice would be $r(x) = x^k$. Then, the bit serial Montgomery multiplication algorithm for GF(2^k) fields can be presented (Algorithm 14).

The Montgomery multiplication algorithm for GF(2^k) fields is similar to the original algorithm of P. Montgomery, however, is more simple because of the GF(2^k) field carry-free logic and the lack of final subtraction. Its space and time complexity is similar to that of the LSB and MSB multipliers [40], and the algorithm can be simplified using special irreducible polynomials [41,42]. However, the removal of the factor x^{-k} in the Montgomery multiplication product in order to get the correct multiplication product adds an extra computational cost in the algorithm. Therefore, the Montgomery multiplication algorithm for GF(2^k) fields is useful in applications that require many multiplications without conversion from Montgomery representation to normal representation.

However, this algorithm is easily made reconfigurable. By padding the unused bits with zeros, a GF(2^k) field Montgomery multiplier can calculate a multiplication product for any GF(2^m) field, where $0 < m < k$. A bit serial GF(2^k) field Montgomery multiplier is shown in Figure 4.7.

4.8.2.1.2.3 Bit Parallel Mastrovito Multiplier

Suppose that $a(x)$, $b(x)$ are polynomials of the GF(2^k) field defined over the irreducible polynomial $f(x)$. Then, as already described, $c(x) = a(x)b(x)$ mod $f(x)$. The polynomials $c(x)$ and $b(x)$ can also be represented as column vectors C and B. Mastrovito in [8] introduced a $k \times k$ Matrix Z with elements $z_{i,j}$ defined as a function of the coefficients f_i and a_i, so that $C = Z \cdot B$.

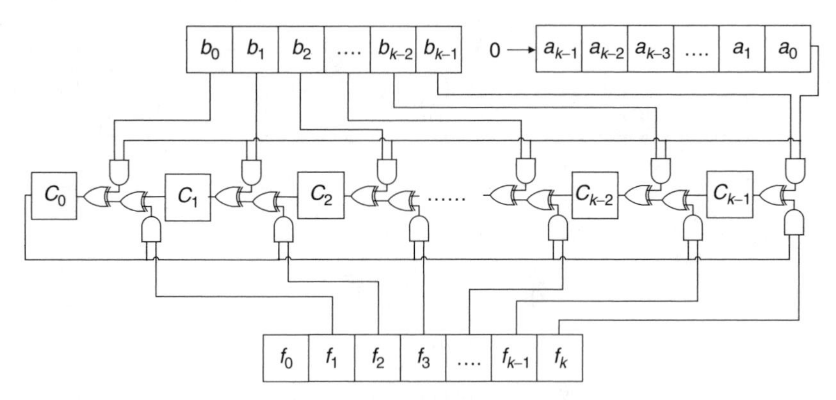

FIGURE 4.7 Bit serial GF(2^k) field Montgomery multiplier architecture.

$$C = Z \cdot B \Rightarrow \begin{bmatrix} c_0 \\ c_1 \\ \cdots \\ c_{k-1} \end{bmatrix} = \begin{bmatrix} z_{0,0} & z_{0,1} & \cdots\cdots & z_{0,k-1} \\ z_{1,0} & z_{1,1} & \cdots\cdots & z_{1,k-1} \\ \cdots & \cdots & \cdots\cdots & \cdots\cdots \\ z_{k-1,0} & z_{k-1,0} & \cdots\cdots & z_{k-1,k-1} \end{bmatrix} \cdot \begin{bmatrix} b_0 \\ b_1 \\ \cdots \\ b_{k-1} \end{bmatrix}.$$

This Z matrix, called product matrix, can be constructed by using the reduction matrix R and the multiplier polynomial $a(x)$, following the formula:

$$z_{i,j} = \begin{cases} a_i & \text{for } i = 0, 1, \ldots, k-1, j = 0, \\ u(i-j) \cdot a_{i-j} + \sum_{t=0}^{j-1} r_{j-1-t,i} \cdot a_{k-1-t} & \text{for } i = 0, 1, \ldots, k-2, j = 1, 2, \ldots, k-1. \end{cases}$$

The function $u(s)$ is defined as

$$u(s) = \begin{cases} 1 & s \geq 0, \\ 0 & s < 0. \end{cases}$$

Using equation $C = Z \cdot B$, each coefficient c_i of the multiplication product $C(x)$ can be written as a linear combination of the coefficients of $b(x)$ and the elements of the product matrix. From a hardware design perspective, each c_i can be calculated as a combination of AND gates and relevant XOR trees. This calculation is completed in one clock cycle. A generic architecture of a Mastrovito multiplier is shown in Figure 4.8.

A close inspection of the product matrix shows that it is highly dependent on the form of the irreducible polynomial defining the GF(2^k) field. Mastrovito was the first to perform a thorough analysis on the effect of $f(x)$ in the resulting design of a bit parallel multiplier in GF(2^k) fields. It has been found that irreducible trinomials, AOPs, and ESPs greatly improve the Mastrovito multiplier. More information can be found in [43].

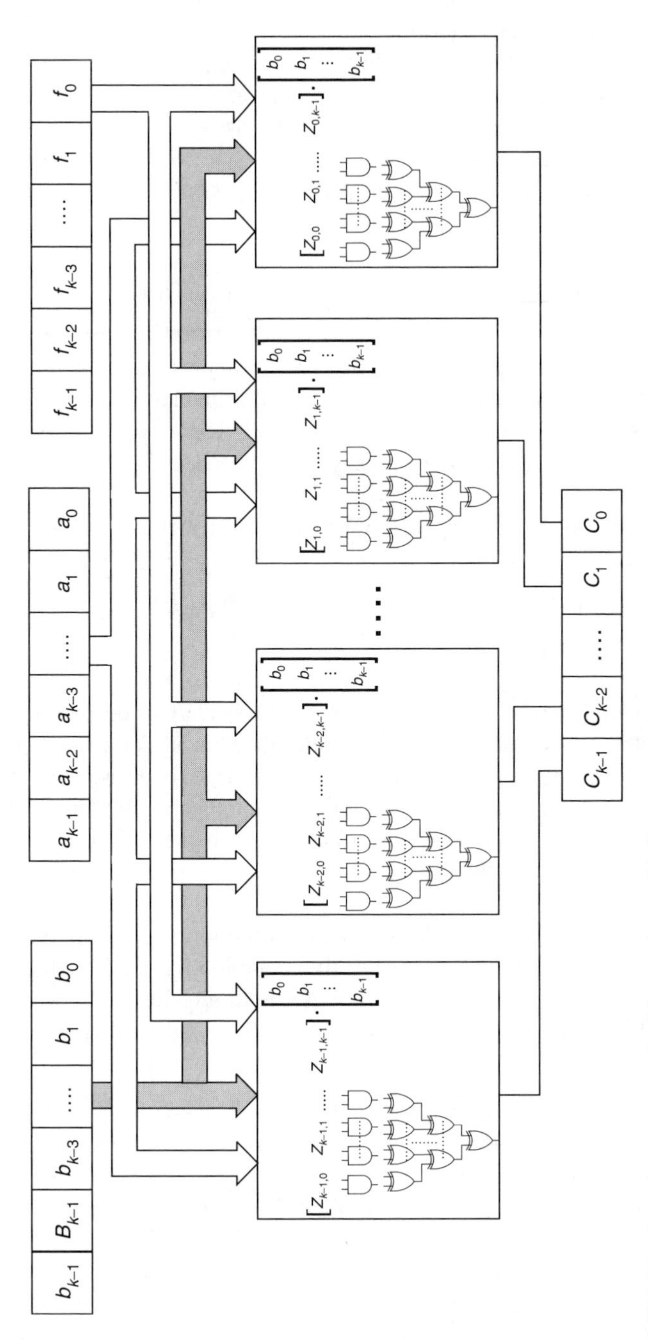

FIGURE 4.8 Generic structure of a Mastrovito multiplier.

4.8.2.1.2.4 Bit Parallel Karatsuba GF(2^k) Field Multiplier

The Karatsuba–Ofman multiplication algorithm applied in GF(p) fields can be successfully utilized for GF(2^k) fields. Suppose that $a(x)$, $b(x)$ are polynomials of a GF(2^k) field defined over an irreducible polynomial $f(x)$, and $k=2m$. We can rewrite $a(x)$, $b(x)$ as $a(x)=A_1(x)x^m+A_2(x)$ and $b(x)=b_1(x)x^m+B_2(x)$ where $A_1(x)$, $A_2(x)$, $B_1(x)$, and $B_2(x)$ are polynomials of $m-1$ degree. In that case, multiplying $a(x)\cdot b(x)$ becomes

$$\begin{aligned} a(x)\cdot b(x) &= (A_1(x)x^m + A_2(x))\cdot(B_1(x)x^m + B_2(x)) \\ &= (A_1(x)\cdot B_1(x))x^{2m} + (A_1(x)\cdot B_2(x) + A_2(x)\cdot B_1(x))x^m + (A_2(x)\cdot B_2(x)). \end{aligned}$$

Instead of performing one multiplication between k-bit polynomials, using Karatsuba–Ofman method three parallel multiplications and four parallel additions of m-bit polynomials are required, where $m=k/2$. We compute $C(x)=A_1(x)\cdot B_1(x)$, $D(x)=A_2(x)\cdot B_2(x)$, and then

$$\begin{aligned} (A_1(x)+A_2(x))\cdot(B_1(x)+B_2(x)) &= A_1(x)\cdot B_2(x) + A_2(x)\cdot B_1(x) \\ &\quad + A_1(x)\cdot B_1(x) + A_2(x)\cdot B_2(x) \Leftrightarrow \\ A_1(x)\cdot B_2(x) + A_2(x)\cdot B_1(x) &= (A_1(x)+A_2(x))\cdot(B_1(x)+B_2(x)) \\ &\quad + C(x) + D(x). \end{aligned}$$

Multiplication between the m degree polynomials can be performed using some other bit parallel or bit serial multiplier but it can also be performed using again the Karatsuba–Ofman method. In the second case, the Karatsuba–Ofman method is used recursively until a multiplication bit length is reached beyond which this method is not affordable. This method works even if k is not even. By padding with zeros, the multiplicand or multiplier can fit to the appropriate bit length.

The output of the Karatsuba–Ofman multiplier is a polynomial of $2k-1$ degree and has to be reduced using the reduction matrix R. Some researchers propose integrating both operations (multiplication and reduction) into one architecture [44], thus improving the speed and hardware resources of the design.

4.8.2.1.2.5 Finite Field Multipliers Based on Composite Finite Fields

Knowing a GF(2^k) field with $k=n\cdot m$, we can create the extension GF($(2^n)^m$) field defined over the irreducible polynomial $f(x)$. This field has polynomial elements $a(x)$ with coefficients a_i, $f_i\in$ GF(2^n) defined over the irreducible polynomial $n(x)$ and is also called composite finite field.

Suppose that we have the $a(x)$, $b(x)$ elements of GF($(2^n)^m$) field, where $a(x)=\sum_{i=0}^{m-1} a_ix^i = a_{m-1}x^{m-1}+a_{m-2}x^{m-2}+\cdots+a_1x+a_0$, $b(x)=\sum_{i=0}^{m-1} b_ix^i = b_{m-1}x^{m-1}+b_{m-2}x^{m-2}+\cdots+b_1x+b_0$ and each a_i, b_i are elements of the GF(2^n) field. Then multiplication between those elements would be

$$
\begin{aligned}
c(x) &= \sum_{i=0}^{m-1} c_i x^i = a(x) \cdot b(x) = a(x) \cdot b(x) \bmod f(x) \\
&= (a(x) \cdot b_{m-1} x^{m-1} + a(x) \cdot b_{m-2} x^{m-2} + \cdots\cdot + a(x) \cdot b_0) \bmod f(x) \\
&= ((a_{m-1} b_{m-1} x^{2m-2} + a_{m-2} b_{m-1} x^{2m-3} + \cdots\cdot + a_1 b_{m-1} x^m + a_0 b_{m-1}) \\
&\quad \cdots + (a_{m-1} b_0 + a_{m-2} b_0 + \cdots\cdot + a_1 b_0 + a_0 b_0)) \bmod f(x).
\end{aligned}
$$

Each product $a_i \cdot b_j$ in GF($(2^n)^m$) fields is not a simple AND operation as in GF(2^k) fields but rather a full GF(2^n) field multiplication, $a_i(x) \cdot b_j(x)$ mod $n(x)$. Therefore, multiplication in GF($(2^n)^m$) fields can be analyzed to multiplication in two different finite fields, a GF(2^m) field defined over $f(x)$ and a GF(2^n) field defined over $n(x)$. However, these multiplication operations of a GF(2^n) field can be carried out in parallel and since $n, m < k$, the overall multiplication delay is reduced.

A usual design approach for multiplication in GF($(2^n)^m$) fields is the choice of an appropriate bit parallel multiplier type for GF(2^n) field operations and a different multiplier type for the overlaid GF(2^m) field. For example, in [45] a Mastrovito multiplier is chosen for GF(2^n) field multiplication and a Karatsuba–Ofman multiplier for GF(2^m) field multiplication.

4.8.2.1.2.6 Digit Serial GF(2^k) Field Multipliers

Suppose that $b(x)$ is an element in digit format of a GF(2^k) field defined over $f(x)$ and assign the digit size as d. Then, the number of digits would be $D = \lceil k/d \rceil$ and the $b(x)$ element in polynomial format would be

$$
b(x) = \sum_{i=0}^{D-1} B_i(x) x^{di} = B_{D-1}(x) x^{d(D-1)} + B_{D-2}(x) x^{d(D-2)} + \cdots + B_1(x) x^d + B_0(x),
$$

where each $B_i(x)$ would be

$$
B_i(x) = \sum_{j=0}^{d-1} b_{Di+j} x^j = b_{Di+d-1} x^{d-1} + b_{Di+d-2} x^{d-2} + \cdots + b_{Di+1} x^1 + b_{Di}.
$$

Suppose that we want to multiply two elements $a(x)$, $b(x)$ of a GF(2^k) field defined over $f(x)$. Then, by representing one or both of those elements in digit format, all the bit serial multipliers presented earlier can be adjusted to process digits instead of bits in each clock cycle. For LSB multiplication we use the following equation:

$$
\begin{aligned}
a(x) \cdot b(x) \bmod f(x) &= \left(a(x) \cdot \sum_{i=0}^{D-1} B_i(x) x^{di}\right) \bmod f(x) = \big((a(x) B_0(x) \\
&\quad + B_1(x)(a(x) x^d \bmod f(x)) + B_2(x)(a(x) x^d \cdot x^d \bmod f(x)) \\
&\quad \cdots + B_{D-1}(a(x) x^{d(D-2)} \cdot x^d \bmod f(x))\big) \bmod f(x).
\end{aligned}
$$

while for MSB multiplication we follow the equation:

$$
\begin{aligned}
a(x) \cdot b(x) \bmod f(x) &= \left(a(x) \cdot \sum_{i=0}^{D-1} B_i(x)x^{di}\right) \bmod f(x) \\
&= \big((((\cdots \cdot (((B_{D-1}(x)a(x) \bmod f(x))x^d + a(x)B_{D-2}(x)) \\
&\qquad \bmod f(x))x^d + \cdots)x^d + B_1(x)a(x)) \bmod f(x))x^d \\
&\qquad + B_0(x)a(x)\big) \bmod f(x).
\end{aligned}
$$

As a result of the above, a digit serial version of LSB, MSB algorithm for GF(2^k) fields can be presented (Algorithm 15 and Algorithm 16).

The digit serial versions of LSB and MSB algorithms involve modular reduction with the irreducible polynomial $f(x)$. Reduction can be performed using the reduction matrix R. This operation is less complex than bit parallel reduction because $d < k$. Several approaches for optimizing this process exist, such as [37,46–48].

Koç and Acar in [39] proposed a word-level version of Montgomery multiplication algorithm for a GF(2^k) field. This version can be considered digit serial, assuming that $a(x)$, $b(x)$, $f(x)$ are in digit format (Algorithm 17).

The above algorithm requires inversion of the irreducible polynomial $f(x)$ and multiplication between d degree polynomials. In hardware design, most of the multiplications between digits can be done in parallel and the inversion can be optimized significantly since the modulus is x^d, a power of the base. However, the algorithm remains more efficient in software designs [39].

4.8.2.1.2.7 Hardware Design of GF(2^k) Field Multipliers

Systolic arrays are widely used in bit serial and digit serial multipliers to increase the multiplication throughput. However, the latency remains unchanged. The use of two-dimensional systolic arrays bears high cost in hardware area resources and can be used in applications where such resources are unimportant. In wireless handheld applications, this cost usually cannot be ignored. Therefore, like in the case of GF(p) fields, one-dimensional systolic arrays are used in the design of GF(2^k) field multipliers.

In the case of bit parallel GF(2^k) field multipliers, the use of special type of irreducible polynomials is inevitable in designing efficient multipliers in speed and hardware resources. Many such designs have been proposed [43,49] giving promising results. Bit parallel versions of Montgomery multiplication have also been proposed for irreducible trinomials [41,42] [50] that manage to achieve results comparable to other types of bit parallel multipliers.

Some researchers have also proposed the use of polynomial residue arithmetic, the polynomial equivalent of RNS. Halbutoğullari in [51] proposes such a multiplier in polynomial residue arithmetic, which uses the MSB multiplication algorithm and lookup table reduction method to speed up the multiplication process. Similarly, Bajard in [52] proposes a Montgomery multiplication algorithm in trinomial residue arithmetic.

An overview of the time delay and area resources required in various GF(2^k) field multipliers is shown in Table 4.2.

TABLE 4.2
Critical Path Delay, Latency, and Gate-MUX-DFF Number of GF(2^k) Field Multiplier Hardware Architectures

GF(2^k) Field Multiplier	Gates: AND	Gates: XOR	2 to 1 MUX	DFF	Latency	Critical Path Delay
Bit Serial Multipliers						
LSB [17]	$2k$	$2k$	0	$3k$	k	$T_A + T_X$
MSB [17]	$2k$	$2k$	0	$3k$	k	$T_A + 2T_X$
Montgomery [53]	k	k	$2k + 1$	$3k$	$2k$	$T_A + T_X$
Systolic–Semisystolic Multipliers						
LSB semisystolic [55]	$2k^2$	$2k^2$	0	$3k^2$	$k + 1$	$T_A + T_X$
MSB systolic [54]	$2k^2$	$2k^2$	0	$7k^2$	$3k$	$T_A + T_X + T_M$
Montgomery general polynomial [53]	$k^2 - k + 1$	$k^2 - 1$	$2k^2 + k - 3$	$2k^2 - k$	$2k$	$T_A + T_X$
Montgomery AOP [50]	$k^2 + k + 1$	$k^2 + k + 1$	0	$3k^2 + 3k + 3$	$k + 1$	$T_A + T_X$
Digit Serial Multipliers						
LSB [46]	$2kd$	$2kd$	k	$2k + d - 1$	$D + 1$	$d(T_A + T_X)$
LSB systolic [48]	$2d^2 + d$	$2d^2$	$2d$	$10d + 1$	$3D$	$T_A + T_X + (d - 1)(T_A + T_X + T_M)$
Bit Parallel Multipliers						
Mastrovito [43]						
General polynomial	k^2	$(k-1)(k+m-1) + \sum_{j \in S*} (2k - 1 - j)$	0	0	1	$T_A + (\lceil \log_2 \lvert S \rvert \rceil + (m-1) + \lceil \log_2 k \rceil) T_X$
EST	k^2	$k^2 - (k/2)$	0	0		$T_A + (1 + \lceil \log_2 k \rceil) T_X$
Trinomials	k^2	$k^2 - 1$	0	0		$T_A + (\lceil k - 1/t \rceil + \lceil \log_2 k \rceil) T_X$
AOP	k^2	$k^2 - 1$	0	0		$T_A + (1 + \lceil \log_2 k \rceil) T_X$
ESP	k^2	$k^2 - t$	0	0		$T_A + (1 + \lceil \log_2 k \rceil) T_X$
Montgomery trinomial [41]	k^2	$k^2 - 1$	0	0	1	$T_A + (2 + \lceil \log_2 (k-2) \rceil) T_X$
Karatsuba [49]	$3\frac{k^2}{4} + 2k + 1$	$3\frac{k^2}{4} + 3k + 1$	0	0		$T_A + (1 + \lceil \log_2 (k+1) \rceil) T_X$

The notation in Table 4.2 is identical to the definitions of the special type of polynomials presented in Section 4.8.2.1.2. The subset S is the set of indices k, $k+t$, $k+t+1,\ldots,2k-3$, $2k-2$, while the subset S^* is $(S-min(S))$. T_A is the delay of an AND gate, T_X is the delay of an XOR gate, and T_M is the delay of an MUX.

4.8.2.1.3 GF(2^k) Field Squaring in Polynomial Basis Representation

Suppose that $a(x)$ is an element of a GF(2^k) field defined over an irreducible polynomial $f(x)$. Then, the square of $a(x)$ would be $A^2=a^2(x) \bmod f(x)$. However, squaring the polynomial $a(x)$ without reduction would give us $a^2(x) = a_{k-1}x^{2k-2} + a_{k-2}x^{2k-4} + \cdots + a_1x^2 + a_0$. The vector of $a^2(x)$ would be $a^2 = \{0, a_{k-1}, 0, a_{k-2}, 0, a_{k-3}, 0, \ldots, a_1, 0, a_0\}$. Therefore, $a^2(x)$ can be created from $a(x)$ element by placing zero values between two consecutive coefficients of $a(x)$. The outcome is a $2k-2$ degree polynomial extended from a $k-1$ degree polynomial. In order to get the correct GF(2^k) field squaring result, a reduction operation is applied to the resulting polynomial using the reduction matrix R. Knowing the structure of $a^2(x)$, the bit serial multiplication algorithms can be greatly simplified and can perform GF(2^k) field squaring with reduced computation cost in comparison with GF(2^k) field multiplication.

4.8.2.1.4 GF(2^k) Field Inversion–Division in Polynomial Basis Representation

Consider a polynomial $a(x)$ of a GF(2^k) field defined over an irreducible polynomial $f(x)$. There exists a polynomial $s(x) \in \mathrm{GF}(2^k)$ so that $a(x) \cdot s(x) \equiv 1(\bmod f(x))$. This polynomial $s(x)$ is denoted as $a^{-1}(x)$ and is called multiplicative inverse of $a(x)$. The process of finding a multiplicative inverse is called inversion. The multiplicative inverse $a^{-1}(x)$ exists as long as $f(x)$ and $a(x)$ are coprime and since $f(x)$ is irreducible, the above constraint is always valid.

Inversion in a GF(2^k) field is performed using algorithms similar to the ones presented for GF(p) fields, adjusted accordingly using carry-free logic. The dominant inversion algorithm is the EEA for GF(2^k) fields, especially in its binary form. Another approach to inversion is the use of consecutive multiplication and squaring operations following Fermat's Little Theorem.

4.8.2.1.4.1 EEA for GF(2^k) Field Inversion

The operation $a(x) \cdot a^{-1}(x) \equiv 1(\bmod f(x))$ can be written as $a(x) \cdot a^{-1}(x) + f(x)\hat{f}(x) = 1$ and can be calculated using the EEA for GF(2^k) fields through proper initialization (Algorithm 18).

In the $EEA(a, f)$, four operations are performed in every round:

- Division of the s and r variables. Its remainder is $s-q \cdot r$ and is used as the r value of the following step and its quotient q is needed for the calculation of $v-q \cdot u$.

- Multiplication for the calculation of $v - q \cdot u$.
- Subtraction operation, which is identical to addition in GF(2^k) fields.
- Swap operation where the values of r and u are exchanged with s and v, respectively.

While the swap and subtraction operations have trivial computation complexity, multiplication and especially division are complex, time-consuming operations. Additionally, the number of repetitions until $r = 0$ (when a result is reached) is not constant and probing the progress of r at every round requires extra computational effort. Therefore, $EEA(a, f)$, in this form, is unsuitable for hardware design because it cannot achieve small critical path delay and high throughput. Techniques like pipelining cannot be efficiently employed and systolic arrays that could dramatically decrease the critical path delay and increase throughput are inapplicable because of the nonconstant loop number.

To avoid these problems, modified versions of the $EEA(a, f)$ have been proposed [56–61]. These algorithms use a different division process in order to find the "remainders" $s - q \cdot r$ and $v - q \cdot u$ without calculating the quotient q. We usually denote those algorithms as MEEA [60] (Algorithm 19). In Algorithm 19, all the values are k degree polynomials and d is an integer.

While $MEEA(a, f)$ is attractive for bit serial or systolic design, Yan et al. [58] have found that it can be further optimized when analyzed bit by bit, resulting in a binary variation of the algorithm (Algorithm 20). In Algorithm 20 all the values are k degree polynomials and the superscripts $^{(i)}$ indicate the current round (round i).

Using this algorithm, an inversion operation can be designed in hardware by one-dimensional (bit serial approach) or two-dimensional systolic arrays, as proposed in [59].

4.8.2.1.4.2 Inversion In GF(2^k) Fields Using Fermat's Little Theorem

For every element a of GF(2^k) field regardless of the field basis representation, the power a^{2^k} can be calculated using Fermat's Little Theorem as $a^{2^k} = a$. In that case, the multiplicative inverse can be found by multiplying both sides of $a^{2^k} = a$ with a^{-2}. Then $a^{2^k-2} = a^{-1}$ and $2^k - 2$ is analyzed into $2^k - 2 = 2 + 2^2 + 2^3 + \cdots + 2^{k-1}$. The multiplicative inverse becomes $a^{-1} = a^{2+2^2+\cdots+2^{k-1}} = a^2 \cdot a^{2^2} \cdot \ldots \cdot a^{2^{k-1}}$ and can be calculated through a process of repeated squarings and multiplications. Such algorithmic processes for performing inversion can be applied to any GF(2^k) field base representation (Algorithm 21).

In polynomial basis representation of a GF(2^k) field the above iterative algorithm is considered slower than other inversion algorithms. The reason is the high number of multiplications and squaring operations. Although squaring is less computationally complex than multiplication, it still employs reduction with the irreducible polynomial. Because of this, this inversion method is not chosen for efficient design of polynomial basis GF(2^k) field inversion.

4.8.2.1.4.3 Other GF(2^k) Field Inversion Methods

Variations of the above inversion algorithms have been proposed by some researchers to reduce the number of computational rounds or increase the hardware efficiency (speed and hardware resources). A well-known variation of the EEA was proposed by Schroeppel et al. in [62]. This method, called almost inverse, achieves similar results and performs inversion in less computation rounds compared with the EEA for GF(2^k) fields.

Another approach in inversion is the Montgomery invert algorithm for GF(2^k) fields as a direct transformation of the Montgomery invert algorithm described in Section 4.8.1.4.2 from GF(p) fields to GF(2^k) fields. In the GF(2^k) field version of the algorithm, addition is identical to subtraction, all the values are polynomials, and the output is the modulus of the irreducible polynomial $f(x)$ instead of p. More on GF(2^k) inversion can be found in [63].

4.8.2.1.4.4 GF(2^k) Field Division

Division in GF(2^k) fields consists of two operations. One inversion and one multiplication are employed to calculate the division output, following the methodology presented for division in GF(p) fields. Such designs include a GF(2^k) systolic inverter concatenated with a GF(2^k) systolic multiplier [58,61].

By proper initialization, the EEA for GF(2^k) fields can be used for division although the resulting hardware architecture does not achieve optimistic results in terms of speed and hardware resources. Another approach is the reusability of functions used for both inversion and multiplication so as to design a reconfigurable architecture that can perform the two operations with small extra cost [64].

4.8.2.2 Normal Basis Representation

Massey and Omura in [65] proposed a new way of representing the elements of a GF(2^k) field. Normal basis (NB) element representation of a GF(2^k) field over GF(2) uses the base $B = \left(x, x^2, x^{2^2}, x^{2^3}, x^{2^4}, \ldots, x^{2^{k-1}}\right)$, where we say that x generates the normal basis or that x is a normal element of GF(2^k) over GF(2).

Every GF(2^k) field has an NB [66,67] and each element A of that GF(2^k) field can be represented using that NB as

$$A = \sum_{i=0}^{k-1} a_i x^{2^i} = a_0 x^{2^0} + a_1 x^{2^1} + a_2 x^{2^2} + \cdots + a_{k-1} x^{2^{k-1}},$$

where $a_i \in$ GF(2) or in vector format as $A = (a_{k-1}, a_{k-2}, \ldots, a_1, a_0)$.

4.8.2.2.1 GF(2^k) Field Addition–Subtraction in Normal Basis Representation

As presented in Section 4.8.2.1.1, addition and subtraction in GF(2^k) fields are identical operations and can both be interpreted as an XOR operation between the GF(2^k) field elements A and B regardless of the representation base used in this field.

4.8.2.2.2 GF(2^k) Field Multiplication in Normal Basis Representation

Suppose that A, B are elements of a GF(2^k) field using normal basis representation. If the result of multiplying those two elements is $C = A \cdot B$, then C would be

$$C = \sum_{m=1}^{k-1} c_m x^{2^m} = A \cdot B = \sum_{i=1}^{k-1} a_i x^{2^i} \cdot \sum_{j=1}^{k-1} b_j x^{2^j}$$
$$= \sum_{i=1}^{k-1} \sum_{j=1}^{k-1} a_i b_j x^{2^i} x^{2^j},$$
$$0 \leq i, j, l \leq k - 1.$$

If we define $t_{i,j}^{(m)} \in \mathrm{GF}(2)$ as $x^{2^i} \cdot x^{2^j} = \sum_{m=0}^{k-1} t_{i,j}^{(m)} x^{2^m}$, then each c_m coefficient of C can be represented as

$$c_m = \sum_{i=1}^{k-1} \sum_{j=1}^{k-1} a_i b_j t_{i,j}^{(m)} = A T_m B^{\mathrm{T}}, \quad 0 \leq m \leq k - 1.$$

where A, B are the multiplier and multiplicand in vector format, B^{T} is the transpose of B and $T_m = (t_{i,j}^{(m)})(^0)$ is a $k \times k$ matrix, called multiplication table matrix, with $t_{i,j}^{(m)}$ elements of a specific m value corresponding to x^{2^m}. The multiplication table matrix can be considered a mapping of all the $x^{2^i} \cdot x^{2^j}$ combinations for a certain m. If a $x^{2^i} \cdot x^{2^j}$ combination is not zero for a given m, then $t_{i,j}^{(m)} = 1$ and since $c_m = A T_m B^{\mathrm{T}}$ the corresponding $a_i b_j$ partial product exists in c_m. The collection of matrixes $\{T_m\}$ is called a multiplication table of a GF(2^k) field over GF(2).

Massey and Omura [65] have found that each coefficient c_m can be calculated from the multiplication table matrix T_0 by rotating m bits the vectors A and B. Each coefficient of the multiplication product C can be found as

$$c_m = \sum_{i=1}^{k-1} \sum_{j=1}^{k-1} a_{i+m} b_{j+m} t_{i,j}^{(0)}, \quad 0 \leq m \leq k - 1,$$

where all the subscripts are considered modulus k.

The number of nonzero elements in the multiplication table matrix T_m of a GF(2^k) field using normal basis representation is called the complexity of the normal basis C_N and is $C_N \geq 2k - 1$. The complexity of a normal basis plays a crucial role in the design of a hardware architecture since each nonzero element of T_m corresponds to a partial product $a_i b_j$ (an AND operation) in c_m.

The normal basis that has the lowest complexity of $C_N = 2k-1$ is called ONB. Those normal bases are special cases of the Gaussian normal basis of type t, for $t = 1$ and $t = 2$ [68]. Therefore, two ONB types exist, Type I ONB (Gaussian normal basis Type 1) and Type II ONB (Gaussian normal basis Type II).

If $k+1$ is prime and 2 is primitive in GF($k+1$) (meaning that 2^s mod($k+1$) is a number in the range $\{0, 1, \ldots, k\}$, where $0 \leq s \leq k-1$), then the nontrivial $(k+1)$ roots of unity form an ONB of GF(2^k) field over GF(2) called Type I ONB.

If $2k+1$ is prime and 2 is primitive over GF($2k+1$) or $2k+1 \equiv 3(mod\ 4)$ and the multiplicative order of 2 in GF($2k+1$) is k, then $x = \gamma + \gamma^{-1}$ generates an ONB of a GF(2^k) field over GF(2), where γ is a primitive $(2k+1)$ root of unity. This ONB is called Type II ONB.

Following the above definitions all $t_{i,j}^{(0)}$ can be found by solving appropriate systems of equation [66]. Whenever an i, j pair satisfying the equation system is found, then $t_{i,j}^{(0)} = 1$. The system of equations for Type I ONB would be

$$2^i + 2^j \equiv 1 \bmod k+1,$$
$$2^i + 2^j \equiv 0 \bmod k+1,$$

and the system of equations for Type II ONB would be

$$2^i + 2^j \equiv \pm 1 \bmod 2k+1,$$
$$2^i - 2^j \equiv \pm 1 \bmod 2k+1.$$

As a result of the above remarks, a multiplier architecture in a normal basis representation GF(2^k) field can be designed by a series of AND operations (one for each a_ib_j partial product of the coefficient c_m of the product C) and a series of XOR operations. The AND operations are performed in parallel and the XOR operations are used for adding all the a_ib_j partial products to calculate each coefficient c_m. The multiplication process depends heavily on the complexity C_N and structure of the multiplication table matrix T_0.

There are various normal basis GF(2^k) field multiplier designs. Bit serial multipliers process the input 1-bit per clock cycle but can give the product C either in parallel after k clock cycles (serial multiplier parallel output, SMPO) or in a serial way by calculating one coefficient c_m per clock cycle (serial multiplier serial output, SMSO). There are also bit parallel normal basis multiplier designs that calculate the multiplication product in one clock cycle at the expense of extra hardware resources. Most of the different NB multipliers are also extended for ONBs, especially bit parallel designs. Those systems achieve far better hardware resource efficiency and speed compared with general normal basis designs.

4.8.2.2.2.1 SMPO Normal Basis Multipliers

Agnew et al. in [69] first proposed the SMPO normal basis design following a principle somewhat different from the classical Massey Omura multiplication approach. Rewriting the Massey Omura equation $c_m = \sum_{i=1}^{k-1} \sum_{j=1}^{k-1} a_{i+m} b_{j+m} t_{i,j}^{(0)}$ we can obtain $c_m = \sum_{j=1}^{k-1} b_{j+m} \sum_{i=1}^{k-1} a_{i+m} t_{i,j}^{(0)} = \sum_{j=1}^{k-1} F_j^{(m)}$, where $F_j^{(m)} = b_{j+m} \sum_{i=1}^{k-1} a_{i+m} t_{i,j}^{(0)}$. Assuming that we work in t clock cycle, the above function can be defined as $F_j^{(m)}(t) = b_{j+m+t} \sum_{i=1}^{k-1} a_{i+m+t} t_{i,j}^{(0)}$. The coefficient c_0 using the above equations would be $c_0 = \sum_{j=1}^{k-1} F_j^{(m)}(-m)$. In each clock cycle t a collection of $F_j^{(m)}(t)$ can be calculated, where $0 \leq m, t \leq k-1$, using A, B rotated by t bits. Each $F_j^{(m)}(t)$ corresponds to a coefficient c_m. The resulting value of each $F_j^{(m)}(t)$ is stored in the $T^{(m)}$ D Flip Flop of register T after adding to it the output of $T^{(m-1)}$. When $t = k-1$ (after k clock cycles), the output of the T register is the multiplication product C. A generic hardware architecture following the design methodology of Agnew is shown in Figure 4.9.

By optimizing the $F_j^{(m)}(t)$ function some researchers [70–72] have managed to reduce the required hardware resources and critical path delay of a resulting multiplier.

SMPO architectures are designed for general normal basis representation. However, if the theory of ONB is applied to SMPO then that type of multiplier becomes very fast and requires reduced hardware resources compared with other designs.

4.8.2.2.2.2 SMSO Normal Basis Multipliers

Massey and Omura in [65] proposed an SMSO multiplier architecture as a direct implementation of equation $c_m = \sum_{i=1}^{k-1} \sum_{j=1}^{k-1} a_{i+m} b_{j+m} t_{i,j}^{(0)}$. To design an architecture so as to calculate the coefficient c_0 of the multiplication product C using $c_0 = \sum_{i=1}^{k-1} \sum_{j=1}^{k-1} a_i b_j t_{i,j}^{(0)}$, a series of AND gates arranged in parallel are required for calculating the partial products $a_i b_j$ and an XOR tree for adding all the resulting partial product results. To find the remaining c_m coefficients of C, cycle shifting of the A, B inputs by m is required for the correct result to reach the output of the XOR Tree. Therefore, after calculating c_0, to find the coefficient c_1, cycle shifting A, B by 1-bit to the left is needed. Using the Massey Omura methodology the resulting multiplier calculates one coefficient of the multiplication product per clock cycle. A general design of the SMSO Massey Omura multiplier is shown in Figure 4.10.

SMSO architectures depend heavily on the complexity C_N of the multiplication table matrix, because they employ an XOR tree for adding all the $a_i b_j$ partial products. They have a very high critical path delay and are relatively slow compared to SMPO designs. To overcome this problem, ONBs are used in the design of SMSO architectures.

4.8.2.2.2.3 Bit Parallel Normal Basis Multipliers

The simpler approach in designing a bit parallel NB multiplier is to use k different Massey Omura multipliers in parallel, fitting equation

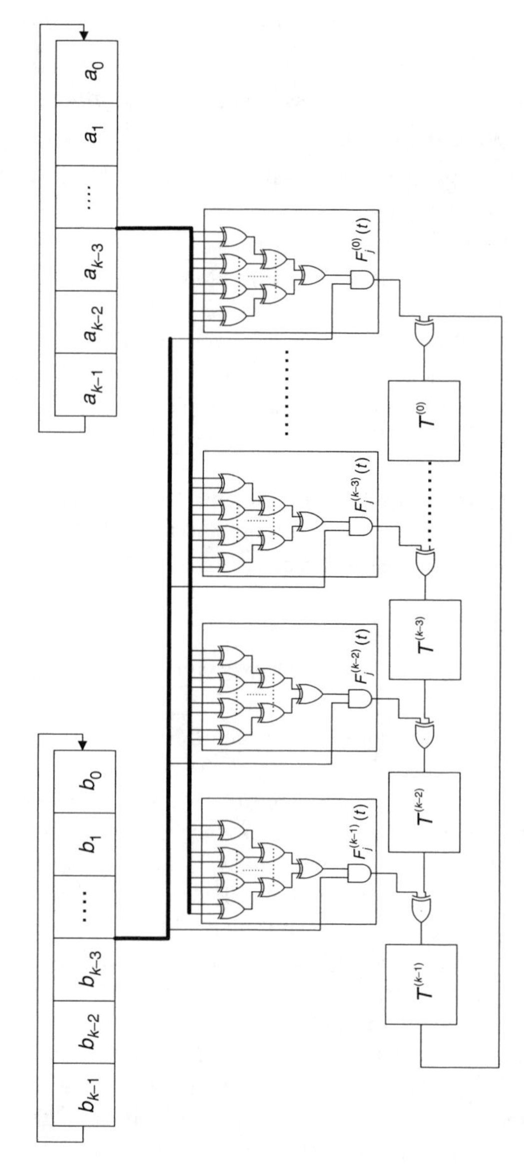

FIGURE 4.9 SMPO normal basis multiplier generic architecture.

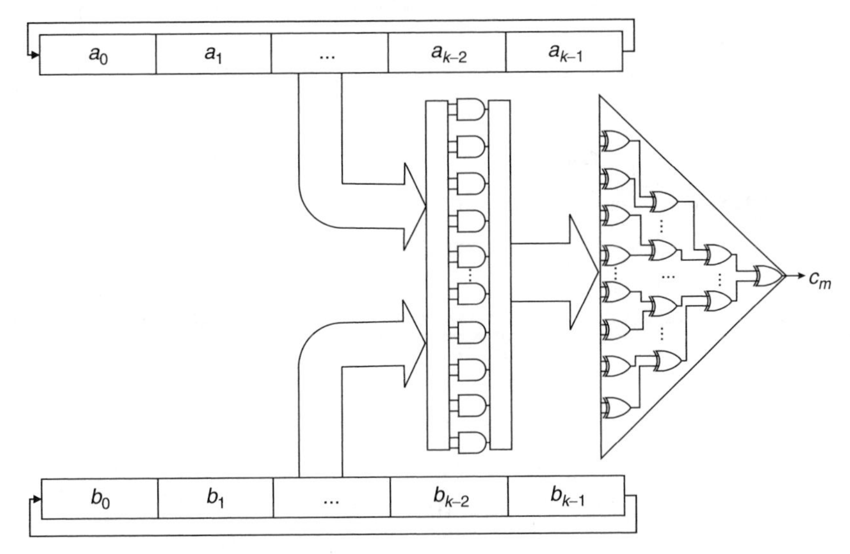

FIGURE 4.10 General SMSO Massey Omura normal basis multiplier architecture.

$c_m = \sum_{i=1}^{k-1}\sum_{j=1}^{k-1} a_{i+m}b_{j+m}t_{i,j}^{(0)}$ for $0 \leq m \leq k-1$ accordingly to every one of them. Such multipliers are called Massey Omura parallel multipliers and are useful only in ONB representation where we have low complexity $C_N = 2k - 1$.

Many researchers propose optimizations that increase the performance of parallel ONB multipliers. In [73,74] the structure of the multiplication table matrix is analyzed, and redundancy is found when ONBs are used (especially Type I ONBs). Some a_ib_j partial products are used more than once in the multiplication process. Using appropriate transformation of the NB multiplication table, this redundancy is removed and the hardware resources along with the critical path delay are reduced.

In other approaches, optimizations consist of an increase in the degree of parallelism by employing composite fields [75] or the definition of the NB as an extension of polynomial basis representation for the same GF(2^k) field [76]. Table 4.3 is an overview of GF(2^k) field ONB multiplication results for each type of multiplier described in the Section 4.8.2.2.2.

4.8.2.2.3 GF(2^k) Field Squaring in Normal Basis Representation

One of the main advantages in the use of normal basis representation for GF(2^k) field elements is the simplicity of the squaring operation in this representation. Suppose that $A = \{a_{k-1}, a_{k-2}, \ldots, a_1, a_0\}$ is an element of a GF(2^k) field. Then, the square of A, denoted as A^2, would be

$$A^2 = \sum_{i=0}^{k-1} (a_i x^{2^i})^2 = a_0(x^{2^0})^2 + a_1(x^{2^1})^2 + a_2(x^{2^2})^2 + \cdots + a_{k-1}(x^{2^{k-1}})^2$$

TABLE 4.3
Critical Path, Latency, and Gate Number of ONB GF(2^k) Field Multipliers

GF(2^k) Field ONB Multiplier	Gates AND	Gates XOR	DFF	Latency	Critical Path
		SMSO ONB Multipliers			
M.O. [65]	$2k-1$	$2k-2$	$2k$	k	$T_{AND} + (1 + \lceil \log_2 (k-1) \rceil) T_{XOR}$
Gao [77]	k	$2k-2$	$2k$	k	$T_{AND} + (1 + \lceil \log_2 (k-1) \rceil) T_{XOR}$
		SMPO ONB Multipliers			
Agnew [69]	k	$2k-1$	$3k$	$k+1$	$T_{AND} + 2T_{XOR}$
Masoleh [70]	$\left\lfloor \frac{k}{2} \right\rfloor + 1$	$2k-1$	$3k$	k	$T_{AND} + 3T_{XOR}$
Yang [71]	k	$\frac{3k-1}{2}$	$3k$	k	$T_{AND} + 2T_{XOR}$
		Bit Parallel ONB Multipliers			
Gao [77]	k^2	$2k^2-2k$	0	1	$T_{AND} + (1 + \lceil \log_2 (k-1) \rceil) T_{XOR}$
Hasan [74]	k^2	k^2-1	0	1	$T_{AND} + (1 + \lceil \log_2 (k-1) \rceil) T_{XOR}$
Sunar & Koc [76]	k^2	k^2-1	0	1	$T_{AND} + (2 + \lceil \log_2 (k-1) \rceil) T_{XOR}$
Masoleh [73]	k^2	k^2-1	0	1	$T_{AND} + (1 + \lceil \log_2 (k-1) \rceil) T_{XOR}$

and since $(x^{2^i})^2 = x^{2^{i+1}}$ from Fermat's Little Theorem ($x^{2^k} = x$), the above equation becomes

$$A^2 = a_{k-1}x^{2^0} + a_0x^{2^1} + a_1x^{2^2} + \cdots + a_{k-2}x^{2^{k-1}}.$$

Therefore, A^2 can be found by shifting each coefficient a_i to the $x^{2^{i+1}}$ base power and rotating the last coefficient to x^{2^0}. The square of A in vector format is $A^2 = (a_{k-2}, a_{k-3}, \ldots, a_0, a_{k-1})$.

4.8.2.2.4 GF(2^k) Field Inversion–Division in Normal Basis Representation

Inversion in GF(2^k) fields using normal basis representation follows the same basic principles as in polynomial basis representation GF(2^k) fields. The algorithms applied to polynomial basis GF(2^k) fields have been proposed for normal basis GF(2^k) fields. However, because of the different ways in which multiplication and squaring operations are performed in NB, some inversion algorithms are more easily applicable than others.

The EEA for GF(2^k) fields can be used for normal basis inversion through an intermediate state conversion to polynomial basis representation, which is applicable only for specific irreducible polynomials, such as AOP, and is usually not affordable in comparison with other normal basis inversion methods. Very recently, a different approach to the Euclidean algorithm for normal basis has been proposed [78], which does not use conversion to and from polynomial basis representation but its efficiency remains to be investigated.

On the other hand, inversion using Fermat's Little Theorem gives optimistic results in terms of speed and required hardware resources because of the trivial cost of squaring operations in normal basis. Itoh and Tsujii [79] proposed several algorithmic solutions for inversion using Fermat's Little Theorem. The general case of those algorithms is called Itoh–Tsujii inversion algorithm.

Suppose that A is an element of a GF(2^k) field in normal basis representation and A^{-1} is its multiplicative inverse, then using Fermat's Little Theorem we can find A^{-1} as

$$A^{-1} = A^{2^k-2} = (A^{2^{k-1}-1})^2.$$

If we write $k-1$ as a sum of powers of 2, meaning $k-1=\sum_{i=1}^{t} 2^{n_i}$ where $n_0 < n_1 < n_2 \cdots < n_t$, then

$$\begin{aligned} A^{-1} &= (A^{2^{k-1}-1})^2 \\ &= \left[(A^{2^{2^{n_t}}-1}) \cdot \left(\left(A^{2^{2^{n_{t-1}}}-1}\right) \cdots \left[\left(A^{2^{2^{n_1}}-1}\right)\left(A^{2^{2^{n_0}}-1}\right)^{2^{2^{n_1}}} \right]^{2^{2^{n_2}}} \cdots \right)^{2^{2^{n_t}}} \right]^2. \end{aligned}$$

From a design point of view, this equation has an important feature. If the quantity $(A^{2^{2^{n_i}}-1})$ is calculated, then all the similar quantities of n_0, $n_1, \ldots, n_{i-1}$ are also calculated. This methodology requires $(\lfloor \log_2 (k-1) \rfloor +$ $Ham \min g\ Weight(k-1) - 1)$ multiplication and $k-1$ squaring operations and can be generalized for any extension finite field (GF(p^k) field) (Algorithm 22).

This version of the algorithm uses calculation in the subfield GF(p). Such calculation can be made using a look up table or the EEA. As shown above, the Itoh–Tsujii algorithm can also be used for polynomial basis representation GF(2^k) fields and composite fields although it is not as efficient as in normal basis representation GF(2^k) fields.

4.9 ELLIPTIC CURVE POINT OPERATIONS

From the analysis of mathematical operations in GF(p) fields and GF(2^k) fields, the overall cost of elliptic curve point operations can be more extensively studied. One obvious result of the finite field analysis is the extensive cost in hardware resources and speed of the inversion–division operation. Inversion algorithms need many rounds to calculate a result and in some cases the number of rounds is not constant, requiring control logic for manipulating the data stream. However, inversion is an essential operation for point addition and doubling as described in Section 4.5, whether we work in GF(2^k) fields or GF(p) fields. A solution to the problem is the use of basic projective geometry principles so as to change the coordinate system of each elliptic curve point from affine coordinates to projective coordinates.

Design problems are also traced in point multiplication, when the algorithm described in Section 4.6 is used. Translating this algorithm in binary form and reducing the required number of point addition and doubling operations can enhance the performance of a resulting design.

In addition, special consideration should be given in efficient design of point operations for cryptographic use so that the performed point operations are undistinguished in an overall cryptographic system to avoid side channel attacks (SCAs). SCA for braking ECC [80] is based on measuring the behavior of an ECC system to identify point addition and doubling operations performed in a period of time in order to extract information about the secret key s.

4.9.1 Point Addition and Point Doubling Using Projective Coordinates

Suppose that we have an elliptic curve E defined over a finite field F. Each EC point is described by two coordinates $x, y \in F$. In this case, we say that the EC points belong to the two-dimensional affine plane $A_F = \{(x, y) \in F \times F\}$. However, there is a mapping between the affine plane A_F^2 and the two-dimensional projective plane $P_F = \{(X: Y: Z) \in F \times F \times F\}$.

The equivalence class in the projective plane is $\{(X: Y: Z) = (\theta^c X, \theta^d Y, \theta^e Z): X, Y, Z, \theta \in F\}$ although we usually choose $e = 0$ so that $\{(X: Y: Z) = (\theta^c X, \theta^d Y, Z): X, Y, Z, \theta \in F\}$. The c, d values are integers. If $Z = 0$ in the projective plane then $(X: Y: 0)$ is the line at infinity, which is identical to the point at infinity in the affine plane. In any other case ($Z \neq 0$) we can map the coordinates (x, y) of the affine plane to the coordinates of the projective plane as $(X, Y, Z) = (x \cdot Z^c, y \cdot Z^d, 1)$ or else $x = X/Z^c$ and $y = Y/Z^d$.

Suppose that E is the equation of the elliptic curve in the affine plane, the equivalent equation E in the projective plane can be found by replacing x, y with their projective coordinate equivalent X/Z^c and Y/Z^d, respectively. According to the values of c and d, various types of projective coordinates can be specified. Among them, the more important variations are the standard projective coordinates ($c = 1$ and $d = 1$), the Jacobian projective coordinates ($c = 2$ and $d = 3$), Chudnovsky projective coordinates ($(X: Y: Z: Z^2: Z^3)$ representation) [81], the Lopez–Dahab projective coordinates ($c = 1$ and $d = 2$) [82], and several different mixes of affine and projective coordinates (mixed affine–projective coordinates). Since the equation E of the elliptic curve in the projective plane is different from the one in the affine plane, the Group Law in its algebraic form will also be different. In the rest of this subsection we analyze the Group Law in the projective plane for GF(p) and GF(2^k) fields using Jacobian projective coordinates.

4.9.1.1 Point Addition–Doubling in Elliptic Curves over GF(p) Fields Using Projective Coordinates

Suppose that we have an elliptic curve defined over GF(p) fields. Then, the short Weierstrass equation of this curve in the affine plane would be E: $y^2 = x^3 + ax + b$ and after applying the projective coordinates transformation it becomes E: $Y^2 = X^3Z^{2d-3c} + aXZ^{2d-c} + bZ^{2d}$ when $2d > 3c$ and E: $Y^2Z^{3c-2d} = X^3 + aXZ^{2c} + bZ^{3c}$ when $2d < 3c$. For $c = 2$ and $d = 1$ (Jacobian projective coordinates), we have E: $Y^2 = X^3 + aXZ^4 + bZ^6$.

Suppose that we have two EC points in the projective plane $P_1 = (X_1, Y_1, Z_1)$ and $P_2 = (X_2, Y_2, Z_2)$. Point addition ($P_3 = (X_3, Y_3, Z_3) = P_1 + P_2$) and point doubling ($P_3 = 2P_1$) can be described as follows.

Each EC point in affine coordinates can be written as $(X/Z^c, Y/Z^d)$, where X, Y, Z are projective coordinates. Then, by replacing x, y accordingly in the equations of the Group Law for affine coordinates we would have when $P_1 \neq P_2$ (point addition)

$$\lambda = \frac{y_2 - y_1}{x_2 - x_1} = \frac{\dfrac{Y_2}{Z_2^d} - \dfrac{Y_1}{Z_1^d}}{\dfrac{X_2}{Z_2^c} - \dfrac{X_1}{Z_1^c}} = \frac{(Y_2Z_1^d - Y_1Z_2^d)Z_2^cZ_1^c}{(X_2Z_1^c - X_1Z_2^c)Z_2^dZ_1^d}$$

for $x_2 \neq x_1$ and the point P_3 would be

$$x_3 = \lambda^2 - x_1 - x_2 = \left(\frac{(Y_2Z_1^d - Y_1Z_2^d)Z_2^cZ_1^c}{(X_2Z_1^c - X_1Z_2^c)Z_2^dZ_1^d}\right)^2 - \frac{(X_1Z_2^c + X_2Z_1^c)}{Z_2^cZ_1^c}$$
$$= \frac{(Y_2Z_1^d - Y_1Z_2^d)^2Z_2^{3c}Z_1^{3c} - (X_1Z_2^c + X_2Z_1^c)(X_2Z_1^c - X_1Z_2^c)^2Z_2^{2d}Z_1^{2d}}{(X_2Z_1^c - X_1Z_2^c)^2Z_2^{2d+c}Z_1^{2d+c}},$$

$$y_3 = \lambda(x_1 - x_3) - y_1 =$$
$$= \frac{(Y_2Z_1^d - Y_1Z_2^d)Z_2^cZ_1^c}{(X_2Z_1^c - X_1Z_2^c)Z_2^dZ_1^d}\left(\frac{X_1}{Z_1^c} - \frac{(Y_2Z_1^d - Y_1Z_2^d)^2Z_2^{3c}Z_1^{3c} - (X_1Z_2^c + X_2Z_1^c)(X_2Z_1^c - X_1Z_2^c)^2Z_2^{2d}Z_1^{2d}}{(X_2Z_1^c - X_1Z_2^c)^2Z_2^{2d+c}Z_1^{2d+c}}\right) - \frac{Y_1}{Z_1^d}$$
$$= \frac{(Y_2Z_1^d - Y_1Z_2^d)Z_2^cZ_1^c}{(X_2Z_1^c - X_1Z_2^c)Z_2^dZ_1^d}\left(\frac{(2X_1Z_2^c + X_2Z_1^c)(X_2Z_1^c - X_1Z_2^c)^2Z_2^{2d}Z_1^{2d} - (Y_2Z_1^d - Y_1Z_2^d)^2Z_2^{3c}Z_1^{3c}}{(X_2Z_1^c - X_1Z_2^c)^2Z_2^{2d+c}Z_1^{2d+c}}\right) - \frac{Y_1}{Z_1^d}$$
$$= \frac{(Y_2Z_1^d - Y_1Z_2^d)(2X_1Z_2^c + X_2Z_1^c)(X_2Z_1^c - X_1Z_2^c)^2Z_2^{2d+c}Z_1^{2d+c} - (Y_2Z_1^d - Y_1Z_2^d)^3Z_2^{4c}Z_1^{4c}}{(X_2Z_1^c - X_1Z_2^c)^3Z_2^{3d+c}Z_1^{3d+c}} - \frac{Y_1}{Z_1^d}$$
$$= \frac{((Y_2Z_1^d - Y_1Z_2^d)(2X_1Z_2^c + X_2Z_1^c) - Y_1Z_2^d(X_2Z_1^c - X_1Z_2^c))(X_2Z_1^c - X_1Z_2^c)^2Z_2^{2d}Z_1^{2d} - (Y_2Z_1^d - Y_1Z_2^d)^3Z_2^{3c}Z_1^{3c}}{(X_2Z_1^c - X_1Z_2^c)^3Z_2^{3d}Z_1^{3d}}.$$

By defining the denominators of x_3 and y_3 as Z_3^c and Z_3^d, X_3 and Y_3 would be the numerators of x_3 and y_3, respectively. When using Jacobian projective coordinates ($c = 2$, $d = 3$), the EC point P_3 would be

$$X_3 = (Y_2Z_1^3 - Y_1Z_2^3)^2 - (X_1Z_2^2 + X_2Z_1^2)(X_2Z_1^2 - X_1Z_2^2)^2,$$
$$Y_3 = ((Y_2Z_1^3 - Y_1Z_2^3)(2X_1Z_2^2 + X_2Z_1^2)$$
$$\quad - Y_1Z_2^3(X_2Z_1^2 - X_1Z_2^2))(X_2Z_1^2 - X_1Z_2^2)^2 - (Y_2Z_1^3 - Y_1Z_2^3)^3,$$
$$Z_3 = (X_2Z_1^2 - X_1Z_2^2)\ Z_2Z_1.$$

When $P_1 \neq P_2$ but $x_2 = x_1$ then $P_3 = P_1 + P_2 = \infty$, where ∞ is the point at infinity of the elliptic curve E in projective coordinates. Its form is dependent on the type of the used projective coordinates (c, d values). For Jacobian projective coordinates, the point at infinity has the form (1: 1: 0).

When $P_1 = P_2$ (point doubling) and $y_1 \neq 0$, by replacing x, y accordingly, $P_3 = P_1 + P_2 = 2P_1 = (X_3, Y_3, Z_3)$ would be

$$\lambda = \frac{3x_1^2 + a}{2y_1} = \frac{3X_1^2 Z_1^d + aZ_1^{2c+d}}{2Z_1^{2c} Y_1},$$

$$x_3 = \lambda^2 - 2x_1 = \frac{(3X_1^2 + aZ_1^{2c})^2 Z_1^{2d}}{4Z_1^{4c} Y_1^2} - 2\frac{X_1}{Z_1^c} = \frac{(3X_1^2 + aZ_1^{2c})^2 Z_1^{2d} - 8Z_1^{3c} X_1 Y_1^2}{4Z_1^{4c} Y_1^2},$$

$$\begin{aligned} y_3 &= \lambda(x_1 - x_3) - y_1 = \lambda(3x_1 - \lambda^2) - y_1 \\ &= \frac{(3X_1^2 + aZ_1^{2c})Z_1^d}{2Z_1^{2c} Y_1} \left(3\frac{X_1}{Z_1^c} - \frac{(3X_1^2 + aZ_1^{2c})^2 Z_1^{2d}}{4Z_1^{4c} Y_1^2} \right) - \frac{Y_1}{Z_1^d} \\ &= \frac{12X_1 Y_1^2 (3X_1^2 + aZ_1^{2c}) Z_1^{3c+2d} - (3X_1^2 + aZ_1^{2c})^3 Z_1^{4d} - 8Z_1^{6c} Y_1^4}{8Z_1^{6c+d} Y_1^3}. \end{aligned}$$

By defining the denominators of x_3 and y_3 as Z_3^c and Z_3^d, X_3 and Y_3 would be the numerators of x_3 and y_3, respectively. When using Jacobian projective coordinates ($c = 2$, $d = 3$), the EC point P_3 for point doubling would be

$$\begin{aligned} X_3 &= (3X_1^2 + aZ_1^4)^2 - 8X_1 Y_1^2, \\ Y_3 &= 12X_1 Y_1^2 (3X_1^2 + aZ_1^4) - (3X_1^2 + aZ_1^4)^3 - 8Y_1^4, \\ Z_3 &= 2Z_1 Y_1. \end{aligned}$$

Point subtraction can be performed by using the point $-P_2$ instead of P_2, where the additive inverse of an EC point (X, Y, Z) has the form $(X{:}-Y{:}\,Z)$ in the projective plane for EC over GF(p) fields.

It can be noted that no inversion–division operation in finite fields is required for calculating EC point P_3 in the projective plane. Only one inversion is needed for moving from the projective plane to the affine plane. Moreover, some intermediate multiplication products and intermediate equations are used more than once in the overall point addition and multiplication process. In a design, such intermediate results can be calculated only once and then be stored in some memory unit or register to reduce the required number of finite field operations (multiplications and additions). For example, the calculation of the EC point $P_3 = 2P_1 = (X_3, Y_3, Z_3)$ can be performed using two intermediate values [14,17].

$$
\begin{aligned}
W &= 3X_1^2 + aZ_1^4, \\
S &= 4X_1Y_1^2, \\
X_3 &= W^2 - 2S, \\
Y_3 &= W(S - X_3) - 8Y_1^4, \\
Z_3 &= 2Z_1Y_1.
\end{aligned}
$$

The overall number of required GF(p) field operations in the above calculations is four multiplications and six squarings.

4.9.1.2 Point Addition–Doubling in Elliptic Curves over GF(2^k) Fields Using Projective Coordinates

A similar methodology can be used for elliptic curves defined over GF(2^k) fields. Suppose that E is the short Weierstrass equation of an elliptic curve defined over a GF(2^k) field as described in Section 4.4. Then, by applying the transformation from affine to projective coordinates ($x = X/Z^c$ and $y = Y/Z^d$) this EC equation E becomes E: $Y^2 + XYZ^{d-c} = X^3Z^{2d-3c} + aX^2Z^{2d-2c} + bZ^{2d}$ when $2d > 3c$ and E: $Y^2Z^{3c-2d} + XYZ^{2c-d} = X^3 + aX^2Z^c + bZ^{3c}$ when $2d < 3c$. For $c = 2$ and $d = 1$ (Jacobian projective coordinates), we have E: $Y^2 + XYZ = X^3 + aX^2Z^2 + bZ^6$.

Suppose that we have two EC points in the projective plane $P_1 = (X_1, Y_1, Z_1)$ and $P_2 = (X_2, Y_2, Z_2)$. The point addition ($P_3 = (X_3, Y_3, Z_3) = P_1 + P_2$) and point doubling ($P_3 = 2P_1$) can be described as follows.

Each EC point in affine coordinates can be written as $(X/Z^c, Y/Z^d)$, where X, Y, and Z are projective coordinates. If we replace x, y accordingly in the equations of the Group Law for affine coordinates, we would have when $P_1 \neq P_2$ (point addition)

$$
\lambda = \frac{y_2 + y_1}{x_2 + x_1} = \frac{(Y_2/Z_2^d) + (Y_1/Z_1^d)}{(X_2/Z_2^c) + (X_1/Z_1^c)} = \frac{(Y_2Z_1^d + Y_1Z_2^d)Z_2^cZ_1^c}{(X_2Z_1^c + X_1Z_2^c)Z_2^dZ_1^d}
$$

for $x_2 \neq x_1$ and the point $P_3 = P_1 + P_2 = (X_3, Y_3, Z_3)$ would be

$$
\begin{aligned}
x_3 &= \lambda^2 + x_1 + x_2 + \lambda + a \\
&= \left(\frac{(Y_2Z_1^d + Y_1Z_2^d)Z_2^cZ_1^c}{(X_2Z_1^c + X_1Z_2^c)Z_2^dZ_1^d}\right)^2 + \frac{(X_1Z_2^c + X_2Z_1^c)}{Z_2^cZ_1^c} + \frac{(Y_2Z_1^d + Y_1Z_2^d)Z_2^cZ_1^c}{(X_2Z_1^c + X_1Z_2^c)Z_2^dZ_1^d} + a \\
&= \frac{(Y_2Z_1^d + Y_1Z_2^d)^2Z_2^{3c}Z_1^{3c} + (X_2Z_1^c + X_1Z_2^c)^3Z_2^{2d}Z_1^{2d} + (X_2Z_1^c + X_1Z_2^c)(Y_2Z_1^d + Y_1Z_2^d)Z_2^{d+2c}Z_1^{d+2c}}{(X_2Z_1^c + X_1Z_2^c)^2Z_2^{2d+c}Z_1^{2d+c}} \\
&\quad + \frac{a(X_2Z_1^c + X_1Z_2^c)^2Z_2^{2d+c}Z_1^{2d+c}}{(X_2Z_1^c + X_1Z_2^c)^2Z_2^{2d+c}Z_1^{2d+c}}.
\end{aligned}
$$

$$
\begin{aligned}
y_3 &= \lambda(x_1+x_3)+x_1+x_3 \\
&= \lambda\left(\frac{(Y_2Z_1^d+Y_1Z_2^d)^2Z_2^{3c}Z_1^{3c}+(X_2Z_1^c+X_1Z_2^c)^3Z_2^{2d}Z_1^{2d}+(X_2Z_1^c+X_1Z_2^c)(Y_2Z_1^d+Y_1Z_2^d)Z_2^{d+2c}Z_1^{d+2c}}{(X_2Z_1^c+X_1Z_2^c)^2Z_2^{2d+c}Z_1^{2d+c}}\right. \\
&\quad \left.+\frac{a(X_2Z_1^c+X_1Z_2^c)^2Z_2^{2d+c}Z_1^{2d+c}++X_1(X_2Z_1^c+X_1Z_2^c)^2Z_2^{2d+c}Z_1^{2d}}{(X_2Z_1^c+X_1Z_2^c)^2Z_2^{2d+c}Z_1^{2d+c}}\right)+\frac{(Y_2Z_1^d+Y_1Z_2^d)^2Z_2^{3c}Z_1^{3c}+(X_2Z_1^c+X_1Z_2^c)^3Z_2^{2d}Z_1^{2d}}{(X_2Z_1^c+X_1Z_2^c)^2Z_2^{2d+c}Z_1^{2d+c}} \\
&\quad +\frac{(X_2Z_1^c+X_1Z_2^c)(Y_2Z_1^d+Y_1Z_2^d)Z_2^{d+2c}Z_1^{d+2c}+a(X_2Z_1^c+X_1Z_2^c)^2Z_2^{2d+c}Z_1^{2d+c}+X_1(X_2Z_1^c+X_1Z_2^c)^2Z_2^{2d+c}Z_1^{2d}}{(X_2Z_1^c+X_1Z_2^c)^2Z_2^{2d+c}Z_1^{2d+c}} \\
&= \left(\frac{(Y_2Z_1^d+Y_1Z_2^d)^3Z_2^{4c}Z_1^{4c}+(Y_2Z_1^d+Y_1Z_2^d)(X_2Z_1^c+X_1Z_2^c)^3Z_2^{2d+c}Z_1^{2d+c}+(X_2Z_1^c+X_1Z_2^c)(Y_2Z_1^d+Y_1Z_2^d)^2Z_2^{d+3c}Z_1^{d+3c}}{(X_2Z_1^c+X_1Z_2^c)^3Z_2^{3d+c}Z_1^{3d+c}}\right. \\
&\quad \left.+\frac{a(Y_2Z_1^d+Y_1Z_2^d)(X_2Z_1^c+X_1Z_2^c)^2Z_2^{2d+2c}Z_1^{2d+2c}+X_1(Y_2Z_1^d+Y_1Z_2^d)(X_2Z_1^c+X_1Z_2^c)^2Z_2^{2d+2c}Z_1^{2d+c}}{(X_2Z_1^c+X_1Z_2^c)^3Z_2^{3d+c}Z_1^{3d+c}}\right) \\
&\quad +\frac{(X_2Z_1^c+X_1Z_2^c)(Y_2Z_1^d+Y_1Z_2^d)^2Z_2^{d+3c}Z_1^{d+3c}+(X_2Z_1^c+X_1Z_2^c)^4Z_2^{3d}Z_1^{3d}+(X_2Z_1^c+X_1Z_2^c)^2(Y_2Z_1^d+Y_1Z_2^d)Z_2^{2d+2c}Z_1^{2d+2c}}{(X_2Z_1^c+X_1Z_2^c)^3Z_2^{3d+c}Z_1^{3d+c}} \\
&\quad +\frac{a(X_2Z_1^c+X_1Z_2^c)^3Z_2^{3d+c}Z_1^{3d+c}+X_1(X_2Z_1^c+X_1Z_2^c)^3Z_2^{3d+c}Z_1^{3d}}{(X_2Z_1^c+X_1Z_2^c)^3Z_2^{3d+c}Z_1^{3d+c}}.
\end{aligned}
$$

By defining the denominators of x_3 and y_3 as Z_3^c and Z_3^d, X_3 and Y_3 would be the numerators of x_3 and y_3, respectively. When using Jacobian projective coordinates ($c=2$, $d=3$), the EC point P_3 would be

$$
\begin{aligned}
X_3 &= (Y_2Z_1^3+Y_1Z_2^3)^2+(X_2Z_1^2+X_1Z_2^2)^3+(X_2Z_1^2+X_1Z_2^2)(Y_2Z_1^3+Y_1Z_2^3)Z_2Z_1 \\
&\quad + a(X_2Z_1^2+X_1Z_2^2)^2Z_2^2Z_1^2, \\
Y_3 &= (Y_2Z_1^3+Y_1Z_2^3)^3+(Y_2Z_1^3+Y_1Z_2^3)(X_2Z_1^2+X_1Z_2^2)^3 \\
&\quad + (X_2Z_1^2+X_1Z_2^2)(Y_2Z_1^3+Y_1Z_2^3)^2Z_2Z_1 \\
&\quad + a(Y_2Z_1^3+Y_1Z_2^3)(X_2Z_1^2+X_1Z_2^2)^2Z_2^2Z_1^2 \\
&\quad + X_1(Y_2Z_1^3+Y_1Z_2^3)(X_2Z_1^2+X_1Z_2^2)^2Z_2^2 \\
&\quad + (X_2Z_1^2+X_1Z_2^c)(Y_2Z_1^3+Y_1Z_2^3)^2Z_2Z_1+(X_2Z_1^2+X_1Z_2^2)^4Z_2Z_1 \\
&\quad + (X_2Z_1^2+X_1Z_2^2)^2(Y_2Z_1^3+Y_1Z_2^3)Z_2^2Z_1^2+a(X_2Z_1^2+X_1Z_2^2)^3Z_2^3Z_1^{3d} \\
&\quad + X_1(X_2Z_1^2+X_1Z_2^2)^3Z_2^3Z_1, \\
Z_3 &= (X_2Z_1^2+X_1Z_2^2)\,Z_2Z_1.
\end{aligned}
$$

When $P_1 \neq P_2$ but $x_2 = x_1$, then $P_3 = P_1 + P_2 = \infty$, where ∞ is the point at infinity of the elliptic curve E in projective coordinates. Its form is dependent on the type of the used projective coordinates (c, d values). For Jacobian projective coordinates, the point at infinity has the form (1:1:0).

When $P_1 = P_2$ (point doubling) and $y_1 \neq 0$, by replacing x, y accordingly, $P_3 = P_1 + P_2 = 2P_1 = (X_3, Y_3, Z_3)$ would be

$$\lambda = x_1 + \frac{y_1}{x_1} = \frac{X_1^2 Z_1^d + Y_1 Z_1^{2c}}{Z_1^{d+c} X_1},$$

$$x_3 = \lambda^2 + \lambda + a = x_1^2 + \frac{b}{x_1^2} = \frac{X_1^4 + bZ_1^{4c}}{X_1^2 Z_1^{2c}} = \frac{(X_1 + \sqrt[4]{b}Z_1^c)^4}{(X_1 Z_1^c)^2},$$

$$y_3 = x_1^2 + \lambda x_3 + x_3 = \frac{X_1^2}{Z_1^{2c}} + \frac{(X_1^2 Z_1^d + Y_1 Z_1^{2c})}{Z_1^{d+c} X_1} \cdot \frac{(X_1^4 + bZ_1^{4c})}{X_1^2 Z_1^{2c}} + \frac{X_1^4 + bZ_1^{4c}}{X_1^2 Z_1^{2c}}$$

$$= \frac{X_1^5 Z_1^{d+c} + (X_1^2 Z_1^d + Y_1 Z_1^{2c})(X_1^4 + bZ_1^{4c}) + X_1^5 Z_1^{d+c} + bX_1 Z_1^{d+5c}}{X_1^3 Z_1^{d+3c}}$$

$$= \frac{X_1^6 Z_1^d + bX_1^2 Z_1^{d+4c} + X_1^4 Y_1 Z_1^{2c} + bY_1 Z_1^{6c} + bX_1 Z_1^{d+5c}}{X_1^3 Z_1^{d+3c}}.$$

By defining the denominators of x_3 and y_3 as Z_3^c and Z_3^d, X_3 and Y_3 would be the numerators of x_3 and y_3, respectively. When using Jacobian projective coordinates ($c=2$, $d=3$), the EC point P_3 for point doubling would be

$$X_3 = (X_1 + \sqrt[4]{b}Z_1^2)^4,$$
$$Y_3 = X_1^6 + bX_1^2 Z_1^8 + X_1^4 Y_1 Z_1 + bY_1 Z_1^9 + bX_1 Z_1^{10},$$
$$Z_3 = X_1 Z_1^2.$$

Point subtraction can be performed by using the point $-P_2$ instead of P_2, where the additive inverse of an EC point (X, Y, Z) has the form $(X: X+Y: Z)$ in the projective plane for EC over GF(2^k) fields.

As in the case of elliptic curves over GF(p) fields, point addition and point doubling in elliptic curves over GF(2^k) fields using projective coordinates can be optimized by storing intermediate results that are used more than once in the calculation process. For example, the calculation of the EC point $P_3 = 2P_1 = (X_3, Y_3, Z_3)$ can be performed using one intermediate value and the reusability of X_3 and Z_3 [14,17].

$$W = Z_3 + X_1^2 + Y_1 Z_1,$$
$$X_3 = (X_1 + \sqrt[4]{b}Z_1^2)^4,$$
$$Y_3 = X_1^4 Z_3 + WX_3,$$
$$Z_3 = X_1 Z_1^2.$$

The overall number of required GF(2^k) field operations in the above calculations is five multiplications, five squarings, and four additions–subtractions.

4.9.1.3 Comparison of EC Point Operations in Affine and Projective Coordinates

Inspecting the equations of the point operation results for the affine and projective coordinates, a tradeoff can be noticed. Using the projective plane

TABLE 4.4
Required Finite Field Operations for EC Point Operations Using Affine and Projective Coordinates (in Inversion (Inv.), Multiplication (Mult.), and Squaring (Sq.) Operations)

	Operation	
Coordinate System	**Point Addition**	**Point Doubling**
	Elliptic curves over GF(p) Fields	
Affine	1 Inv. + 2 Mult. + 1 Sq.	1 Inv. + 2 Mult. + 2 Sq.
Jacobian projective	12 Mult. + 4 Sq.	4 Mult. + 6 Sq.
Standard projective	12 Mult. + 2 Sq.	7 Mult. + 5 Sq.
Chudnovsky projective	11 Mult. + 3 Sq	5 Mult. + 6 Sq.
	Elliptic curves over GF(2^k) Fields	
Affine	1 Inv. + 2 Mult. + 1 Sq.	1 Inv. + 2 Mult. + 1 Sq.
Jacobian projective	15 Mult. + 5 Sq.	5 Mult. + 5 Sq.
Standard projective	15 Mult. + 5 Sq.	

we have managed to exchange finite field inversions with a number of finite field multiplications. In Table 4.4, the cost in finite field operations for each coordinate system using the presented optimizations is shown.

4.9.1.4 Design Issues for Elliptic Curve Point Addition and Point Squaring

In IEEE 1363 Draft [14] along with [17], there is an analysis of the equations in projective coordinates for calculating point addition and doubling results, by breaking those equations in small reusable intermediate values. Moreover, algorithms that take advantage of this reusability to increase the parallelism of calculations are presented.

Resulting designs of those algorithms have an increased degree of parallelism, meaning that they can perform several algorithmic steps in the same clock cycle. For this reason intermediate storage elements (registers) are needed. Pipelining is also used in this design methodology to increase the throughput speed of a point addition or point doubling architecture. An example of a point addition and doubling architecture for an elliptic curve over GF(2^k) fields (with $b = 1$ of the EC equation E) [83], along with possible pipeline stages, is presented in Figure 4.11.

However, considering the cost of a single finite field multiplier (bit parallel architecture) the design described here requires a considerable amount of hardware resources, such as power consumption, chip covered area, and storage cells, and can be considered unaffordable especially for wireless applications. The use of bit serial finite field architectures can

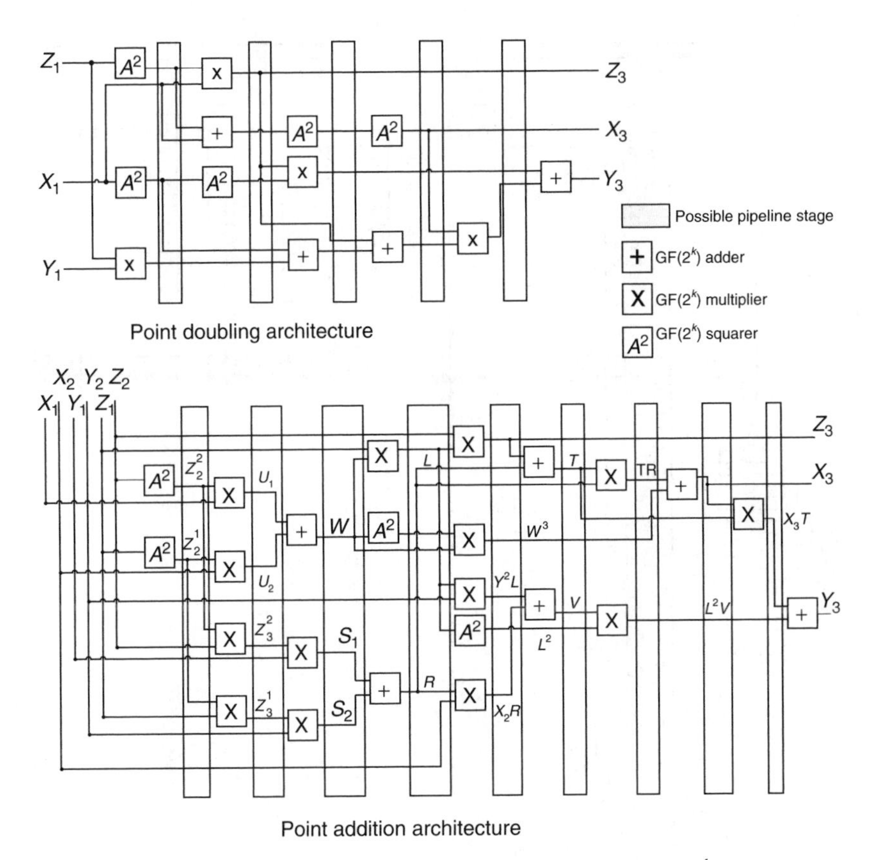

FIGURE 4.11 Point addition–doubling architecture for EC over GF(2^k) fields.

minimize the described problem but will lead to a dramatic increase in the required clock cycle number for one point operation. Another solution could be the use of time multiplexing by appropriate input–output registers managed by a control unit. In that case, the maximum number of each GF(2^k) field operations that can be performed in parallel in each point operation is estimated and designed. The outputs pass through a controllable register series, which use feedback to reinsert those outputs as inputs to the architecture to perform the next round of parallel GF(2^k) field operations correctly or store them for future use. Such a design is shown in Figure 4.12 for the architectures in Figure 4.11.

4.9.2 Point Multiplication Design Issues

Point multiplication, as presented in Section 4.6, is the most complex of the elliptic curve point operations. One point multiplication, $Q = sP$, requires

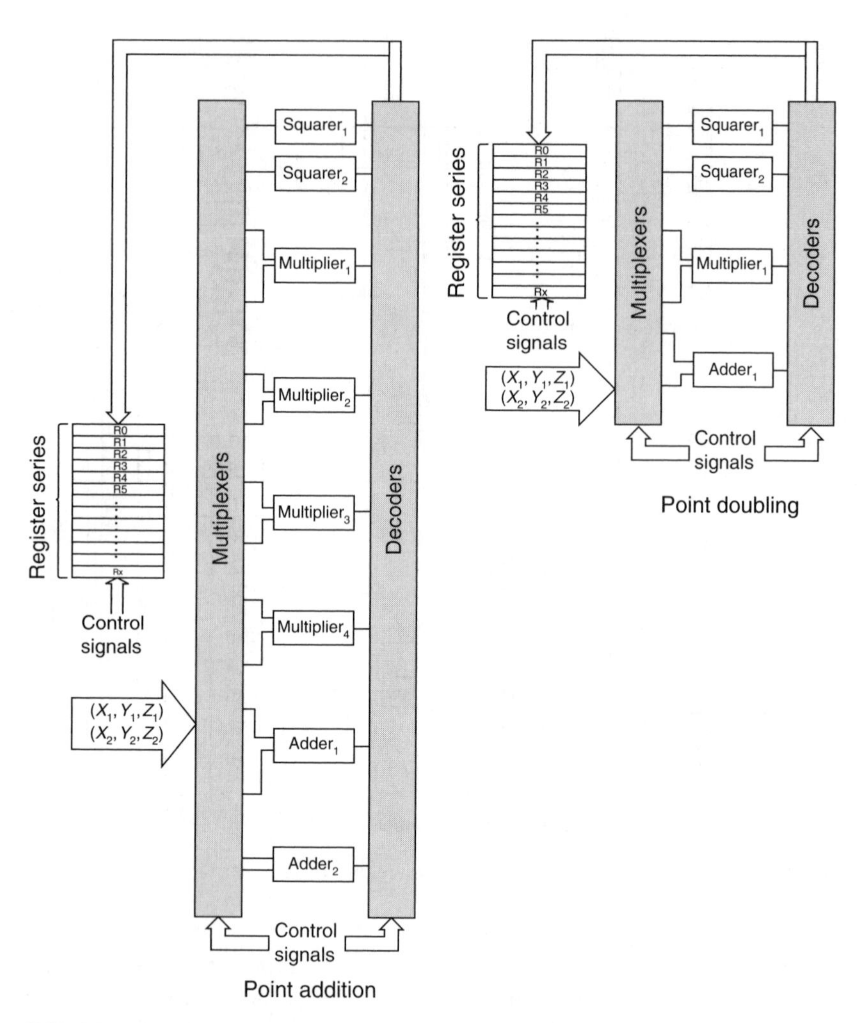

FIGURE 4.12 Time multiplexed point addition–doubling architecture for EC over GF(2^k) fields.

many point doubling and additions depending on the integer s. Expressing s in binary format highlights this dependency. The number of zero and nonzero bits, their place in the binary vector s, and the bit length of s can lead to different number of point addition and doubling operations used in one point multiplication. This is shown in the binary version of the point multiplication algorithm (Algorithm 23).

The Hamming Weight of s (HW(s)) determines the number of point addition operations and the bit length of s determines the number of point doubling operations. Therefore, in the binary point multiplication algorithm there are

HW(s) point additions (pA) and t point doublings (pD). Assuming that the bit length of s is close to k and that HW(s) $= k/2$ at average, meaning that the number of zeros is approximately the same as the number of ones, the overall point operations required for $Q = sP$ would be $k/2\ pA + k\ pD$.

A more general approach to point multiplication concerning the form of the integer s is to process w bits of s at a time. Such a method is called window method and is found in several forms [17] such as the fixed window method or the sliding window method.

The main idea of the window methodology is the use of w bits of the multiplier s in each clock cycle to reduce the overall number of point addition and doubling operations. Those w bits are called a window of s. Window methods use a precomputation procedure to calculate $P_j = jP$, where j takes odd values from 1 to $2^{w-1} - 1$. Those values are stored in order to be used in the main multiplication process that varies according to the window methodology employed. In the fixed window method, the multiplier s is split into fixed w-bit length windows with pinpointed bit beginning and end. The sliding window method uses a window with w-bit length at most and arbitrary beginning or end. This arbitrary window slides from the right to the left of the s-bit vector, skipping consecutive zero s_i bits after a nonzero s_i-bit is processed. The sliding window method is faster than the fixed window method overcoming several problems of the second [17] (Algorithm 24).

Another approach to point multiplication derives from the properties of the elliptic curve point addition. It can be noted that point subtraction and point addition require approximately the same number of finite field operations; thus, they have the same computational and hardware cost. This is true both for elliptic curves over GF(p) fields, where the additive inverse of $P = (x, y) \in E(\mathrm{GF}(p))$ is $-P = (x, -y)$ and for elliptic curves over GF(2^k) fields, where the additive inverse of $P = (x, y) \in E(\mathrm{GF}(p))$ is $-P = (x, x + y)$.

Taking advantage of this, signed digit representation of s in sP point multiplication can be introduced to reduce the required number of point addition and doubling operations. In signed digit representation, the value s would be $s' = \sum_{i=0}^{t} s_i' 2^i$, where $s_i' \in \{0, \pm 1\}$. If $s_{t-1}' \neq 0$ and no two consecutive digits $s_i(s_{i-1}' s_i')$ are nonzero, the signed digit representation of s is called nonadjacent form (NAF) and has some interesting properties that can be used for point multiplication. More specifically, NAF representation of s is unique, its bit length is at most $t + 1$, and a value in NAF representation has fewer nonzero digits ($t/3$ at average) than in binary representation.

NAF representation construction of an integer s, denoted as NAF(s) $= s' = \sum_{i=0}^{t} s_i' 2^i$, can be done by repeatedly dividing s by 2, allowing remainders of 0 or ± 1. If s is odd, then the remainder $r \in \{-1, 1\}$ is chosen so that the quotient $(s - r)/2$ is even, ensuring that the next NAF digit is 0 (Algorithm 25). NAF representation can be introduced to all point multiplication algorithms. The binary NAF point multiplication algorithm as a direct realization of the binary point multiplication algorithm is presented in Algorithm 26.

Assuming that the bit length of s is close to k ($t \approx k$) and that the cost of point addition is the same as that of point subtraction, then the HW(s') $= k/3$ at average and the overall point operations required for $Q = sP$ using the NAF binary point multiplication algorithm would be $k/3\ pA + k\ pD$. This cost is smaller than the one without NAF representation. There are several variations of NAF point multiplication methods, where window or sliding window techniques are used. More on the subject can be found in [14,17,84].

The number of point operations can be further reduced if the multiplicand point P is known, remaining fixed for many point multiplications. In that case, some point operations can be precomputed once and stored in a storage element (memory and registers) if the bit length of s is known. Their result can be added at the end of each sP point multiplication. Applying this methodology to the binary point multiplication algorithm (in NAF format or not) leads to elimination of all point doubling operations during the algorithm's execution. Similar optimizations can be made in other point multiplication algorithms and additional techniques are introduced, such as point multiplication comb methods that are more applicable to software designs. More on the subject can be found in [17].

Fixed point multiplication techniques are useful for EC cryptographic algorithms that use the same point P for many point multiplications, like the ECDSA signature generation scheme.

4.9.2.1 Point Multiplication Design

Point multiplication is an operation that employs point addition and point doubling. Therefore, by designing a point multiplication architecture, we manage to complete an elliptic curve arithmetic unit that can support all elliptic curve operations. Such an arithmetic unit can be used by an ECC algorithm and can be extended to a fully functional elliptic curve coprocessor.

A point multiplication architecture consists of a point addition unit and a point doubling unit that are connected to a control logic. Each point operation unit consists of finite field adders, squarers, multipliers, and inverters interconnected according to the coordinate system used. The point addition and doubling units can be constructed using time multiplexing techniques such as the ones presented in Figure 4.12. An abstract design of a point multiplication architecture is shown in Figure 4.13a.

There can be several other approaches to the design of point multiplication. Reusability of the finite field operations can be employed so that the same circuitry through proper input adjustment and control can perform both point addition and subtraction [85–87]. The point multiplication steps can be microcoded on a general purpose processor and a finite field arithmetic unit for both point addition and doubling can be implemented in hardware [88]. In Figure 4.13b, a generic point multiplication architecture, taking into consideration the above propositions, is shown.

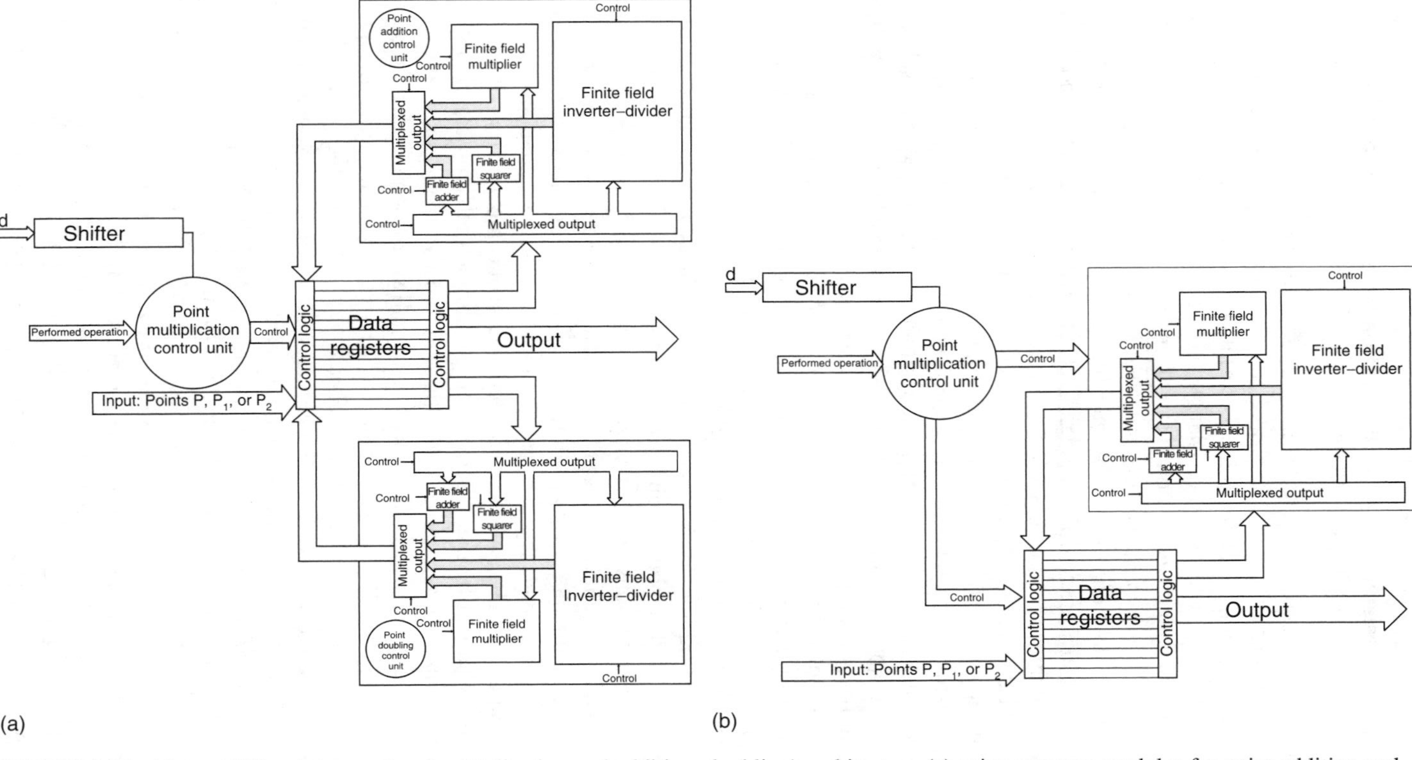

FIGURE 4.13 Abstract EC point operation (multiplication and addition–doubling) architecture (a) using separate modules for point addition and doubling (b) with reusable module for both point operations.

4.9.2.2 Designing Point Multiplication for SCA Resistant Elliptic Curve Cryptosystems

SCAs are considered some of the most fruitful cryptanalytic methods. They exploit information that can "leak" from a cryptographic algorithm's implementation during this algorithm's execution. Such information may include computation time or power consumption traces and can be used to characterize specific computations performed at a given time in the cryptographic execution process. SCA cryptanalytic methods have been proposed for ECC [80], including simple power analysis (SPA) and differential power analysis (DPA) models [89,90], which are considered a significant cryptographic threat. In those attacks, point addition and doubling operations are identified during the point multiplication calculations by probing the power or calculation time of the system and analyzing a specific value in the point multiplication process statistically. This information along with the knowledge of the point multiplication algorithm (that is not considered secret) is enough to determine integer s thus solving the ECDLP problem easily. It must be noted, however, that SCAs are not applicable to all possible cryptographic systems, since there must be an easy way to take time or power consumption measurements. When a device operates in potentially hostile, not trusted environments, SCA can be a serious threat. Such unprotected devices could be smart cards, Radio frequency identification cards (RFIDs), or other wireless handheld devices.

To avoid the SCA threat, point multiplication algorithms should be restated so that point addition operations are indistinguishable from point doubling operations to the external environment. Coren in [80] introduced the SPA and DPA attack on elliptic curve cryptosystems and Okeya in [91] set specific requirements for avoiding such attacks. SPA attacks can be avoided by using independency of secret information and computation procedures. DPA attacks can be avoided by randomization of computing objects. The SPA requirement can be met by performing both point addition and doubling in every round of the point multiplication process (point- and always-add method) at the expense of an increase in hardware resources and a reduction in speed. The DPA requirement can be met by randomizing the private exponent s of $Q = sP$ point multiplication, by blinding the point P and by using randomized projective coordinates. In order to randomize the value s in $Q = sP$, we choose a random number d (~20 bits) and calculate $d' = s + d \cdot \#E(F)$, where $\#E(F)$ is the number of the EC points. Then, we calculate $Q' = d'P$, which is identical to $Q = sP$, since $\#E(F)P = \infty$. Blinding the point P involves adding to it a secret point R with known sR outcome. We perform point multiplication with s on the addition result $(P + R)$ and at the end of the calculation we subtract the point sR from the outcome.

However, the most promising technique proposed in [80] is the use of randomized projective coordinates. Using this method, we can represent the point P as $P = (X, Y, Z)$ and after the first point operation (addition, doubling, or multiplication) change the coordinate system using the equivalent class of the elliptic curve $\{(X : Y : Z) = (\theta^c X, \theta^d Y, \theta^e Z) : X, Y, Z, \theta \in F\}$ by appointing a random θ. Several papers exist on the appropriate choice of c, d, e values to obtain fast point operation designs. Coron in [80] uses $c = d = e = 1$, whereas Izu et al. [95] favor $c = 2$, $d = 3$, $e = 1$. Similar coordinate randomization can also be used for the affine plane using isomorphism, as proposed in [92].

In order to reduce the extra cost introduced in the point- and always-add method, a methodology introduced by Montgomery in [93], is used (Algorithm 27). This algorithm possesses some interesting advantages. The difference P–Q in each iteration of this algorithm is known and equal to the initial P point. Montgomery also observed that the x coordinate of the sum of two points with constant difference can be computed using only the x coordinates of those involved points. Using the above remarks, designs for computing point multiplication using specific coordinate types were proposed. Initially, this methodology was analyzed for elliptic curves over GF(2^k) fields and point multiplication, point addition, and point doubling were parameterized using projective coordinates [82]. Additionally, optimistic results were given when using a special type of elliptic curve equation of the form E: $by^2 = x^3 + ax^2 + x$ called Montgomery equation [93,94]. The method has also been proposed for elliptic curves over GF(p) fields in projective [95] and affine coordinates [92]. By combining randomization of computing objects for DPA resistance and Montgomery's technique described earlier for SPA resistance, a very secure cryptosystem can be designed, which is fully protected against SCAs.

The methodology of Montgomery for elliptic curve point multiplication and its realization in accordance to the used coordinate system, resulting in appropriate algorithms for point addition–doubling and multiplication, is advantageous against the similar algorithms presented in Section 9.1 and Section 9.2 [82,96].

4.10 ELLIPTIC CURVE CRYPTOGRAPHIC ALGORITHMS FOR SECURE WIRELESS SYSTEMS

As explained in the introduction of this chapter, the security of wireless devices is dependent on key pair generation and management along with digital signature schemes. In this section, we analyze the key generation procedure, the Elliptic Curve Diffie–Hellman (ECDH) key exchange–establishment used in ECC and we present ECDSA digital signature that has become a standard by many international organizations like IEEE [14], ANSI [97]. This digital signature scheme is the only one used so far in some wireless security protocols (WAP–WTLS [3]) and is the most promising to be adopted by future wireless protocols employing ECC.

Before we proceed in analyzing the above issues, the elliptic curve has to be generated and firmly described for cryptographic use through a set of parameters. Such parameters are called domain parameters. Suppose that we have a finite field F and a created elliptic curve $E(F)$ using this field. Then, the domain parameters consist of

- Finite field order: $Order(F)$
- Finite field type, meaning whether it is a GF(p) or GF(2^k) field, along with the element representation of this finite field: F_R
- Value S called seed, if the elliptic curve was randomly generated
- Two coefficients a, b used in the equation E of the elliptic curve $E(F)$
- Point $P = (x, y)$ of prime order, caller base point
- Order n of the point P
- Value $h = \#E(F)n$ called cofactor

Generation of the domain parameters involves the use of an elliptic curve generation algorithm (finding S, a, b) such as the ones described in [17,84], computation of the number of elliptic curve points $\#E(F)$, verification that $\#E(F)$ is divisible by a large prime n ($n \neq Order(F)$ and not divisible by $Order(F)^k - 1$, where $1 \leq k \leq 20$) and calculation of h and point $P \neq \infty$.

The domain parameters are used for generating EC public key pairs. The key generation process with inputs of the domain parameters ($Order(F)$, F_R, S, a, b, P, n, h) consists of

- Selecting an integer d, where $1 \leq d \leq n - 1$
- Computing $Q = dP$

The public key is the point Q, while the private key is the value d.

The generated keys are used for the ECDH key exchange and establishment protocols. In ECDH key exchange–establishment protocols an entity A generates a key pair and sends the public key to an entity B. Similarly, entity B generates a different key pair and sends the public key to entity A. Each entity possessing two public keys multiplies them to get a session key K, which can be used for encryption in symmetric key algorithms or message authentication. Each entity can validate that the public key is indeed a legitimately created point Q on the elliptic curve using the domain parameters [2].

Another use of public key cryptography in wireless systems is for certification. Certification protocols require digital signature schemes. A digital signature scheme uses the domain parameters and key pairs for the procedure of digital signature generation and digital signature verification. Both procedures involve hash functions $H(G)$, where a hash function is a transformation that takes a variable-size input G and returns a fixed-size string $H(G)$. The most widely used digital signature scheme is ECDSA and is presented in Algorithm 28 and Algorithm 29.

ALGORITHMS

Algorithm 1. Point Multiplication Algorithm (abstract form)

Input: P, s
Output: $Q = sP$
1. $Q = \infty$
2. While $s \neq 0$ do
    2.1. If s is even then $s = s/2$ and $P = 2P$
    2.2. If s is odd then $s = s - 1$ and $Q = Q + P$
3. Return Q.

Algorithm 2. Barrett's Reduction Algorithm

Input: $x = \{x_{2n-1}, \ldots, x_2, x_1, x_0\}_b$, $p = \{p_{n-1}, \ldots, p_1, p_0\}_b$, $\mu = \left\lfloor \frac{b^{2n}}{p} \right\rfloor$
Output: $x \bmod p$
1. $p' = \left\lfloor \left\lfloor \frac{x}{b^{n-1}} \right\rfloor \cdot \frac{\mu}{b^{n+1}} \right\rfloor$
2. $x' = (x \bmod b^{n+1}) - (p' \cdot p \bmod b^{n+1})$
2. If $x' < 0$ then $x' = x' + b^{n+1}$
3. While $x' \geq p$ then $x' = x' - p$
4. Return x'.

Algorithm 3. Montgomery Modular Reduction Algorithm
($MontR(x, p)$ *Function*)

Input: $x = \{x_{2n-1}, \ldots, x_2, x_1, x_0\}_b$, $p = \{p_{n-1}, \ldots, p_1, p_0\}_b$, $r = b^n$, $p' = -p^{-1}$ mod b
Output: $c = x \cdot r^{-1} \bmod p$
1. $c = x$
2. For $i = 0$ to $n-1$ do
    2.1 $q = c_i \cdot p' \bmod b$
    2.2 $c = c + q \cdot p \cdot b^i$
3. $c = c/b^n$
4. If $c \geq p$ then $c = c - p$
5. Return c.

Algorithm 4. Montgomery Modular Multiplication
($MontM(x, y, p)$ *Function*)

Input: $x = \{x_{n-1}, \ldots, x_2, x_1, x_0\}_b < p$, $y = \{y_{n-1}, \ldots, y_2, y_1, y_0\}_b < p$,
$p = \{p_{n-1}, \ldots, p_1, p_0\}_b$, $r = b^n$,
$p' = -p^{-1} \bmod b$
Output: $c = x \cdot y \cdot r^{-1} \bmod p$
1. $c = 0$

2. For $i=0$ to $n-1$ do
   2.1 $q=(c_0+x_i\cdot y_0)\,p' \bmod b$
   2.2 $c=(c+x_i\cdot y+q\cdot p)/b$
3. If $c\geq p$ then $c=c-p$
4. Return c.

Algorithm 5. Binary Montgomery Modular Multiplication
(MontMb(x, y, p) Function)

Input: $x=\{x_{n-1},\ldots,x_2,x_1,x_0\}_2<p$, $y=\{y_{n-1},\ldots,y_2,y_1,y_0\}_2<p$,
$p=\{p_{n-1},\ldots,p_1,p_0\}_2$, $r=2^n$,
$p'=-p^{-1} \bmod 2=1$
Output: $c=x\cdot y\cdot 2^{-n} \bmod p$
1. $c=0$
2. For $i=0$ to $n-1$ do
   2.1 $q=(c_0+x_i\cdot y_0) \bmod 2$
   2.2 $c=(c+x_i\cdot y+q\cdot p)/2$
3. If $c\geq p$ then $c=c-p$
4. Return c.

Algorithm 6. Extended Euclidean Algorithm for Inversion
(ExEucl(a, p) Function)

Input: $a=\{a_{n-1},\ldots,a_2,a_1,a_0\}_2<p$, $p=\{p_{n-1},\ldots,p_1,p_0\}_2$
Output: $a^{-1} \bmod p$
1. $r_{i-1}=p$, $r_{i-2}=a$, $u_{i-2}=1$, $u_{i-1}=0$, $i=0$
2. While $r_i\neq 0$ do
   2.1 $q_i=\left\lfloor \frac{r_{i-1}}{r_{i-2}} \right\rfloor$
   2.2 $r_i=r_{i-1}-q_i\cdot r_{i-2}$
   2.3 $u_i=q_i\cdot u_{i-1}+u_{i-2}$
   2.4 $i=i+1$
3. Return u_i.

Algorithm 7. Binary Extended Euclidean Algorithm for Inversion
(ExEuclB(a, p) Function)

Input: $a=\{a_{n-1},\ldots,a_2,a_1,a_0\}_2<p$, $p=\{p_{n-1},\ldots,p_1,p_0\}_2$
Output: $a^{-1} \bmod p$
1. $s=p$, $r=a$, $u=1$, $v=0$
2. While ($r\neq 0$ and $s\neq 0$) do
   2.1 While $r_0=0$ do
      2.1.1 $r=r/2$
      2.1.2 If $u_0=0$ then $u=u/2$ else $u=(u+p)/2$
   2.2 While $s_0=0$ do
      2.2.1 $s=s/2$

2.2.2 If $v_0 = 0$ then $v = u/2$ else $v = (v + p)/2$
2.3 If $r \geq s$ then ($r = r - s$ and $u = u - v$) else ($s = s - r$ and $v = v - u$)
3. If $r = 1$ then return u mod p else return v mod p.

Algorithm 8. Phase I: Montgomery Almost Inverse Algorithm
(*MontAI(a, p) Function*)

Input: $a = \{a_{n-1}, \ldots, a_2, a_1, a_0\}_2 < p$, $p = \{p_{n-1}, \ldots, p_1, p_0\}_2$
Output: $c = a^{-1} 2^k$ mod p and k
1. $s = p$, $r = a$, $u = 1$, $v = 0$, $k = 0$
2. While ($s > 0$) do
2.1 If $s_0 = 0$ then $s = s/2$, $u = 2u$
else if $r_0 = 0$ then $r = r/2$, $v = 2v$
else if $s > r$ then $s = (s - r)/2$, $v = v + u$, $u = 2u$
else if $s \leq r$ then $r = (r - s)/2$, $u = v + u$, $v = 2v$
2.2 $k = k + 1$
3. If $v \geq p$ then $v = v - p$
4. Return $c = v = p - v$ and k.

Algorithm 9. Phase II: Montgomery Inverse Correction Algorithm
(*MontIcor(c, k, p) Function*)

Input: $c = \{c_{n-1}, \ldots, c_2, c_1, c_0\}_2$, $p = \{p_{n-1}, \ldots, p_1, p_0\}_2$, k
Output: $a^{-1} 2^n$ mod p
1. For $i = 0$ to $(k - n)$ do
1.1 If $c_0 = 0$ then $c = c/2$ else $c = (c + p)/2$
2. Return c.

Algorithm 10. Phase II: Modified Montgomery Inverse Correction Algorithm
(*MontIcor(c, k, p) Function*)

Input: $c = \{c_{n-1}, \ldots, c_2, c_1, c_0\}_2$, $p = \{p_{n-1}, \ldots, p_1, p_0\}_2$, k
Output: a^{-1} mod p
1. For $i = 1$ to k do
1.1 If $c_0 = 0$ then $c = c/2$ else $c = (c + p)/2$
2. Return c.

Algorithm 11. Binary Extended Euclidean Algorithm for Modular Division
(*ExEuclBdiv(a, b, p) Function*)

Input: $a = \{a_{n-1}, \ldots, a_2, a_1, a_0\}_2$, $a = \{b_{n-1}, \ldots, b_2, b_1, b_0\}_2$, $p = \{p_{n-1}, \ldots, p_1, p_0\}_2$
Output: $\frac{a}{b}$ mod p

1. $s = p$, $r = b$, $u = a$, $v = 0$, $z = 0$
2. While $(r > 0)$ do
   - 2.1 While $r_0 = 0$ do
     - 2.1.1 $r = r/2$, $u = u/2 \bmod p$, $z = z - 1$
   - 2.2 If $z < 0$ then $(r \leftrightarrow s, u \leftrightarrow v, z = -z)$
   - 2.3 If $((s + r) \bmod 4 = 0)$ then
     - 2.3.1 $r = (r + s)/2$, $u = (u + v)/2 \bmod p$

       else
     - 2.3.2 $r = (r - s)/2$, $u = (u - v)/2 \bmod p$
3. If $s = 1$ then return v else return $p - v$.

Algorithm 12.
Bit Serial LSB
Multiplication Algorithm

Input: $a = \{a_{k-1}, \ldots, a_2, a_1, a_0\}$,
$b = \{b_{k-1}, \ldots, b_1, b_0\}$,
$f(x) = x^k + \sum_{i=0}^{k-1} f_i x^i = x^k + r(x)$,
$r = \{f_{k-1}, \ldots, f_1, f_0\}$
Output: $c = a \cdot b$, $c(x) = a(x)b(x) \bmod f(x)$

1. $c = 0$
2. For $i = 0$ to $k - 1$ do
   - 2.1 $c = c + b_i \cdot a$
   - 2.2 $a = x \cdot a + a_{k-1} \cdot r$
3. Return c.

Algorithm 13.
Bit Serial MSB
Multiplication Algorithm

Input: $a = \{a_{k-1}, \ldots, a_2, a_1, a_0\}$,
$b = \{b_{k-1}, \ldots, b_1, b_0\}$,
$f(x) = x^k + \sum_{i=0}^{k-1} f_i x^i = x^k + r(x)$,
$r = \{f_{k-1}, \ldots, f_1, f_0\}$
Output: $c = a \cdot b$, $c(x) = a(x)b(x) \bmod f(x)$

1. $c = 0$
2. For $i = k - 1$ to 0 do
   - 2.1 $c = x \cdot c + c_{k-1} \cdot r$
   - 2.2 $c = c + b_i \cdot a$
3. Return c.

Algorithm 14. Bit Serial Montgomery Multiplication Algorithm for GF(2^k) Fields

---

Input $a(x)$, $b(x)$, $f(x)$
Output $c(x) = a(x)b(x)x^{-k} \bmod f(x)$

1. $c(x) = 0$
2. For $i = 0$ to $k - 1$ do
   - 2.1 $c(x) = c(x) + a_i \cdot b(x)$
   - 2.2 $c(x) = c(x) + c_0 \cdot f(x)$
   - 2.3 $c(x) = c(x)/x$
3. Return $c(x)$.

Algorithm 15. Digit Serial LSB Multiplication Algorithm

Input: $a = \{a_{k-1}, \ldots, a_2, a_1, a_0\}$,
$b = \{B_{D-1}, \ldots, B_1, B_0\}$,
$B_i = \{b_{Di+d-1}, b_{Di+d-2}, \ldots, b_{Di+1}, b_{Di}\}$
$f(x) = x^k + \sum_{i=0}^{k-1} f_i x^i = x^k + r(x)$,
$r = \{f_{k-1}, \ldots, f_1, f_0\}$
Output: $c = a \cdot b$,
$c(x) = a(x)b(x) \bmod f(x)$
1. $c = 0$
2. For $i = 0$ to $D - 1$ do
    2.1 $c = c + B_i \cdot a$
    2.2 $a = (x^d \cdot a) \bmod f(x)$
3. Return c.

Algorithm 16. Digit Serial MSB Multiplication Algorithm

Input: $a = \{a_{k-1}, \ldots, a_2, a_1, a_0\}$,
$b = \{B_{D-1}, \ldots, B_1, B_0\}$,
$B_i = \{b_{Di+d-1}, b_{Di+d-2}, \ldots, b_{Di+1}, b_{Di}\}$
$f(x) = x^k + \sum_{i=0}^{k-1} f_i x^i = x^k + r(x)$,
$r = \{f_{k-1}, \ldots, f_1, f_0\}$
Output: $c = a \cdot b$,
$c(x) = a(x)b(x) \bmod f(x)$
1. $c = 0$
2. For $i = D - 1$ to 0 do
    2.1 $c = (x^d \cdot c) \bmod f(x)$
    2.2 $c = c + B_i \cdot a$
3. Return c.

Algorithm 17. Digit Serial Montgomery Multiplication Algorithm for GF(2^k) Fields

Input: $a(x)$, $b(x)$, $f(x)$, $\hat{f}(x) = f^{-1}(x) \bmod x^d$
Output: $c(x) = a(x)b(x)x^{-k} \bmod f(x)$
1. $c(x) = 0$
2. For $i = 0$ to $D - 1$ do
    2.1 $c(x) = c(x) + A_i(x) \cdot b(x)$
    2.2 $m(x) = (C_0(x) \cdot \hat{F}_0(x)) \bmod x^d$
    2.2 $c(x) = c(x) + m(x) \cdot f(x)$
    2.3 $c(x) = c(x)/x^d$
3. Return $c(x)$.

Algorithm 18. Extended Euclidean Algorithm for GF(2^k) Field Inversion *(EEA(a, f) Function)*

Input: $f(x)$, $a(x)$
Output: $v = a^{-1}(x) \bmod f(x)$
1. $s^{(-1)} = f(x)$, $r^{(-1)} = a(x)$, $u^{(-1)} = 1$, $v^{(-1)} = 0$, $i = 0$
2. While $r^{(i)} \neq 0$ repeat
    2.1 $q = \left\lfloor \frac{s^{(i-1)}}{r^{(i-1)}} \right\rfloor$
    2.2 $r^{(i)} = s^{(i-1)} - q \cdot r^{(i-1)}$
    2.3 $u^{(i)} = v^{(i-1)} - q \cdot u^{(i-1)}$
    2.4 $s^{(i)} = r^{(i-1)}$, $v^{(i)} = u^{(i-1)}$
    2.5 $i = i + 1$
3. Return v.

Algorithm 19. Modified Extended Euclidean Algorithm for GF(2^k) Field Inversion
(MEEA(a, f) Function)

Input: $f(x)$, $a(x)$
Output: $a^{-1}(x) \bmod f(x)$
1. $s(x) = f(x)$, $r(x) = xa(x)$, $u(x) = 1$, $v(x) = 0$, $d = 0$
2. For $i = 1$ to $2k$ do
    2.1 if $r_k = 0$ then
        2.1.1 $r(x) = x \cdot r(x)$
        2.1.2 $u(x) = x \cdot u(x) \bmod f(x)$
        2.1.3 $d = d + 1$
    else
        2.1.4 if $s_k = 1$ then
            2.1.4.1 $s(x) = s(x) - r(x)$
            2.1.4.2 $v(x) = v(x) - u(x)$
    2.2 $s(x) = x \cdot s(x)$
    2.3 if $d = 0$ then
        2.3.1 $r(x) \leftrightarrow s(x)$ (exchange $r(x)$ with $s(x)$)
        2.3.2 $u(x) \leftrightarrow v(x)$ (exchange $u(x)$ with $v(x)$)
        2.3.3 $u(x) = x \cdot u(x) \bmod f(x)$
        2.3.4 $d = 1$
    else
        2.3.5 $u(x) = u(x)/x \bmod f(x)$
        2.3.6 $d = d - 1$
3. Return $V(x)$.

Algorithm 20. Binary Modified Extended Euclidean Algorithm for GF(2^k) Field Inversion
(bMEEA(a, f) Function)

Input: $a(x), f(x)$
Output: $a^{-1}(x) \bmod f(x)$
1. $s^{(0)}(x) = f(x), r^{(0)}(x) = a_{k-1}x^k + a_{k-2}x^{k-1} + \cdots + a_0x, u^{(0)}(x) = 1, v^{(0)}(x) = 0,$
   $e^{(0)} = -1$, $\text{sign}^{(0)} = 1$
2. For $i = 1$ to $2k - 1$
    2.1 $r^{(i)}(x) = x \cdot r^{(i-1)}(x) + x \cdot r^{(i-1)}(x) \cdot r_k^{(i-1)}$
    2.2 $u^{(i)}(x) = u^{(i-1)}(x) + v^{(i-1)}(x) \cdot r_k^{(i-1)}$
    2.3 If $r_k^{(i-1)} \cdot \text{sign}^{(i-1)}$ then
        2.3.1 $s^{(i)}(x) = s^{(i-1)}(x)$
        2.3.2 $v^{(i)}(x) = x \cdot v^{(i-1)}(x)$
        2.3.3 $e^{(i)} = e^{(i-1)} - 1$
    else
        2.3.4 $s^{(i)}(x) = r^{(i-1)}(x)$

2.3.5 $v^{(i)}(x) = x \cdot u^{(i-1)}(x)$
2.3.6 $e^{(i)} = -e^{(i-1)} - 1$
3. Return $a^{-1}(x) = \sum_{j=0}^{k-1} t_{k-i}^{(2k-1)} x^j$.

Algorithm 21. Square and Multiply Inversion Algorithm

Input: $a(x), f(x)$
Output: $a^{-1}(x) \bmod f(x)$
1. $X = Y = a^2(x) \bmod f(x)$
2. For $i = 1$ to $k - 1$
 2.1 $Y = X^2 \cdot Y$
 2.2 $X = X^2$
3. Return Y.

Algorithm 22. Itoh–Tsujii Algorithm for GF(p^k) Fields

Input: $A \in \mathrm{GF}(p^k)$
Output: A^{-1}
1. $r = (p^{k-1})/(p-1)$
2. A^{r-1} in GF(p^k)
3. $A^r = A^{r-1} \cdot A$
4. $(A^r)^{-1}$ in GF(p)
6. $A^{-1} = (A^r)^{-1} \cdot A^{r-1}$
7. Return A^{-1}.

Algorithm 23. Binary Point Multiplication Algorithm (*Point and Add Method*)

Input: $s = (s_{t-1}, \ldots, s_1, s_0)_2$, $P \in E(F)$
Output: $Q = sP$
1. $Q = \infty$
2. For i from 0 to $t - 1$ do
 2.1 If $s_i = 1$ then $Q = Q + P$
 2.2 $P = 2P$
3. Return Q.

Algorithm 24. Sliding Window Point Multiplication Algorithm

Input: $s = \sum_{i=0}^{t-1} s_i 2^i : (s_{t-1}, \ldots, s_1, s_0)_2$, $P \in E(F)$
Output: $Q = sP$
Precomputation Phase (calculation of $P_e = eP$, where $e \leq 2^w - 1$ is an odd integer)
1. $P_1 = P$, $Q = \infty$, $r = t - 1$
2. For $j = 1$ to $2^{w-1} - 1$
 2.1 $P_{2j+1} = P_{2j-1} + 2P$

Main Calculation Phase

1. While $r \geq 0$ do
   1.1 If $s_r = 0$ then
      1.1.1 $Q = 2Q$
      1.1.2 $r = r - 1$
   else
      1.1.3 $v = w$
      1.1.4 While $s_{r-v+1} = 0$ do $v = v - 1$ (finding the largest integer v for odd u)
      1.1.5 $u = \{s_r, s_{r-1}, \ldots, s_{r-v+1}\}$ (u is an odd integer)
      1.1.6 $Q = 2^v Q + P_u$
      1.1.7 $r = r - v$
2. Return Q.

Algorithm 25. NAF Construction Algorithm

Input: $s = \sum_{i=0}^{t-1} s_i 2^i : (s_{t-1}, \ldots, s_1, s_0)_2$
Output: $NAF(s) = s' = \sum_{i=0}^{t} s_i' 2^i$

1. $g = 0$
2. For $i = 0$ to k
   2.1 $g_{i+1} = \left\lfloor \frac{s_i + s_{i+1} + g_i}{2} \right\rfloor$
   2.2 $s_i' = s_i + g_i - 2g_{i+1}$
3. Return s'.

Algorithm 26. Binary NAF Point Multiplication Algorithm (*NAF Point and Add Method*)

Input: $\text{NAF}(s) = s'$, $P \in E(F)$
Output: $Q = s' P$

1. $Q = \infty$
2. For $i = 0$ to t
   2.1 If $s_i' = 1$ then $Q = Q + P$
   2.2 If $s_i' = -1$ then $Q = Q - P$
   2.3 $P = 2P$
3. Return Q.

Algorithm 27. Binary Montgomery Point Multiplication

Input: $s = (s_{t-1}, \ldots, s_1, s_0)_2$, $P \in E(F)$
Output: $Q = sP$

1. $Q = \infty$
2. For $i = t - 2$ to 0
   2.1 If $s_i = 1$ then
      2.1.1 $Q = Q + P$ and $P = 2P$
   else
      2.1.2 $P = Q + P$ and $Q = 2Q$
3. Return Q.

Algorithm 28. ECDSA Signature Generation

Input: Domain parameters (*Order*(*F*), *F_R*, *S*, *a*, *b*, *P*, *n*, *h*), private key *d*, message *m*.
Output: Signature (*r*, *s*).
1. Select k, where $1 \leq k \leq n-1$
2. $e = H(m)$
3. $kP = (x_1, y_1)$
4. Represent $x_1 \in F$ as an integer $x_1' \in Z$
5. $r = x_1' \bmod n$. If $r = 0$ then go to step 1.
6. $s = k^{-1}(e + d \cdot r) \bmod n$. If $s = 0$ then go to step 1.
7. Return (r, s).

Algorithm 29. ECDSA Signature Verification

Input: Domain parameters *D* (*Order*(*F*), *F_R*, *S*, *a*, *b*, *P*, *n*, *h*), public key *Q*, message *m*, signature (*r*, *s*).
Output: Valid or invalid signature.
1. If ($1 \leq r, s \leq n-1$ and r, s are integers) is not true then return (Invalid signature) else
    1.2 $e = H(m)$.
    1.3 $w = s^{-1} \bmod n$.
    1.4 $u_1 = e \cdot w \bmod n$ and $u_2 = r \cdot w \bmod n$.
    1.5 $X = u_1P + u_2Q = (x_1, y_1)$.
    1.6 If $X = \infty$ then return ("Invalid signature");
    1.7 Represent $x_1 \in F$ as an integer $x_1' \in Z$
    1.8 $v = x_1 \bmod n$.
    1.8 If $v = r$ then return ("Valid signature") else return ("Invalid signature").

REFERENCES

1. Carter B. and Shumway R. *Wireless Security-End to End*, Wiley Publishing Indianapolis, 2002.
2. Aydos M. Efficient Wireless Security Protocols Based on Elliptic Curve Cryptography, PhD thesis, Oregon State University, June 2001.
3. WAP Forum. Wireless Transport Layer Security, WAP-261–WTLS-20010406-a, www.wapforum.org.
4. Koblitz N. Elliptic curve cryptosystems, *Mathematics of Computation*, 48, 20, 1987.
5. Miller V. Use of elliptic curves in cryptography. In *Proceedings of the Advances in Cryptology—CRYPTO'85*, Springer-Verlag, LNCS 218, 1986, 417.
6. Schneier B. *Applied Cryptography—Protocols, Algorithms and Source Code in C*, 2nd ed., John Wiley & Sons, New York, 1996.
7. Menezes A., Van Oorschot P.C., and Vanstone S.A. *Handbook of Applied Cryptography*, CRC Press, Boca Raton, FL, 1997.

8. Mastrovito E.D. VLSI Architectures for Computations in Galois Fields, PhD thesis, Linköping University, Sweden, 1991.
9. Washington L.C. *Elliptic Curves: Number Theory and Cryptography*, Chapman & Hall/CRC, New York, 2003.
10. Menezes A., Okamoto T., and Vanstone S. Reducing elliptic curve logarithms to logarithms in a finite field, *IEEE Transactions on Information Theory* 39, 1639, 1993.
11. Behrooz P. *Computer Arithmetic: Algorithms and Hardware Designs*, Oxford University Press, New York, 2000.
12. Rabaey J., Chandrakasan A., and Nikolic B. *Digital Integrated Circuits: A Design Perspective*, 2nd ed., Prentice Hall, Englewood Cliffs, NJ, 2003.
13. FIPS 186-2. Digital Signature Standard (DSS). Federal Information Processing Standards Publication 186-2, National Institute of Standards and Technology (NIST), 2000.
14. IEEE STD 1363–2000. IEEE Standard Specifications for Public-Key Cryptography, 2000.
15. Karatsuba A. and Ofman Y. Multiplication of multidigit numbers on automata, *Soviet Physics—Doklady*, 7, 595, 1963.
16. Montgomery P.L. Modular multiplication without trial division, *Mathematics of Computation*, 44(170), 519, 1985.
17. Hankerson D., Menezes A., and Vanstone S. *Guide to Elliptic Curve Cryptography*, Springer-Verlag & Hall/CRC, New York, 2004.
18. Eldridge S. and Walter C.D. Hardware implementation of Montgomery's modular multiplication algorithm, *IEEE Transactions on Computers*, 42(6), 693, 1993.
19. Tenca A.F. and Koç Ç. A scalable architecture for modular multiplication based on Montgomery's algorithm, *IEEE Transactions on Computers*, 52(9), 1215, 2003.
20. Hong J.-H. and Wu C.-W. Cellular-array modular multiplier for fast RSA public-key cryptosystem based on modified Booth's algorithm, *IEEE Transactions on Very Large Scale Integration (VLSI) Systems*, 11(3), 2003.
21. Fournaris A.P. and Koufopavlou O. Montgomery modular multiplier architectures and hardware implementations for an RSA cryptosystem. In *Proceedings of 46th IEEE Midwest Symposium on Circuits and Systems '03 (MWSCAS 2003)*, Egypt, 2003.
22. Fournaris A.P. and Koufopavlou O. A new RSA encryption architecture and hardware implementation based on optimized Montgomery multiplication. In *Proceedings of the IEEE International Symposium on Circuits and Systems (ISCAS 2005)*, Japan, 2005.
23. McIvor C., McLoone M., McCanny J.V., Daly A., and Marnane W. Fast Montgomery modular multiplication and RSA cryptographic processor architectures. In *Proceedings of 37th Annual Alisomar Conference on Signals Systems and Computers*, CA November, 2003.
24. Cilardo A., Mazzeo A., Romano L., and Saggese G.P. Carry-Save Montgomery exponentiation on reconfigurable hardware. In *Proceedings of Design, Automation and Test in Europe Conference and Exhibition*, Paris, France, February, 2004.
25. Wu X., Chen H., Sun Y., and Gai W. A fully pipeline linear systolic architecture for modular multiplier in public-key crypto-systems, *Journal of VLSI Signal Processing*, 33, 191, 2003.

26. Kornerup P. High radix modular multiplication for cryptosystems. In *Proceedings of 11th Symposium on Computer Arithmetic*, Canada, July 1993.
27. Orup H. Simplifying quotient determination in high-radix modular multiplication. In *Proceedings of 12th Symposium on Computer Arithmetic (ARITH'95)*, Bath, England, July 1995.
28. Schinianakis D.M., Fournaris A.P., Kakarountas A.P., and Stouraitis T. An RNS architecture of an elliptic curve point multiplier. In *Proceedings of International Symposium on Circuits and Systems 2006 (ISCAS'06)*, Koc-Greece, May 2006.
29. Bajard J.C., Didier L.S., and Kornerup P. An RNS Montgomery's modular multiplication, *IEEE Transaction on Computers*, 19(2), 167, 1998.
30. Bajard J.C., Didier L.S., and Kornerup P. Modular multiplication and base extensions in residue number systems. In *Proceedings of IEEE Symposium on Computer Arithmetic 2001* Vail, CO, 2001.
31. Kaliski B.S. The Montgomery inverse and its applications, *IEEE Transactions on Computers*, 44(8), 1064, 1995.
32. Kaihara M.E. and Takagi N. A hardware algorithm for modular multiplication/division, *IEEE Transactions on Computers*, 54(1), 12–21, 2005.
33. Takagi N. A VLSI algorithm for modular division based on the binary GCD algorithm, *IEICE Trans. on Fundamentals and Electronics*, E81-A(5), 724, 1998.
34. Gutub A.A. and Tenca A.F. Efficient scalable VLSI architecture for Montgomery inversion in GF(p), *Integration, the VLSI Journal*, 37(2), 103, 2004.
35. Daly A., Marnane W., and Popovici E. Fast modular inversion in the Montgomery domain on reconfigurable logic. In *Proceedings of the Irish Signals and Systems Conference (ISSC 2003)*, Limerick, Ireland, July 2003.
36. Savas E. and Koç Ç.K. The Montgomery modular inverse-revisited, *IEEE Transactions on Computers*, 49(7), 763, 2000.
37. Hütter M., Großschädl J., and Kamendje G.-A. A versatile and scalable digit-serial/parallel multiplier architecture for finite fields GF(2^m). In *Proceedings of the International Conference on Information Technology: Computers and Communications (ITCC 03)*, Las Vegas, NV, 2003.
38. Berlekamp E.R. *Algebraic Coding Theory*. McGraw-Hill, New York, 1968.
39. Koç Ç.K. and Acar T. Montgomery multiplication in GF(2^k), *Designs, Codes and Cryptography*, 14(1), 57, 1998.
40. Lee W.H., Lee K.J., and Yoo K.Y. Design of a linear systolic array for computing modular multiplication and squaring in GF(2^m), *Computer and Mathematics with Applications*, 42, 231, 2001.
41. Wu H., Montgomery multiplier and squarer for a class of finite fields, *IEEE Transactions on Computers*, 51(5), 521, 2002.
42. Fournaris A.P. and Koufopavlou O. A systolic trinomial GF(2^k) multiplier based on the Montgomery multiplication algorithm. In *Proceedings of 12th IEEE International Conference on Electronics, Circuits and Systems (ICECS 2005)*, Gammarth, Tunisia, 2005.
43. Halbutogullari A. and Koç Ç.K. Mastrovito multiplier for general irreducible polynomials, *IEEE Transactions on Computers*, 49(5), 503–518, 2000.
44. Ernst M., Jung M., Madlener F., Huss S., and Blümel R.A. Reconfigurable system on chip implementation for elliptic curve cryptography over GF(2^n), *In Proceedings of 2002 Workshop on Cryptographic Hardware and Embedded Systems (CHES 2002)*, Kaliski B.S. Jr. et al. (eds), Springer-Verlag, LNCS 2523, 2003, 381.

45. Paar C. A new architecture for a parallel finite field multiplier with low complexity based on composite fields, *IEEE Transactions on Computers*, 45(7), 856, 1996.
46. Song L. and Parhi K.K. Low-energy digit-serial/parallel finite field multipliers, *Journal of VLSI Signal Processing Systems*, 19(2), 149, 1998.
47. Guo J.-H. and Wang C.-L. Digit-serial systolic multiplier for finite fields GF(2^m), *IEE Proceedings-Computers and Digital Techniques*, 145(2), 143, 1998.
48. Kim C.H., Hong C.P., and Kwon S. A digit-serial multiplier for finite field GF(2^m), *IEEE Transactions on Very Large Scale Integration (VLSI) Systems*, 13(4), 2005.
49. Chang K.-Y., Hong D., and Cho H.-S. Low complexity bit-parallel multiplier for GF(2^m) defined by all-one polynomials using redundant representation, *IEEE Transactions on Computers*, 54(12), 2005.
50. Lee C.Y., Horng J.-S., Jou I.-C., and Lu E.-H. Low-complexity bit-parallel systolic Montgomery multipliers for special classes of GF(2^m), *IEEE Transactions on Computers*, 54(9), 2005.
51. Halbutoğullari A. and Koç Ç.K. Parallel multiplication in GF(2^k) using polynomial residue arithmetic, *Designs, Codes and Cryptography*, 20(2), 155, 2000.
52. Bajard J.-C., Imbert L., and Jullien G.A. Parallel Montgomery multiplication in GF (2^k) using trinomial residue arithmetic. In *Proceedings of 17th IEEE Symposium on Computer Arithmetic (ARITH'05)*, Cape Cod, MA, 2005.
53. Fournaris A.P. and Koufopavlou O. GF(2^k) Multipliers based on the Montgomery multiplication algorithm. In *Proceedings of the 2004 IEEE International Symposium on Circuits and Systems (ISCAS 2004)*, Canada, 2004.
54. Wang C.L. and Lin J.L. Systolic array implementation of multipliers for finite fields GF(2^k), *IEEE Transactions on Circuits and Systems*, 38(7), 796, 1991.
55. Jain S.K., Song L., and Parhi K.K. Efficient semisystolic architectures for finite-field arithmetic, *IEEE Transactions on Very Large Scale Integration (VLSI) Systems*, 6(1), 101, 1998.
56. Guo J.-H. and Wang C.-L. Hardware efficient systolic architecture for inversion and division in GF(2^k), *IEE Proceedings on Computer and Digital Techniques*, 272, 1998.
57. Wu C.-H., Wu C.-M., Shieh M.-D., and Hwang Y.-T. Systolic VLSI realization of a novel iterative division algorithm over GF(2^k): a high-speed, low-complexity design. In *Proceedings of International Symposium on Circuits and Systems (ISCAS'01)*, Sydney, Australia, May 2001.
58. Yan Z. and Sarwate D.V. New systolic architectures for inversion and division in GF(2^k), *IEEE Transactions on Computers*, 52(11), 1514, 2003.
59. Yan Z., Sarwate D.V., and Liu Z. High-speed systolic architectures for finite field inversion, *Integration, the VLSI Journal*, 38 (3), 382, 2005.
60. Brunner H., Curiger A., and Hofstetter M. On computing multiplicative inverses in GF(2^k), *IEEE Transactions on Computers*, 42(8), 1010, 1993.
61. Guo J.-H. and Wang C.-L. Systolic array implementation of Euclid's algorithm for inversion and division in GF(2^m), *IEEE Transactions on Computers*, 47(10), 1161, 1998.
62. Schroeppel R., Orman H., O'Malley S., and Spatscheck O. Fast key exchange with elliptic curve systems. In *Proceedings Advances in Cryptology (CRYPTO '95)*, Coppersmith D. (ed.), Springer-Verlag, LNCS 963, 1995, 43.
63. Olofsson M. VLSI Aspects on Inversion in Finite Fields, PhD thesis, Linköping University, Sweden, 2002.

64. Fournaris A.P. and Koufopavlou O. A novel systolic GF(2^k) field multiplication-inversion arithmetic unit. In *Proceedings of 12th IEEE International Conference on Electronics, Circuits and Systems (ICECS 2005)*, Gammarth, Tunisia, December 2005.
65. Massey J.L. and Omura J.K. Computational method and apparatus for finite field arithmetic, US patent #4,587,627, May 1986.
66. Mullin R.C., Onyszchuk I.M., Vanstone S.A., and Wilson R.M. Optimal normal bases in GF(p^n), *Discrete Applied Mathematics*, 22, 149, 1988/89.
67. Gao S. Normal Bases over Finite Fields, PhD dissertation in Combinatorics and Optimization, University of Waterloo, Canada, 1993.
68. Ash D., Blake I., and Vanstone S. Low complexity normal bases, *Discrete Applied Mathematics*, 25, 191, 1989.
69. Agnew G.B., Mullin R.C., Onyszchuk I.M., and Vanstone S.A. An implementation for a fast public-key cryptosystem, *Journal of Cryptology*, 3, 63, 1991.
70. Masoleh A.R. and Hasan M.A. Low complexity sequential normal basis multipliers over GF(2^m). In *Proceedings of 16th Symposium on Computer Arithmetic*, New York, 2003, 8.
71. Yang D.-J., Kim C.-H, Park Y., Kim Y., and Lim J. Modified sequential normal basis multipliers for type II optimal normal bases. In *Proceedings of ICCSA 2005*, Gervasi O. et al. (eds), Springer-Verlag, LNCS 3481, 2005, 647.
72. Kwon S., Gaj K., Kim C.-H., and Hong C.-P. Efficient linear array for multiplication in GF(2^m) using a normal basis for elliptic curve cryptography. In *Proceedings of 2004 Workshop on Cryptographic Hardware and Embedded Systems (CHES 2004)*, Joye M. and Quisquater J.-J. (eds), Springer-Verlag, LNCS 3156, 2004, 76.
73. Masoleh A.R. and Hasan M.A. A new construction of Massey-Omura parallel multiplier over GF(2^m), *IEEE Transactions on Computers*, 51(5), 511, 2002.
74. Hasan M.A., Wang M.Z., and Bhargava V.K. A modified Massey-Omura parallel multiplier for a class of finite fields, *IEEE Transactions on Computers*, 42(10), 1278, 1993.
75. Masoleh A.R. and Hasan M.A. Fast normal basis multiplication using general purpose processors, *IEEE Transactions on Computers*, 52(11), 1379, 2003.
76. Koç Ç.K. and Sunar B. Low-complexity bit-parallel canonical and normal basis multipliers for a class of finite fields, *IEEE Transactions on Computers*, 47(3), 353, 1998.
77. Gao L. and Sobelman G.E. Improved VLSI designs for multiplication and inversion in GF(2^m). In *Proceedings of 13th Annual IEEE International ASIC/SOC Conference*, Washington, 2000, 97.
78. Sunar B. A Euclidean algorithm for normal bases, *Special Issue on Finite Fields: Applications and Implementations, Acta Applicandae Mathematicae*, Editor J.L. Imana, 93(1–3), 57, 2006.
79. Itoh T. and Tsujii S. A fast algorithm for computing multiplicative inverses in GF(2^m) using normal basis, *Information and Computing*, 78, 171, 1988.
80. Coron J.S. Resistance against differential power analysis for elliptic curve cryptosystems. In *Proceedings of 1999 Workshop on Cryptographic Hardware and Embedded Systems (CHES 1999)*, Koç Ç.K. and Paar C. (eds), Springer-Verlag, LNCS 1717, 1999, 292.

81. Chudnovsky D.V. and Chudnovsky G.V. Sequences of numbers generated by addition in formal groups and new primality and factorizations tests, *Advances in Applied Mathematics*, 7, 385, 1987.
82. Lopez J. and Dahab R. Fast multiplication on elliptic curves over GF(2^m) without precomputation. In *Proceedings of 1999 Workshop on Cryptographic Hardware and Embedded Systems (CHES 1999)* Koç Ç.K. and Paar C. (eds), Springer-Verlag, LNCS 1717, 1999, 316.
83. Karandeinos G. Cryptographic Applications on Elliptic Curves: Design and Implementation of an Elliptic Curve Arithmetic Unit Defined over GF(2^n) Fields, Diploma thesis, University of Patras, Patras, July 2005.
84. Blake I., Seroussi G., and Smart N. *Elliptic Curves in Cryptography*, Ist ed., Cambridge University Press, Cambridge, UK, 1999.
85. Gao L., Shrivastava S., and Sobelman G.E. Elliptic curve scalar multiplier design using FPGAs. In *Proceedings of 1999 Workshop on Cryptographic Hardware and Embedded Systems (CHES 1999)* Koç Ç.K. and Paar C. (eds), Springer-Verlag, LNCS 1717, 1999, 257.
86. Orlando G. and Paar C. A high performance elliptic curve processor for GF(2^m). In *Proceedings of 2000 Workshop on Cryptographic Hardware and Embedded Systems (CHES 2000)*, Springer-Verlag, LNCS 1965, 2000.
87. Orlando G. and Paar C. A scalable GF(p) elliptic curve processor architecture for programmable hardware. In *Proceedings of 2001 Workshop on Cryptographic Hardware and Embedded Systems (CHES 2001)* Koç Ç.K., Naccache D., and Paar C. (eds), Springer-Verlag, LNCS 2162, 2001, 348.
88. Leong P.H.W. and Leung I.K.H. A microcoded elliptic curve processor using FPGA technology, *IEEE Transactions on Very Large Scale Integration (VLSI) Systems*, 10(5), 2002.
89. Kocher P. Timing attacks on implementations of Diffie–Hellman, RSA, DSS and other systems, *Advances in Cryptology, Proceedings of CRYPTO '96*, Koblitz N. (ed.), Springer-Verlag, LNCS 1109, 1996, 104.
90. Kocher P., Jaffe J., and Jun B. Differential power analysis, *Advances in Cryptology, Proceedings of CRYPTO '99*, Wiener M. (ed.), Springer-Verlag, LNCS 1666, 1999, 388.
91. Okeya, K. and Sakurai, K. Power analysis breaks elliptic curve cryptosystems even secure against the timing attack. In *Proceedings of Progress in Cryptology–INDOCRYPT 2000*, Springer-Verlag, LNCS 1977, 2000, 178.
92. Joye M. and Tymen C. Protections against differential analysis for elliptic curve cryptography. In *Proceedings of 2001 Workshop on Cryptographic Hardware and Embedded Systems (CHES 2001)*, Koç Ç.K., Naccache D., and Paar C. (eds), Springer-Verlag, LNCS 2162, 2001, 377.
93. Montgomery, P.L. Speeding the pollard and elliptic curve methods of factorizations, *Mathematics of Computation*, 48, 243, 1987.
94. Okeya K., Miyazaki K., and Sakurai K. A fast scalar multiplication method with randomized projective coordinates on a Montgomery-form elliptic curve secure against side channel attacks. In *Proceedings of ICICS 2001*, Kim K. (ed.), Springer-Verlag, LNCS 2288, 428.
95. Izu T. and Takagi T. A fast parallel elliptic curve multiplication resistant against side channel attacks. In *Proceedings of PKC 2002*, Naccache D. and Paillier P. (eds), Springer-Verlag, LNCS 2274, 2002, 280.

96. Joye M. and Yen S.-M. The Montgomery powering ladder. In *Proceedings of 2002 Workshop on Cryptographic Hardware and Embedded Systems (CHES 2002)*, Kaliski B.S. Jr. et al. (eds), Springer-Verlag, LNCS 2523, 2003, 291.
97. ANSI X9.62–1998, Public Key Cryptography for the Financial Services Industry: The Elliptic Curve Digital Signature Algorithm (ECDSA).

5 Efficient Elliptic Curve Cryptographic Hardware Design for Wireless Security

Lo'ai A. Tawalbeh and Çetin Kaya Koç

CONTENTS

The spread of wired and wireless communications, the continuous growth of the Internet, and the E-commerce transactions increased the necessity for security in applications that involve sharing or exchange of secret or private information. Public-key cryptography is widely used in establishing secure communication channels between the users on the Internet and in wireless communication networks.

5.1 INTRODUCTION

A small set of public-key cryptosystems are used extensively, which includes ElGamal cryptosystem [1], Diffie–Hellman (DH) key exchange algorithm [2], the digital signature algorithm (DSA) [3], and elliptic curve cryptography–based algorithms such as EC–ElGamal and ECDSA. Elliptic curve cryptography (ECC), which was introduced by Miller [4] and Koblitz [5], is based on a more difficult mathematical problem to solve than the one used in traditional public-key algorithms. Thus, ECC stands out from this crowd of algorithms because of its unique property of providing the highest degree of security with the smallest key sizes. For example, an elliptic curve system with 313-bits can replace a certain 4096-bit key size conventional system [6]. Using smaller key sizes to gain the same level of security leads to a big reduction in hardware resources used in implementations.

In this chapter, we mainly concentrate on efficient hardware realization of elliptic curve cryptography for wireless applications. Elliptic curve cryptography involves huge arithmetic operations performed over finite fields (most commonly used fields are the prime extension fields, GF(p), and the binary extension fields, GF(2^n)), and therefore, an efficient ECC system requires efficient hardware implementations of finite field operations. Once realized, similar hardware can also be used to support other public-key cryptographic functions. Furthermore, long-term deployment of public-key cryptography hardware requires flexibility in key size as better cryptanalytic techniques are developed.

Recently, two important developments took place in this area. The first one is called *scalability* which refers to the ability of the hardware to reconfigure itself to support longer key sizes, limited only by the amount of available input, output, and scratch memory space. The second one is about designing a single hardware to support all kinds of elliptic curves based on finite fields of different characteristics. This property of hardware is called *unified* or *dual field*.

Our research starts from these premises and moves on to create better algorithms to support long-term, efficient, scalable, and unified hardware implementations. We address and provide solutions for dual-field Montgomery multipliers, modular dividers, and unified dividers and inverters. Particularly, we introduce a novel algorithm suitable for hardware design which computes division (inverse) and multiplication in a very efficient way for

GF(p) and GF(2^n) fields. The new algorithm is called the unified division/multiplication algorithm (UDMA). In addition, we propose the hardware architecture that efficiently supports all operations in the UDMA and uses carry-save unified adders for reduced critical path delay, making the proposed architecture faster than other previously proposed designs. We present example designs of our algorithms using field programmable gate arrays (FPGAs) and the benchmark results of our implementations.

At the end of this chapter, we introduce an elliptic curve crypto-processor (ECCP) architecture over GF(2^n) that is based on the efficient UDMA hardware implementation. The scalability feature of the proposed crypto-processor allows the adjustment of the word size used in the datapath to meet area and performance requirements. On the other hand, the processor allows the user to choose the value of the field parameter (n). Finally, the experimental results obtained for the ECCP are analyzed and compared with other proposed designs.

5.2 ELLIPTIC CURVE THEORY

In the mid-1980s, Niel Koblitz and Victor Miller proposed the elliptic curve cryptography (ECC) [4,5]. It is based on the discrete logarithm (DL) problem over the points on an elliptic curve (EC). Recently, the elliptic curve crypto-systems started to replace many known conventional public-key cryptography algorithms. This is due to the high level of security they provide and their fast and compact size implementations over finite fields.

Data in an ECC are represented as points on an elliptic curve. They are called elliptic because they arose historically from the problem of computing the solutions for an equation of an ellipse. These curves have special characteristics and provide the base for particular arithmetic operations.

In cryptography, we are interested in the elliptic curves defined over finite fields. In other words, the coefficients of the defining equation ($F(x,y) = 0$) are elements of GF(q), and the points on the curve are of the form $P = (x,y)$, where x and y are the elements of GF(q) that satisfy the equation. The general form for an elliptic curve equation is

$$y^2 + axy + by = x^3 + cx^2 + dx + e.$$

A point at infinity (O) is also defined [7]. O plays a role similar to zero in ordinary addition. It is computed as the sum of three points that lie on a straight line on the EC.

The complexity of elliptic curve arithmetic operations that includes rules used to add two points (point addition) or add a point to itself (point doubling) on the elliptic curves, depends on the finite field (GF(p) or GF(2^n)) and on the coordinate system (affine or projective) that is used. Moreover, choosing the suitable representation for the elements of the finite field may lead to more efficient implementations of the field arithmetic in hardware or in software.

The core operation on ECC is the scalar point multiplication, which consists of a certain number of point additions. When a point P defined on the curve is added to itself k times, it is very difficult to find what was P without knowing k. That is the characteristic that provides security to ECC:

$$Q = kP = \underbrace{P + P + \cdots + P}_{k \text{ times}}. \tag{5.1}$$

In the following subsections, we discuss the elliptic curves defined over GF(p) and GF(2^n) and the arithmetic algorithms defined in each field.

5.2.1 Elliptic Curves Defined over GF(p)

The elements of the field GF(p) are the integers in the set $\{0, 1, 2, \ldots, p-1\}$, where p is an n-bit prime modulus in the range of $2^{n-1} < p < 2^n$. The basic arithmetic operations defined in this field are

- *Addition modulo p*. The addition of elements in a prime field is a conventional integer addition with modulo reduction (mod p). For example, let $X, Y, R \in \text{GF}(p)$, then $R = X + Y \bmod p$, where R is the remainder of $(X + Y)$ divided by p.
- *Multiplication modulo p*. Let $M = X \cdot Y$, where $X, Y, M \in \text{GF}(p)$, M is the remainder of $X \cdot Y$ divided by p.
- *Squaring*. If $X \in \text{GF}(p)$, then $X^2 = X \cdot X$ is the remainder of X^2 divided by p.
- *Inversion modulo p*. Inversion is defined for a nonzero element $X \in \text{GF}(p)$ as X^{-1} to be the unique integer $W \in \text{GF}(p)$, such that $X \cdot W \equiv 1 \bmod p$.

The elliptic curves defined over GF(p) satisfy the following equation:

$$y_2 = x_3 + ax + b \bmod p,$$

where $p > 3$, $4a^3 + 27b^2 \neq 0$ and $x, y, a, b \in \text{GF}(p)$. As mentioned earlier, the point at infinity O plays a role similar to zero in the integer domain. But, there are some addition rules for O in this field. Assume that (x, y) is a point on an EC, then

1. $(x, y) + O = (x, y)$.
2. $(x, y) + (x, -y) = O$.
3. $O = -O$.

The points on the curves can be represented using affine or projective coordinates. A brief description of each coordinate is given in the following sections.

5.2.1.1 Affine Coordinates

To add two points on an elliptic curve represented in affine coordinates as $P_1 = (x_1, y_1)$ and $P_2 = (x_2, y_2)$, we compute $P_3 = (x_3, y_3) = P_1 + P_2$ and $P_1 \neq P_2$ According to the addition rules,

$$\alpha = \frac{y_2 - y_1}{x_2 - x_1},$$
$$x_3 = \alpha^2 - x_1 - x_2,$$
$$y_3 = \alpha(x_1 - x_3) - y_1,$$

and when $P_1 = P_2$ (point doubling $P_3 = 2P_1$ and $P_1 \neq 0$), the addition rules are

$$\alpha = \frac{3x_1^2 + a}{2y_1},$$
$$x_3 = \alpha^2 - 2x_1,$$
$$y_3 = \alpha(x_1 - x_3) - y_1.$$

If we assumed that the squaring calculation is equivalent to a multiplication, then the addition of two different points in GF(p) requires: six additions, one inversion, and three multiplication operations. On the other hand, to add a point to itself (point doubling) a total of four additions, one inversion, and four multiplications are required [8].

5.2.1.2 Projective Coordinates

Adding or doubling points represented in affine coordinates involve modular inversion calculations. The inversion is considered a time-consuming operation. The projective coordinates are used to almost eliminate the need for performing inversion [8].

The elliptic point, $P_1 = (x, y)$ defined over GF(p), is represented in the projective coordinates as (X, Y, Z), where $x = X/Z^2$ and $y = Y/Z^3$. This transformation is performed at the beginning to represent the point in projective coordinates. After performing the point addition operation, this transformation is carried out again to get the point back in affine coordinates. Algorithm 1 is used to add two points $(P + Q, P \neq Q)$ in projective coordinates:

$$
\begin{aligned}
P = (X_1, Y_1, Z_1);\ Q &= (X_2, Y_2, Z_2);\ P + Q = (X_3, Y_3, Z_3) \\
(x, y) &= (X/Z^2, Y/Z^3), \\
T_1 &= X_1 Z_2^2, \\
T_2 &= X_2 Z_1^2, \\
T_3 &= T_1 - T_2, \\
T_4 &= Y_1 Z_2^3, \\
T_5 &= Y_2 Z_1^3, \\
T_6 &= T_4 - T_5, \\
T_7 &= T_1 + T_2, \\
T_8 &= T_4 + T_5, \\
Z_3 &= Z_1 Z_2 T_3, \\
X_3 &= T_6^2 - T_7 T_3^2, \\
T_9 &= T_7 T_3^2 - 2X_3, \\
Y_3 &= \frac{T_9 T_6 - T_8 T_3^3}{2}.
\end{aligned}
$$

The doubling point algorithm $(P + P)$ in projective coordinates is given by

$$
\begin{aligned}
P = (X_1, Y_1, Z_1);\ P + P &= (X_3, Y_3, Z_3) \\
(x, y) &= (X/Z^2, Y/Z^3), \\
T_1 &= T_3 X_1^2 + a Z_1^4, \\
Z_3 &= 2Y_1 Z_1, \\
T_2 &= 4X_1 Y_1^2, \\
X_3 &= T_1^2 - 2T_2, \\
T_3 &= 8Y_1^4, \\
T_4 &= T_2 - X_3, \\
Y_3 &= T_1 T_4 - X_3.
\end{aligned}
$$

From these algorithms, we found that the number of multiplication operations needed to add 2 points is 16, whereas the number of multiplications for doubling a point is found to be only 10 [8].

5.2.2 Elliptic Curves Defined over GF(2^n)

The elliptic curves defined over GF(2^n) satisfy the equation

$$E\colon y^2 + xy = x^3 + ax^2 + b,$$

where $a,b \in GF(2^n)$ and $b \neq 0$. The addition law for two points in affine coordinates involves multiplication, division, and squaring in the underlying finite field.

5.2.2.1 Affine Coordinates

Adding two points in the affine coordinates can be achieved as follows: let $P_1 = (x_1, y_1)$ and $P_2 = (x_2, y_2)$ be two points defined on the curve; then $P_3 = (x_3, y_3) = P_1 + P_2$ is defined when $P_1 \neq P_2$ as

$$\begin{aligned} \alpha &= \frac{y_1 + y_2}{x_1 + x_2}, \\ x_3 &= \alpha^2 + \alpha + x_1 + x_2 + a, \\ y_3 &= (x_1 + x_3)\alpha + x_3 + y_1, \end{aligned}$$

and when $P_1 = P_2$ (point doubling) as

$$\begin{aligned} \alpha &= x_1 + \frac{y_1}{x_1}, \\ x_3 &= \alpha^2 + \alpha + a, \\ y_3 &= (x_1 + x_3)\alpha + x_3 + y_1. \end{aligned}$$

5.2.2.2 Projective Coordinates

To eliminate the need for performing inversion in $GF(2^n)$, the affine coordinates (x, y) are projected to (X, Y, Z), where $x = X/Z^2$ and $y = Y/Z^3$ [8]. The point doubling algorithm $(P + P)$ in projective coordinates is given by

$$\begin{aligned} P &= (X_1, Y_1, Z_1);\ P + P = (X_3, Y_3, Z_3), \\ Z_3 &= X_1 Z_1^2, \\ X_3 &= (X_1 + bZ_1^2)^4, \\ T &= Z_3 + X_1^2 + Y_1 Z_1, \\ Y_3 &= X_1^4 Z_3 + T X_3. \end{aligned}$$

On the other hand, the point addition of two elliptic curve points $(P + Q)$, where $P \neq Q$, is given by

$$\begin{aligned} P &= (X_1, Y_1, Z_1);\ Q = (X_2, Y_2, Z_2);\ P + Q = (X_3, Y_3, Z_3), \\ (x,y) &= (X/Z^2, Y/Z^3), \\ T_1 &= X_1 Z_2^2, \end{aligned}$$

$$
\begin{aligned}
T_2 &= X_2Z_1^2, \\
T_3 &= T_1 + T_2, \\
T_4 &= Y_1Z_2^3, \\
T_5 &= Y_2Z_1^3, \\
T_6 &= T_4 + T_5, \\
T_7 &= Z_1T_3, \\
T_8 &= T_6X_2 + T_7Y_2, \\
Z_3 &= T_7Z_2, \\
T_9 &= T_6 + Z_3, \\
X_3 &= aZ_3^2 + T_6T_9 + T_3^3, \\
Y_3 &= T_9X_3 + T_8T_7^2.
\end{aligned}
$$

When using GF(2^n), the number of multiplication processes for adding 2 points is found to be 20, whereas it is found to be 10 for doubling a point.

5.2.3 Arithmetic Complexity of Affine and Projective Coordinates

A research was carried out by Gutub [8] to evaluate the complexity of performing arithmetic operations in affine and projective coordinates, and in both finite fields (GF(p) and GF(2^n)). The research was based on using the binary algorithm to compute kP from a given point P on the elliptic curve. Assuming that k is n-bits, then the algorithm performs exactly n point doubling. To evaluate the average point additions, we assume that k has half ones and half zeros. This results in $n/2$ point additions.

Table 5.1 shows the total number of multiplications and inversions for both GF(p) and GF(2^n) needed to perform n point doubling and $n/2$ point additions. The table indicates that for an affine coordinates system to be faster than a projective system, the time to compute $1.5n$ inversions and $5.5n$ multiplications should be less than $18n$, GF(p) multiplications or $20n$, GF(2^n) multiplications. But, it is worth mentioning that even using projective coordinates did not eliminate the inversion step completely. It is still required at the end of the computations to convert the result back to affine coordinates. This fact motivates the research for efficient hardware implementations for the inverse operation.

TABLE 5.1
Comparison between Affine and Projective Coordinates

Finite Field	Affine Coordinates Operations	Projective Coordinates Operations
GF(p)	$1.5n$ inversions, $5.5n$ multiplications	$18n$ multiplications
GF(2^n)	$1.5n$ inversions, $5.5n$ multiplications	$20n$ multiplications

5.3 ELLIPTIC CURVE CRYPTOSYSTEMS

Computing kP from the point p can be carried out easily using the algorithms mentioned in the earlier sections, based on which field and coordinates are used. Now, computing the value of k from the points kP and P is very hard. This fact is used to build many elliptic curve–based cryptosystems and techniques.

To change conventional systems that are based on DL problem [9] into an elliptic curve system, the following two rules are applied:

- Any modular multiplication operation defined in the conventional system is replaced by the addition of points on the elliptic curve version.
- Any modular exponentiation operation is replaced by point multiplication on the elliptic curve version of the conventional system.

There are many conventional systems that can be transferred to elliptic curve systems. As an example, we mention the elliptic curve digital signature algorithm (ECDSA) and the elliptic curve ElGamal cryptosystem (ECEC).

5.3.1 Elliptic Curve Digital Signature Algorithm

The process of ECDSA is composed of three main steps: key generation, signature generation, and signature verification. Each step is described as follows.

5.3.1.1 ECDSA Key Generation

The following procedure shows how the users should generate the public and the private keys:

1. Choose an elliptic curve E over a finite field, GF(p), for example. Assume that n is a large prime, then the number of points on E should be divisible by n.
2. Choose a point $P=(x,y) \in \mathrm{GF}(p)$ of order n (see [6] for more information about the order).
3. Choose randomly an integer $d \in [1, n-1]$.
4. Compute $Q = dP$.
5. The public keys for the users are (Q, n, P, E), and the private key is d.

5.3.1.2 ECDSA Signature Generation

The following steps describe how to generate a signature for a certain message m:

1. Choose k to be a random integer $\in [1, n-1]$.
2. Compute $kP = (x_1, y_1)$, and set $x_1 \bmod n = r$. If r is zero then go back to step 1.

3. Compute k^{-1} mod n.
4. Compute $s = k^{-1}$ $(H(m) + dr)$ mod n, where $H(m)$ is the hash value of the message m obtained using a suitable hash function.
5. If $s = 0$, go to step 1. This is because s^{-1} mod n does not exist, and the signature cannot be verified.
6. The pair of integers (s, r) is included in the message m as a signature.

5.3.1.3 ECDSA Signature Verification

The last step is to verify the signature (s, r) on the message m, which is executed as follows:

1. Obtain an authentic copy of the public key (Q, n, P, E).
2. Make sure that the integers r and $s \in [1, n-1]$.
3. Compute $w = s^{-1}$ mod n and $H(m)$.
4. Compute $u_1 = H(m) \cdot w$ mod n and $u_2 = r \cdot w$ mod n.
5. Compute $u_2 Q + u_1 P = (x_0, y_0)$ and $v = x_0$ mod n.
6. If $r = v$, the signature is accepted, otherwise it is not verified.

To reduce the public-key size (Q, n, P, E), the users can agree on a fixed curve E and a base point P as system parameters, instead of generating different E and P for each user. After that, each user defines only the point Q.

5.3.2 Elliptic Curve ElGamal Cryptosystem

First, we describe the conventional version of the ElGamal algorithm introduced by ElGamal [1]. If Alice has to send a message m to Bob, Bob needs to have both public and private keys. Bob selects a large prime p, an integer ι mod p, and a secret integer a. He computes $v = \iota^a$ mod p. The public key for Bob consists of (p, ι, v), whereas his private key is a. Now, to encrypt the message m, Alice chooses a random integer n and computes x_B, y_B such that

$$x_B \equiv \iota^n, \; y_B \equiv mv^n (\text{mod } p).$$

After that, x_B and y_B are sent to Bob to be decrypted. The decryption process is carried out by computing

$$m \equiv y_B x_B^a (\text{mod } p).$$

On the other hand, the ElGamal elliptic curve version can be described as follows: first, Bob selects an elliptic curve E mod p, a point ι on E, and a secret integer a. He computes

$$v = a\iota.$$

The public key consists of the two points ι and v. The secret key is the integer a. The message m is translated into a point on E by Alice. Then she chooses a random integer n and computes

$$x_{\mathrm{B}} = n\iota \text{ and } y_{\mathrm{B}} = m + nv.$$

Then she sends x_{B} and y_{B} to Bob. Finally, the decryption is done by computing

$$m = y_{\mathrm{B}} - ax_{\mathrm{B}}.$$

5.4 SCALABLE HARDWARE DESIGN FOR ELLIPTIC CURVE CRYPTOGRAPHY

The main operation in elliptic curve cryptography is to compute the point multiplication that consists of point additions and point doubling. As discussed earlier, computing point multiplication involves huge arithmetic operations done over the finite fields (mostly GF(p) and GF(2^n)), and therefore, an efficient ECC system requires efficient hardware implementations of finite field operations. The main two operations are modular multiplication and modular division (inverse). The proposed elliptic curve hardware design has the following two features:

1. computing point multiplication based on efficient implementation of UDMA and
2. meeting the most required two features of any efficient hardware design: being scalable and unified.

In the following subsections, UDMA and its hardware implementation are proposed.

5.4.1 Unified Division/Multiplication Algorithm

We use a novel algorithm (UDMA) [10] to compute Montgomery modular multiplication (proved to be a very efficient modular multiplication method) and modular division in GF(p) and GF(2^n) finite fields. UDMA is presented in Figure 5.1.

The UDMA mode of operation is controlled by input *Op* (*div* or *mult*), and the finite field is controlled by the input field (GF(p) or GF(2^n)). For simplicity, the polynomials $X(x)$, $Y(x)$, and $p(x)$ are denoted as X, Y, and p, respectively, which correspond to the bit-vector representation of these polynomials.

Most of the arithmetic operations in the algorithm are common to both modes of operation. The initialization of variables depends on the operation.

```
Function: Modular Division and Multiplication in GF(p) and GF(2^n)
Inputs: 0 ≤ X < p, 0 < Y < p, 2^(n-1) < p < 2^n, Field, Op, n
Output: Z = XY2^(-n) mod p when Op = mult, Z = X/Y mod p when Op = div.
Algorithm:
C = Y.

  IF Op = mult THEN                    /* Multiplication Mode */
    D = 0, U = 0, W = X, δ = n
  ELSE                                 /* Division Mode */
    D = p, U = X, W = 0,  δ = 0
  END IF;
  WHILE [(C ≠ 0 AND Op = div) OR (δ ≠ 0 AND Op = mult)]
    IF c0 = 0 THEN
      C := C >> 1
      δ := δ - 1        /* Integer Operation */
    ELSE
      k = 1
      IF (Op = div) THEN
        IF δ < 0 THEN C ⇔ D, U ⇔ W,  δ := -δ  END IF; /*Swapping */
        IF((C + D) mod 4 ≠ 0 AND Field = GF(p))THEN k = -1
        ELSE   δ :=  δ - 1   END IF;
      ELSE    /*Op = mult */
        δ :=  δ -1
      END IF;
      C := (C + k * D) >> 1, U := (U + k *W)
    END IF;
    U := (U + u0 * p) >> 1
  END WHILE;
  IF Op = div THEN Z := W     ELSE Z := U
  END IF;
```

FIGURE 5.1 Unified modular division/multiplication algorithm (UDMA) for GF(p) and GF(2^n).

For a given field, all the additions or subtractions are done in the field, besides the arithmetic operations on δ (subtractions and change of sign) which are always integer operations.

The algorithm integrates the extended binary GCD algorithm and the Montgomery multiplication algorithm and it was verified using Maple. To compute Montgomery multiplication using an n-bit modulus p, UDMA performs n iterations. The counter δ is initialized with value n, and in each iteration it is decremented by 1. The variables used in the algorithm are initialized as $C=Y$, $D=0$, $U=0$, and $W=X$. The result is ready ($Z=U$), when $\delta=0$. The partial product U is reduced mod p in each iteration. In both fields, while multiplying, addition is used in the operations that update C and $D(k=1)$. The $\gg$ operator indicates a 1-bit right shift operation.

UDMA computes modular division using the same structure used by the extended binary GCD algorithm for modular division [11]. The variables are initialized as $C=Y$, $D=p$, $U=X$, $W=0$, and $\delta=0$. If the division is computed in GF(p), UDMA tests the least significant 2-bits of C and D ($(C+D)$ mod $4 \neq 0$) to conditionally subtract C from D (set $k=-1$). Otherwise, C is always added to D in both fields. The division is completed when $C=0$, and the final result is available in W. For more details about the operation of UDMA, the reader is referred to [10,12].

5.4.2 Top Level Hardware Architecture Implementing UDMA

Figure 5.2 shows the top level architecture of the unified modular divider or multiplier (let us call it UMDM) that implements UDMA. The main functional blocks are Register file, Datapath, and Control.

5.4.2.1 Register File

The register file has five registers ($R1$ to $R5$). As the computations are done in carry-save form, each intermediate variable (C, U, D, W) is represented in two vectors (sum, carry). Therefore, the registers inside the register file are designed to store two n-bit vectors. In other words, the ith register R_i is represented as $R_i = (\text{sum}, \text{carry}) = (R_i s, R_i c)$.

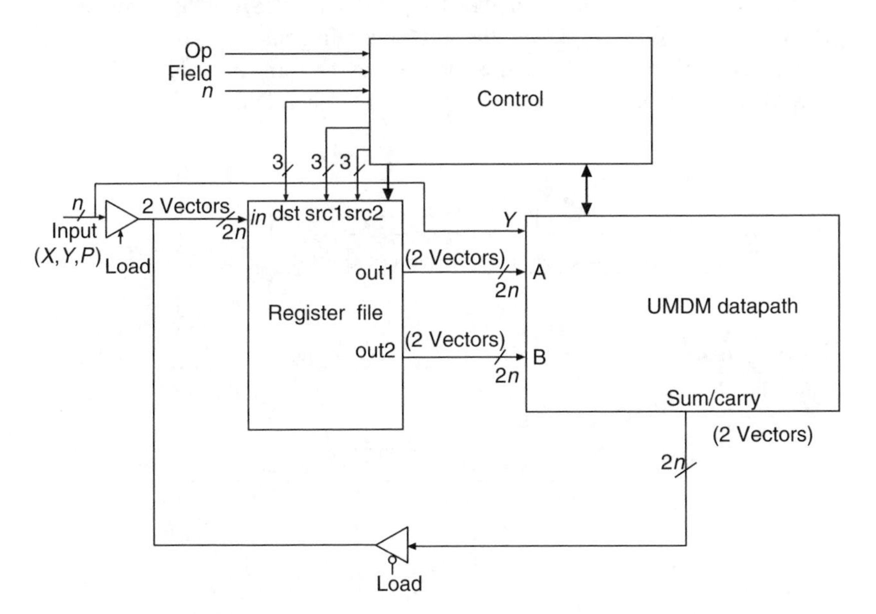

FIGURE 5.2 Top level hardware architecture of the unified modular divider/multiplier (UMDM).

The register file has one input and two output ports. The control block provides the register file with the signals necessary to perform reading or writing operations. The 3-bit signal dst determines the destination register to be written. The signals src1 and src2 (3-bits each) specify the registers to be read at output ports out1 and out2, respectively.

5.4.2.2 Datapath

The n-bit datapath implementing UDMA is shown in Figure 5.3. Each iteration of the algorithm is implemented in one clock cycle for multiplication mode, three clock cycles for division if C is odd, and two clock cycles if C is even, as explained later.

The proposed datapath has two inputs represented in carry-save form as $\mathrm{A} = (\mathrm{As}, \mathrm{Ac})$ and $\mathrm{B} = (\mathrm{Bs}, \mathrm{Bc})$, which receive their values from the register file ports out1 and out2, respectively. The main components of the datapath are two (3–2) unified carry-save adders (UCSAs), which are similar in complexity to full-adders [13]. The unified adders can perform bit addition with or without carry depending on the input FSEL (Field Select).

The unified adder may be used to implement a redundant or nonredundant adder. The use of nonredundant form of the operands and results reduces the register area but increases the addition time (because of carry propagation). We decided to use carry-save adders to make the addition time constant and independent of the operand's precision.

UCSAs. The first adder in the datapath is a UCSA with complement (UCSA1). Figure 5.4a shows the bit slice diagram for this adder and Figure 5.4b shows the connection of n slices to form an n-bit adder. The UCSA1 outputs are $(\mathrm{sum}, \mathrm{carry}) = a + b + c$, when $\mathrm{NEG} = C\ in = 0$, and

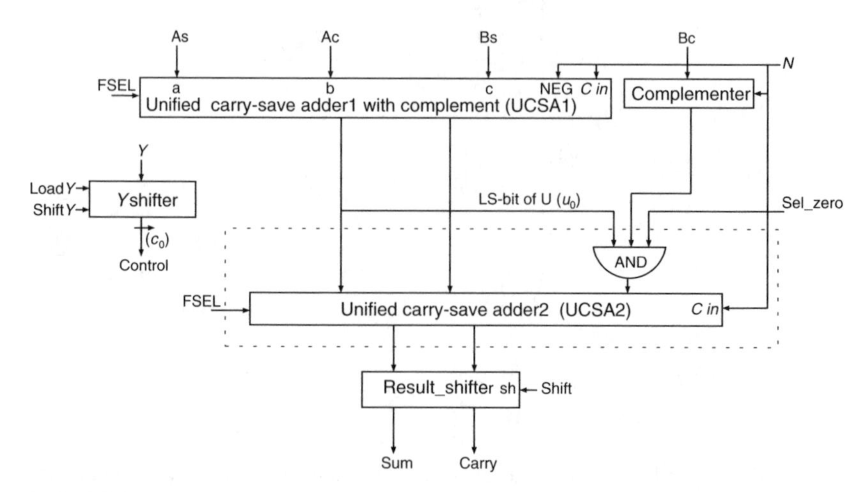

FIGURE 5.3 Unified datapath of the modular divider/multiplier (UMDM datapath).

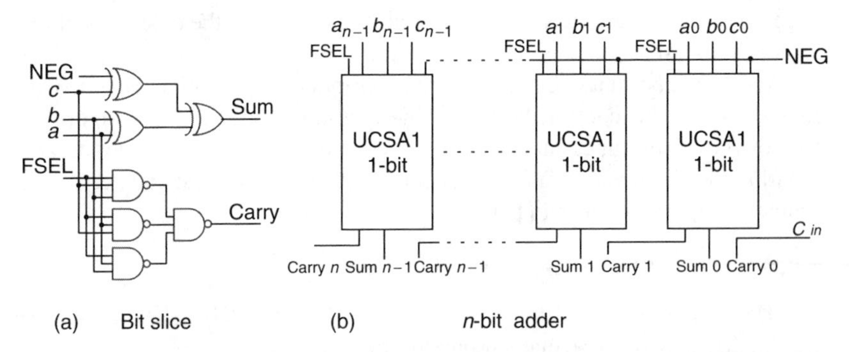

FIGURE 5.4 Unified carry-save adder with complement (UCSA1) for 1-bit and n-bit precision.

(sum, carry) $= a + b - c$, when NEG $= C\ in = 1$. Addition and subtraction in GF(2^n) are the same.

The delay of the two UCSAs, and the delay of the result_shifter ($2t_{\text{MUX}} \simeq 2t_{\text{XOR}}$), mainly determines the delay of the UMDM datapath (t_{datapath}). The delay of the AND gate is not considered because it was integrated with the second adder (shown in dashed box in Figure 5.3). As each UCSA has a delay of a full adder ($t_{\text{FA}} = 2t_{\text{XOR}}$), we get

$$t_{\text{datapath}} = t_{\text{USCA1}} + t_{\text{UCSA2}} + t_{\text{result\_shifter}} = 4t_{\text{XOR}} + t_{\text{MUX}} = 5t_{\text{XOR}}.$$

The Yshifter shown in Figure 5.3 is a shift register used to implement. The operation ($C \gg 1$) in the multiplication mode is implemented by the shift register Yshifter shown in Figure 5.3. The least significant bit of the shifted C goes to the control section to be tested ($c_0 = 0$).

The datapath outputs (sum, carry) are shifted right 1-bit by correct wiring using the result_shifter at the output of the UCSA2.

5.4.2.3 Control Block

The control block provides the necessary signals to control the flow of the operations in the system. The major component in the control unit is a finite state machine that was implemented using a hardwired control methodology. With the intention to design a robust and reliable control unit, the state machine was coded as a Moore machine in which the output signals depend solely on the present state, minimizing or eliminating glitches. More implementation details can be found in [10].

The algorithm's swap functions ($C \Leftrightarrow D$ and $U \Leftrightarrow W$) are accomplished within control unit to avoid actual data transfer between registers. An actual data transfer would be costly in terms of time, especially for a system with large precision. Thus, the swap is performed, exchanging the addresses of the register in question, inside the control unit.

Another important component of the control unit is the delta counter. This counter is used to control the swapping operation and the major algorithm control flow. The functionality for delta counter includes decrementing and negating the count value. With the goal of implementing a fast counter, a ring counter design was chosen [14].

5.4.3 Experimental Results for UMDM

The UMDM design was implemented in ASIC and FPGAs. Therefore, we present two sets of experimental data in this section.

5.4.3.1 ASIC Results for the UMDM Scalable Design

The experimental data presented in this section were generated using Mentor Graphics CAD tools. The target technology was set to AMI05_fast auto (0.5 μm CMOS with hierarchy preserved) provided in the ASIC Design Kit (ADK) from the same company [15].

The UMDM architecture was described in VHDL and simulated in ModelSim for functional correctness. It was synthesized using Leonardo synthesis tool for the mentioned technology.

Figure 5.5 shows the critical path delays (in nanoseconds) of the UMDM for the precision range from 128 to 512-bits. The maximum delay at 512-bits is around 12.8 ns.

Table 5.2 shows the total number of gates for the UMDM design as a function of operand size. The area for the UMDM design was extracted from the experimental data presented in Table 5.2 as

$$A_{\text{UMDM}} = 236.12 * n + 180 = O(n) \text{ gates.}$$

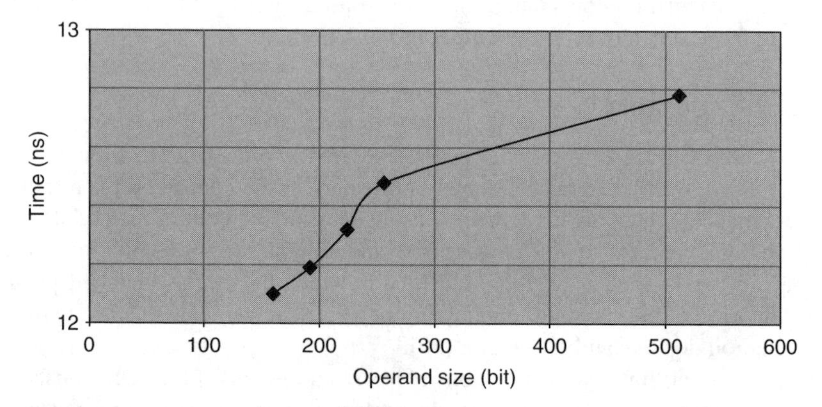

FIGURE 5.5 Critical path delays of the UMDM in nanoseconds (operand size from 160 to 512-bits).

TABLE 5.2
Area of the UMDM Design in Gates for Different Operand Sizes

Operand Size (Bits)	Area (Gates)
128-bits	30,403
160-bits	37,059
192-bits	45,513
224-bits	53,075
256-bits	60,629
512-bits	121,070

The integration of Montgomery multiplication and modular division in one design adds extra gates when compared with a dedicated divider. In the design proposed in this work, Montgomery multiplication is computed in almost the same time and complexity of a separate multiplication unit. In addition to that, this design allows the ability to compute division in the same unit with the flexibility to choose the required finite field.

5.4.3.2 FPGA Results for the UMDM Scalable Design

The scalable divider or multiplier design was synthesized for the FPGAs VertixII chip. The technology was set to $xc2vp50-7ff148$. The following paragraphs present the area and the critical path delay results obtained for the design.

Figure 5.6 shows the area synthesis results (in number of slices) of the scalable UMDM. The area is presented as function of the operand size (n)

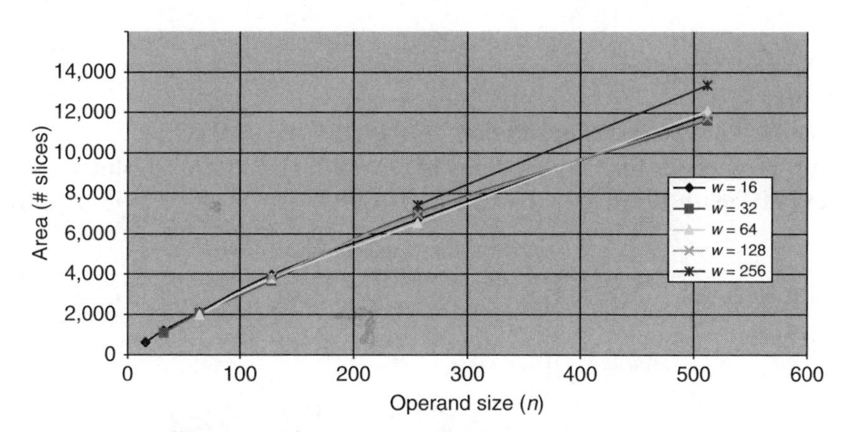

FIGURE 5.6 Area (FPGA technology) of the scalable UMDM in number of slices for combinations of operand size (n) from 16 to 512-bits and datapath word size (w) from 16 to 256-bits.

with different combinations of the datapath word sizes (w). The area results were obtained for the operand size in the range from 16 to 512-bits. The datapath word size was in the range from 16 to 256-bits. The reason why we did not use larger operand sizes is because the machines we are using could not handle operand size greater than 512-bits.

From the figure, we note that the area increases linearly as the operand size increases. There is a little difference in the number of slices when using different datapath word sizes for the same operand size.

The area for the scalable UMDM design was extracted from the experimental data presented in Figure 5.6 approximately as

$$A_{\text{scUMDM}} = 28 * n + 275 = O(n).$$

The same as in the area results, the experimental data for the critical path delay were obtained for the operand size (n) in the range from 16 to 512-bits, and the datapath word size (w) range from 16 to 256-bits. Table 5.3 shows the critical path delay (clock period) for all the possible combinations of the operand size and the datapath word size. The symbol—indicates that the combination is not possible.

The operating frequency of the UMDM design can be found by taking the reciprocal of the clock period at any point. From the table, the lowest clock period (19.83 ns) is at $n = 16$ and $w = 16$, and therefore, the maximum operating frequency is around 50 MHz.

The question now is how to choose the best design points, or in other words, the (n, w) combinations that give the lowest delay. By looking at Table 5.3, we note that at a given operand size n, the minimum delay happens at the datapath word size $w = n$. For example, the best combination at the operand size $n = 256$ happens when the word size $w = 256$ also, with a minimum delay equal to 28.4 ns.

TABLE 5.3
Critical Path Delay (Clock Period) of the Scalable UMDM in Nanoseconds for Combinations of Operand Size (16 to 512-bits) and Datapath Word Size from 16 to 256-Bits

	Datapath Word Size (w)				
Operand size (n)	**16**	**32**	**64**	**128**	**256**
16	19.83	—	—	—	—
32	24.55	22.13	—	—	—
64	25	26.55	24.7	—	—
128	32	31	27.9	25.4	—
256	34.7	37.3	34.3	31.9	28.4
512	47.15	38.71	38.5	37.4	35.4

5.5 ELLIPTIC CURVE CRYPTO-PROCESSOR OVER GF(2^n)

After introducing UDMA and its efficient hardware implementation, we propose an ECCP over the binary extension field GF(2^n) to compute the point multiplication operation kP. The ECCP architecture is based on the UDMA hardware implementation shown in the previous sections, with some simplifications applied in GF(2^n).

5.5.1 ECCP Hardware Architecture

Figure 5.7 shows the top level diagram of the ECCP. Its components are the arithmetic unit (AU) data section and control, and the main control block. The AU unit represents the UDMA architecture. The main control block interacts with the user to get the scalar multiple (k) and the point to be multiplied (P), passing them to the AU.

The details of the main blocks in the ECCP are similar to that presented in the previous sections, taking into consideration the simplifications applied to the algorithm and its implementation due to the use of GF(2^n).

The scalability feature of the proposed crypto-processor allows the adjustment of the word size used in the datapath to meet area and performance requirements. On the other hand, the processor allows the user to choose the value of the field parameter (n).

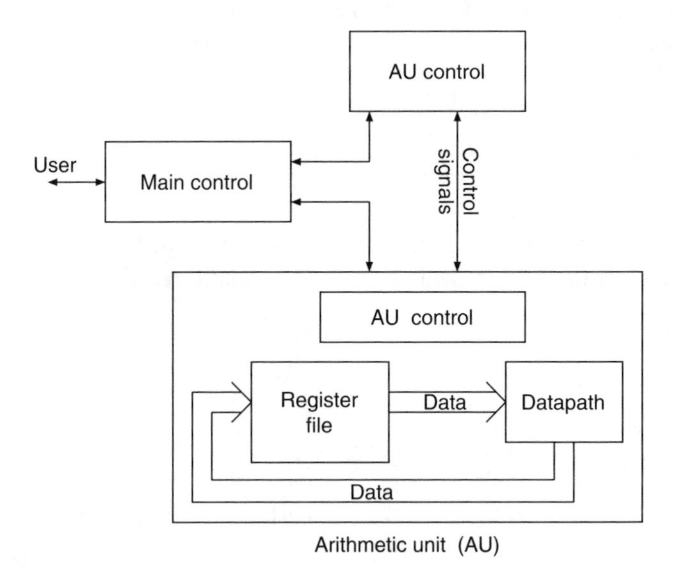

FIGURE 5.7 Top level diagram of the elliptic curve crypto-processor (ECCP).

5.5.2 Experimental Results and Analysis for GF(2^n) ECCP

As performed for the UDMA design, the experimental data presented in this section were generated using Mentor Graphics CAD tools with the target technology set to AMI05_fast auto (0.5 μm CMOS with hierarchy preserved) provided in the ADK from the same company [15]. The scalable architecture of the ECCP was described in VHDL and simulated in ModelSim to validate functional correctness. It was synthesized using Leonardo synthesis tool for the available technology.

Table 5.4 shows the critical path delays (in nanoseconds) of the ECCP for the precision range from 16 to 512-bits at different combinations of the datapath word size (from 16 to 512-bits).

We can see in the table that the minimum delay happens when the datapath word size is 16. When the word size increases, the delay increases slightly for a fixed operand precision, and the delay increases as the number of bits increases and it saturates at higher precision.

The ECCP architecture based on UDMA performs one iteration of the algorithm in each clock cycle when computing Montgomery multiplication. This means that we need n cycles to compute Montgomery modular multiplication. The ECCP has no dedicated hardware for squaring (x^2), and therefore the multiplication algorithm is used for squaring.

On the other hand, it takes a maximum of 2 iterations/bit and on an average 1.5 iterations/bit to compute the modular inverse in GF(2^n) using the simplified algorithm. The ECCP architecture performs each iteration of the algorithm in two clock cycles on an average, one to compute ($C + c_0 \cdot D$) and another to compute $U + W$ with the modulus reduction. Therefore, the GF(2^n) inversion by the simplified algorithm takes on an average of $1.5 \times 2 = 3$ cycles for each bit.

TABLE 5.4
Critical Path Delay of the ECCP in Nanoseconds for Operand Precision 16 to 512-bits and Different Datapath Word Sizes

Delay (ns)	Datapath Word Size (w)					
Precision (bits)	16	32	64	128	256	512
16-bits	17.2	—	—	—	—	—
32-bits	17.6	17.8	—	—	—	—
64-bits	17.6	19.2	20.4	—	—	—
128-bits	17.5	19.2	20.8	20	—	—
256-bits	16.5	19.1	20.7	20.4	19	—
512-bits	16.7	18.2	20.7	20.5	19.5	20.2

In computing kP using the double-add method [7], where $P=(X_1, Y_1, Z_1)$, $Q=(X_2, Y_2, Z_2)$ are the points on the curve in the projective coordinates, we can assume that $Z_2=1$, computing point addition ($P \neq Q$) requires 13 field multiplications and computing point doubling ($P=Q$) requires 7 field multiplications [16]. To compute the scalar point multiplication (kP) using Equation 5.1, n point doubling operations are needed (n is the order of the field), and $\sim n/2$ point additions are needed (given that the number of ones in the binary expansion of k is $0.5n$).

Let the total average computation time of a given design to compute multiplication or division be T_{design}, which is given by

$$T_{\text{design}} = (\text{cycles/bit}) * n * \text{clock period}.$$

At operand precision of $n=512$-bits, the time required to compute one multiplication by the ECCP is $T_{\text{mult}} = 1 * 512 * 20.2 * 10^{-9} = 10.3\ \mu\text{s}$. Then, at $n=512$-bits, the ECCP computes point addition in

$$T_{\text{P Add}} = 13 * T_{\text{mult}} \approx 134\ \mu\text{s},$$

and half of that time is required to compute point doubling $T_{\text{P Double}}=0.5 * T_{\text{P Add}}$. To compute the scalar point multiplication (kP), an inversion operation is required to transform back the result from the projective to the affine coordinates. The total time to compute the modular division (inverse) by the ECCP is $T_{\text{inv}} = 3 * T_{\text{mult}} \approx 31\ \mu\text{s}$. Then, the total time to compute kP by the proposed ECCP is

$$\begin{aligned} T_{kP} &= 0.5n * T_{\text{P Add}} + n * T_{\text{P Double}} + T_{\text{inv}} = 13/2n * T_{\text{mult}} + 7n * T_{\text{mult}} + 3T_{\text{mult}} \\ &= (13.5n+3) * T_{\text{mult}} = (13.5n+3)(n * \text{clock period}) \\ &= (13.5n^2+3n) * \text{clock period}. \end{aligned}$$

At precision $n=512$-bits, $T_{kP}=71$ ms. The proposed ECCP computes the kP faster than previously proposed elliptic curve architectures. As an example, the FPGA implementation of the elliptic curve processor presented in [17] computes the scalar point multiplication in 80.3 ms at operand size of 163-bits. In addition, the ECCP has an advantage over other designs by its scalablity (i.e., the user can choose the word size to achieve the required performance).

Table 5.5 shows the total area (in number of gates) for the ECCP design as a function of operand precision and different datapath word sizes. The area for the ECCP design was extracted from the experimental data presented in Table 5.5 as

$$A_{\text{ECCP}} = 236.12 * n + 180 = O(n) \text{ gates}.$$

TABLE 5.5
Total Area of ECCP in Gates for Operand Precision 16 to 512-bits and Different Datapath Word Sizes

Area (Gates)	Datapath Word Size (w)					
Precision (bits)	**16**	**32**	**64**	**128**	**256**	**512**
16-bits	3,857	—	—	—	—	—
32-bits	4,251	5,567	—	—	—	—
64-bits	5,012	6,267	8,945	—	—	—
128-bits	6,389	7,664	10,265	15,727	—	—
256-bits	8,212	10,310	12,928	18,274	29,434	—
512-bits	12,602	13,861	18,109	23,473	34,458	56,570

From Table 5.5, we can see that the proposed ECCP has area complexity of $O(n)$ at a given datapath word size. These results are compatible with many other designs [18,19].

REFERENCES

1. T. ElGamal, A public key cryptosystem and signature scheme based on discrete logarithms, *IEEE Transactions on Information Theory*, Vol. IT-31, No. 4, pp. 469–472, July 1998.
2. M.E. Hellman and W. Diffie, New directions on cryptography, *IEEE Transactions on Information Theory*, Vol. 22, pp. 644–654, November 1976.
3. National Institute for Standards and Technology, Digital Signature Standard (DSS), Technical Report 168–2, FIPS PUB, January 2000.
4. V. Miller, Elliptic curves in cryptography, in *Advances in Cryptology CRYPTO '85*. Editor H.C. Williams, Lecture Notes in Computer Science, No. 218, pp. 417–426, Springer 1985.
5. N. Koblitz, Elliptic curve cryptosystems, *Mathematics of Computation*, Vol. 48, No. 177, pp. 203–209, January 1987.
6. G. Seroussi, I. Blake, and N. Smart, *Elliptic Curves in Cryptography*, Cambridge University Press, UK, 1st ed., 1999.
7. P1363, Standard specifications for public key cryptography (draft version 13), IEEE, November 1999.
8. Adnan Abdul-Aziz Gutub, New Hardware Algorithms and Designs for Montgomery Modular Inverse Computation in Galois Fields GF(p) and GF(2^n), Ph.D. thesis, Oregon State University, Oregon, USA, June 2002.
9. W. Trappe and L.C. Washington, *Introduction to Cryptography with Coding Theory*, Prentice Hall, Englewood Cliffs, NJ, 2002.
10. L.A. Tawalbeh, A Novel Unified Algorithm and Hardware Architecture for Integrated Modular Division and Multiplication in GF(p) and GF (2^n) Suitable for Public-Key Cryptography, Ph.D. thesis, Oregon State University, Oregon, USA, October 2004.

11. A.F. Tenca and L.A. Tawalbeh, An algorithm for unified modular division in *GF*(*p*) and *GF*(2^n) suitable for cryptographic hardware, *IEE Electronics Letters*, Vol. 40, No. 5, pp. 304–306, March 2004.
12. L.A. Tawalbeh and A.F. Tenca, An algorithm and hardware architecture for integrated modular division and multiplication in *GF*(*p*) and *GF*(2^n), in *The IEEE 15th International Conference on Application-Specific Systems, Architecture, and Processors (ASAP)*. September 27–29, 2004, pp. 247–257, ieeecs.
13. E. Savas, A.F. Tenca, and Ç.K. Koç, A scalable and unified multiplier architecture for finite fields *GF*(*p*) and *GF*(2^m), in *Cryptographic Hardware and Embedded Systems—CHES 2000*, Ç.K. Koç and C. Paar, Eds. 2000, Lecture Notes in Computer Science, No. 1717, pp. 281–296, Springer, Berlin, Germany.
14. M. Stan, A. Tenca, and M. Ercegovac, Long and fast up/down counters, *IEEE Transactions on Computers*, Vol. 47, No. 7, pp. 722–734, July 1998.
15. ASIC Design Kit. Mentor Graphics Co, http://www.mentor.com/partners/hep/AsicDesignKit/dsheet/ami05databook.html
16. G.B. Agnew, R.C. Mullin, and S.A. Vanstone, An implementation of elliptic curve cryptosystems over *GF*(2^{155}), *IEEE Journal on Selected Areas in Communications*, Vol. 11, No. 5, pp. 804–813, 1993.
17. G. Orlando and C. Paar, Implementation of elliptic curve cryptographic Coprocessor over *GF*(2^m) on an FPGA, in *Cryptographic Hardware and Embedded Systems—CHES 2000*, Ç.K. Koç and C. Paar, Eds. 2000, Lecture Notes in Computer Science, No. 2162, pp. 25–40, Springer, Berlin, Germany.
18. J. Goodman and A.P. Chandrakasan, An energy-efficient reconfigurable public-key cryptography processor, *IEEE Journal of Solid-State Circuits*, Vol. 36, No. 11, pp. 1808–1820, November 2001.
19. J. Wolkerstorfer, Dual-field arithmetic unit for *GF*(*P*) and *GF*(2^n), in *Cryptographic Hardware and Embedded Systems—CHES 2002*, B.S. Kaliski Jr. et al., Eds. 2003, Lecture Notes in Computer Science, No. 2523, pp. 484–499, Springer, Berlin, Germany.

6 Cryptographic Algorithms in Constrained Environments

Vincent Rijmen and Norbert Pramstaller

CONTENTS

6.1 INTRODUCTION AND MOTIVATION

Building information technology applications is not possible without ensuring security. Cryptographic algorithms are an essential part of a modern security layer. In real-world applications, security measures have to be balanced against their cost, as well as the value of the protected data and the probability of attacks. In wireless applications it happens quite often that in at least one of the devices there are severe limitations on the available computing power, memory, chip area or code size, and electrical power or energy. This influences the choices a designer has to make when deciding on the trade-off between security and cost.

We begin this chapter with a discussion of cryptographic primitives and the security services they can deliver. We argue that by using only a block cipher it is possible to deliver a wide range of security services. Additionally, a hash function can be included to increase the performance. Subsequently, we discuss the implementation of the Advanced Encryption Standard (AES), used for symmetric encryption and authentication, and Whirlpool, a dedicated hash function standardized in ISO/IEC 10118-3. Interest in Whirlpool has increased significantly after the recent attacks on MD5 and SHA-1.

6.2 CRYPTOGRAPHIC PRIMITIVES FOR CONSTRAINED ENVIRONMENTS

6.2.1 SECURITY SERVICES, MECHANISMS, AND PRIMITIVES

Before we start discussing the implementation of cryptographic primitives, we need to determine which primitives we need. This finally depends on the security services we want to provide. In a constrained environment, we usually want to provide all necessary security services using a set of primitives as small as possible. The ISO 7498-2 standard distinguishes 5 types of security services and 13 types of security mechanisms [1]. The five types of security services are data confidentiality, data integrity, (entity) authentication, access control, and nonrepudiation. Examples of security mechanisms are cryptographic techniques like encryption, data integrity mechanisms, and digital signatures, and also other techniques like padding and routing control. The cryptographic security mechanisms are implemented by using one or more cryptographic primitives.

On the highest level, we distinguish between symmetric primitives and asymmetric primitives. Most asymmetric primitives can be used to provide data confidentiality services by means of asymmetric encryption and data integrity services by means of digital signatures. They are also used in some key exchange protocols. The nonrepudiation security service can be provided only by using asymmetric primitives. The implementation of asymmetric primitives invariably requires significantly more resources than those required for the implementation of symmetric primitives. They are therefore not suited for the most constrained environments. In this chapter, we concentrate on symmetric primitives.

There are four types of symmetric primitives: block ciphers, stream ciphers, hash functions, and message authentication codes (MACs). They can be used to provide confidentiality and data integrity services. Confidentiality is provided by encryption, a mechanism that is typically implemented using stream ciphers or block ciphers. Data integrity is provided by means of data integrity mechanisms, which can be implemented using hash functions or MACs.

However, it is also possible to implement encryption by means of hash functions or MACs. Likewise, block ciphers and some stream ciphers can be used to implement a data integrity mechanism. Another way to describe this multipurpose nature is to state that symmetric primitives can be used to construct other symmetric primitives. For instance, when a block cipher is used to implement a data integrity mechanism, we can also say that we use a hash function that is constructed from a block cipher, instead of a dedicated hash function. Figure 6.1 illustrates which symmetric primitives can be constructed from which other primitives. It can also be observed that

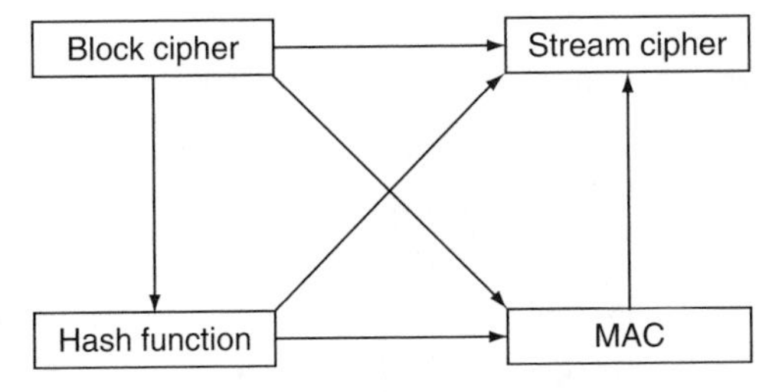

FIGURE 6.1 Possibilities to use symmetric primitives to construct other symmetric primitives.

dedicated MAC algorithms are rare; they are usually replaced by block cipher–based constructions.

6.2.2 Types of Primitives to Include

We discuss here the types of symmetric primitives that should be preferred to be implemented in constrained environments. The choice depends on the security mechanisms that are used to implement the desired security services.

6.2.3 First Primitive: Block Ciphers

Two block ciphers are very popular nowadays: the Data Encryption Standard (DES or 3-DES) and the AES. We discuss here only the latter, because it is the most future-oriented choice. The AES design makes extensive use of finite field arithmetic, a type of arithmetic that is less intuitive than ordinary integer arithmetic, but suited well for hardware implementations. An n-bit block cipher with a h-bit key can be defined formally as a family of 2^h permutations in the space of n-bit vectors. Every value of the key defines one permutation in the family.

Block ciphers are the most versatile symmetric primitives. They can be used to provide confidentiality, data integrity, and authentication services. Block ciphers can be considered as fundamental cryptographic building blocks, which can be used to construct a stream cipher, a hash function, and a MAC.

6.2.3.1 Confidentiality

Block ciphers can be used to provide confidentiality by means of encryption. Different modes of operation have been standardized [2]. In the *Electronic Code Book* (ECB) mode, the message is divided into n-bit blocks, and every message block m_i is replaced by $c_i = E(k;\ m_i)$. The ECB mode has several shortcomings. The most obvious one is that repeating message blocks results in repeating ciphertext blocks

$$m_i = m_j \ \Leftrightarrow \ c_i = c_j.$$

In the cipher block chaining (CBC) mode, patterns occurring in the plaintext are hidden by introducing feedback

$$c_i = E(k;\ m_i \oplus c_{i-1}), \tag{6.1}$$

where c_{i-1} is also called the initial value (IV), which does not have to be secret, but should not be predictable by an attacker. When more than $2^{n/2}$ blocks are encrypted under the same key, the CBC mode starts leaking information in the same way as the ECB mode. A recently standardized mode of operation is the counter (CTR) mode. Here, the ciphertext is produced by encrypting a counter z and XORing the output with the message $c_i = m_i \oplus E(k; z_i)$, with $z_i = z_0 + i$. Observe that a block cipher in this mode of operation resembles closely a synchronous stream cipher. Other popular block cipher modes of operation resemble self-synchronizing stream ciphers. Hence, we conclude that the reason for the popularity of block cipher is not that they would have superior qualities as encryption primitives, but rather the fact that they are so versatile. The performance of block ciphers is often inferior to the performance of dedicated stream ciphers, hash functions, or MAC algorithms, but still acceptable in many applications.

6.2.3.2 Data Integrity

Data integrity is usually not implied by data confidentiality. For many encryption mechanisms, it is possible for an attacker to alter even encrypted messages in such a way that the legitimate receiver of the messages cannot detect the modifications. When data integrity needs to be provided together with data confidentiality, it is possible to provide the data integrity by using a hash function. Hashing is also used as a preprocessing step when data integrity is provided by means of asymmetric cryptography (digital signatures).

Hash functions can easily be constructed from block ciphers. The most popular constructions can be divided into two groups. Single-length constructions result in hash functions with output size equal to the block length of the block cipher. These constructions resist second preimage attacks, but are typically not sufficiently resistant to brute-force collision attacks. Double-length constructions result in hash functions with output size equal to twice the block length. They resist collision attacks, but all known constructions are much slower than the single-length constructions. An overview of hash function constructions using block ciphers is given in [3].

When the data integrity service needs to be independent of the data confidentiality service, a hash function alone is not sufficient to provide the service. The solution is to use a MAC. There are very few standardized dedicated MACs. A very popular class of data integrity mechanisms is based on the block cipher–based CBC–MAC construction. In the simplest configuration, the data are first processed (encrypted) using the CBC encryption mode (1.1). Only the last ciphertext block is returned, and possibly even that block is truncated [4]. A number of variants to the basic configuration have been defined to remediate weaknesses that occur when the basic scheme is instantiated with a weak cipher like the DES. This approach has proven to be error prone, see for example, the overview of attacks presented in [5]. When AES is used as the underlying block cipher, the basic configuration is secure and there is no need to implement any of the variants.

6.2.3.3 Entity Authentication

Entity authentication or identification is a process whereby a party proves that it possesses a certain secret key. The simplest identification method is by using a (static) password. In many settings, passwords are vulnerable to eavesdropping and replay attacks. Replay attacks are countered by using dynamic passwords; if a password can be used only once, then eavesdropping becomes useless. A very simple authentication protocol consists of encrypting the current time with a block cipher and using the resulting ciphertext as password. The verifier decrypts the received password and checks whether the resulting plaintext is a valid time, not too far in the past. Note that this protocol becomes trivially breakable if the block cipher would be replaced by a synchronous stream cipher, as is illustrated by some attacks on the wired equivalent privacy (WEP) protocol [6].

6.2.4 Additional Primitives

We discuss here which type of primitive is most useful when added to a block cipher.

6.2.4.1 Stream Ciphers

A stream cipher computes the ciphertext symbols by adding to the plaintext symbols the output of a pseudo-random number generator. If the pseudo-random number generator operates independently of the message symbols and the (previous) ciphertext symbols, then the stream cipher is called a synchronous stream cipher. In a self-synchronizing stream cipher, the pseudo-random number generator takes as input a number of previous ciphertext symbols also besides the key.

The main advantage of stream ciphers compared with block ciphers is that they often provide a higher performance at a lower cost, especially in hardware. If only confidentiality is needed, then stream ciphers are sometimes preferred over block ciphers. On the other hand, for many proposals, the increase in performance comes at the cost of a reduced security level. While it proves to be already difficult to combine a high performance with a high level of confidentiality, this appears to be even more the case when data integrity also is required; see for instance several results available from [7]. Synchronous stream ciphers can provide only confidentiality. Because of these limitations, we think that for many applications a stream cipher is not the best choice.

6.2.4.2 Hash Functions

A hash function takes inputs of a variable length and produces outputs of a fixed length. Hash functions are mainly used in data integrity mechanisms, where their most important advantages over block cipher–based constructions are higher performance and larger output length, which increases the resistance against brute-force collision attacks.

The dedicated hash function widely used today is SHA-1. However, recently there have been several breakthroughs in cryptanalysis, indicating that the security level of SHA-1 is borderline. It can be expected that it will soon be known how to construct collisions for SHA-1. Therefore, we discuss here a completely different hash function, Whirlpool. It is based on similar design principles as the AES, and its implementation can be done using similar strategies as for the AES.

Hash functions can be used to construct MACs, for instance the HMAC construction [8]. However, here the advantages over block cipher–based constructions are more limited. First, the larger output length does not lead to an increase in security and might be unwanted in applications where the amount of bits to be stored should be minimized. Second, the intrinsic higher performance is handicapped by a relatively long setup. Finally, the inclusion of a dedicated hash function algorithm can offer benefits in some applications, but it is recommended to do a careful evaluation of the expected improvements in performance and security. For many applications, it is better to replace the dedicated hash function by block cipher–based constructions [9].

6.3 ON OPTIMIZATION OF HARDWARE IMPLEMENTATIONS

The aim of this section is to give the reader a basic understanding of the considerations under which one can optimize hardware implementations. We present ideas for the optimization of the block cipher AES and the hash function Whirlpool when implemented on field programable gate arrays (FPGAs). Advanced readers may skip this section.

6.3.1 Optimization Targets

We discuss in general optimization targets such as area requirements, throughput (speed), and power or energy consumption. These three optimization targets cannot be considered to be independent of each other but are rather closely related. Without going into details of VLSI design, we show this relation on a fictive example. Assume we have implemented an algorithm in hardware. This implementation requires a certain chip area, has a certain power consumption, and achieves a certain throughput for a fixed clock frequency. A possible way to increase the throughput (throughput optimization) is to do more computations of the algorithm in parallel. To do so, we need more chip area and, therefore, the power consumption increases as well. This is due to the fact that now we have more computations running in parallel. So, roughly speaking, we can say that with more area resources, we can achieve higher throughput rates but also the power consumption increases. Another possibility to increase the throughput is, for instance, to increase the clock frequency (assuming that this is possible). This leads again to an increase of the throughput but also the power consumption increases. It is easy to see that if we consider one optimization target, other parameters get modified too. The ideal hardware design should be the smallest, the fastest, and the most power-efficient one. Based on the short introduction given already, it is easy to see that this is not realistic. This makes optimization a challenging process in hardware design.

Before we concentrate on the optimization targets area and speed, we briefly consider the power consumption of hardware implementations. Power consumption becomes more and more important for emerging technologies such as radio frequency identification (RFID). Let us consider two possible target devices: FPGAs and application specific integrated circuits (ASICs). On the one side, we have a lot of freedom for optimization strategies to reduce the power consumption of ASIC implementations (for instance clock gating, operand isolation, etc.). On the other side, it is difficult to control the power consumption of FPGAs. In general, the structure of an FPGA is given and designers can only control few parameters. In the past years, power consumption of FPGAs has become more and more important, and basic strategies to reduce power consumption have been proposed. Power consumption of FPGAs is an ongoing research topic and is one of the main factors that determine the future of FPGAs.

In this section, we focus on optimization at the algorithmic level, i.e., optimizations that mainly focus on the algorithm we want to implement. The reason for this is that optimizations at this level can in general be done with only little knowledge of the target device and technology. For the first step in the design phase, it is enough to know whether we want to use an FPGA or we want to design an ASIC. For an easier and clearer demonstration of optimization techniques for the two algorithms described in this chapter, we focus only on hardware implementations on FPGAs. This has the advantage that we do not require a deep understanding of ASIC design. Since we use FPGAs, we omit optimizations regarding the power consumption and only focus on area and speed.

6.3.2 Optimization Techniques

We describe possible optimization techniques for the two algorithms presented in this chapter: the block cipher AES and the hash function Whirlpool. We focus on hardware implementations on FPGAs and show how to optimize in direction area and speed at the algorithmic level. To keep the discussion general, we do not consider any optimization strategies that are provided by design tools.

If we want to optimize a hardware implementation with respect to throughput (speed), the following parameters are important: the maximum possible frequency and the required number of cycles to perform a certain number of iterations. The higher the frequency and the lower the required number of cycles, the higher the throughput becomes. The maximum frequency is determined by the maximum combinatorial path, i.e., depth of logic gates between registers. The number of cycles is determined by how many operations we can do within one clock cycle. Both of these requirements for high-speed designs require a lot of hardware resources. To achieve a high frequency, we have to introduce registers to divide the combinatorial path into several shorter paths. This increases the area requirements. Since these additional registers also increase the required number of cycles, it is important to find a trade-off between these two parameters. If we want to do more operations in parallel to reduce the number of cycles, we need additional area since we have to implement additional components for these computations.

If we look at the two cryptographic algorithms discussed in this chapter, we can basically divide them into two parts: the state that serves as storage for initial and intermediate data and the transformations that are applied to the state to compute the result of the algorithm. Let us first consider the state. The size of the state is determined by the algorithm. Therefore, it is not possible to implement a smaller state. For an n-bit state we need an n-bit storage. For now, it is left open how the state is implemented. This is discussed in more detail when we present the hardware implementation of AES and Whirlpool. Since the size of the state is fixed, we now consider the transformations that are applied to the state. In general, the transformations are defined for smaller bit sizes than the bit size of the state. Therefore, one and the same transformation is applied several times to different parts of the state. Here it becomes immediately clear that if we want to achieve high throughput rates then it is necessary that we apply the transformations in parallel. This increases the throughput and also the area requirements. If we want to be area efficient, then it is better to implement the transformation only once and apply it several times to the state. This decreases the area but also the throughput, since more cycles are then required to compute the result of the algorithm. So we see that regarding the transformations, we already have several possibilities to optimize a hardware implementation either for speed or area efficiency. Another important decision is how to implement the transformations. For the two discussed algorithms, the transformations have a very nice mathematical structure. This gives the designer the freedom to choose whether the transformations are implemented in such a way that only few hardware resources are needed or whether he wants the transformations with a small combinatorial delay. Again, the smaller the slower and the bigger the faster.

Another possibility to increase the throughput is to use pipelining, where for instance one iteration of the algorithm is computed within one clock cycle. To achieve this, techniques such as loop-unrolling are used. Once the pipeline has been filled, a new result is produced in each cycle. It is clear that this increases the throughput but it also requires more hardware resources for the implementation. For pipelining strategies, one has to consider the target application of the hardware implementation. If the algorithm is for instance used in a certain mode of operation that does not support pipelining, then the whole effort of the hardware implementation is lost. In the following sections, we discuss the different optimization techniques on the hardware implementations of AES and Whirlpool.

6.4 ADVANCED ENCRYPTION STANDARD

6.4.1 DESCRIPTION

The AES encrypts plaintext blocks of 128 bits with a 128-bit, 192-bit, or 256-bit key [10]. A plaintext block is represented as a 4×4 array of bytes, called the state. The AES repeatedly applies a so-called round transformation to the state. The number of iterations (Nr) depends on the key length: 10 rounds for 128-bit keys, 12 rounds for 192-bit keys, and 14 rounds for 256-bit keys. The round transformation consists of the following steps:

- Nonlinear layer SubBytes, where a nonlinear S-Box is applied to each byte of the state individually. SubBytes is a multiplicative inversion in GF(2^8), followed by an affine transformation.
- Cyclical permutation ShiftRows, where the bytes of row i are rotated to the left by i positions.
- Linear diffusion layer MixColumns, where each column of the state is multiplied by a matrix with coefficients that are elements of GF(2^8).
- Key addition AddRoundKey, where the round key is added to the state. Addition corresponds to the XOR operation.

Figure 6.2 depicts the AES dataflow and the steps of the round transformation (for encryption). As can be seen in Figure 6.2, MixColumns is omitted in the final round. Each round uses a different round key. The round keys are

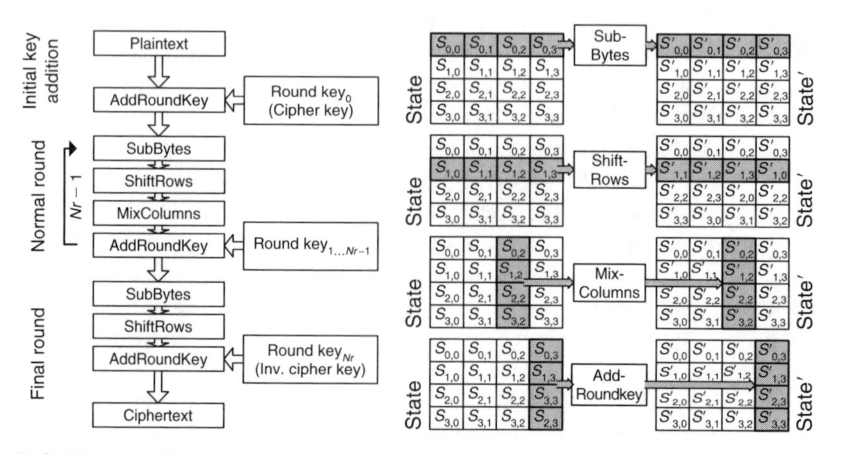

FIGURE 6.2 AES dataflow (left) and steps of the round transformation (right).

derived from the cipher key by applying the key schedule. The key schedule needs the SubBytes transformation and simple XOR operations. Obtaining the initial round key requires no transformations: the first 128 bits of the cipher key is used for the initial key addition. All subsequent round keys are derived from their respective predecessors.

Decryption is done by inverting the process of encryption; the round iterations are executed in the reverse order. This requires generating round keys in reverse order too. Additionally, the sequence of the round functions SubBytes, ShiftRows, MixColumns, and AddRoundKey is reversed and their inverse functions are applied: (Inv)SubBytes), (Inv)ShiftRows, and (Inv)MixColumns. AddRoundKey requires no extra inverse function because the XOR operation is its own inverse.

6.4.2 Overview of Implementations

Numerous AES hardware implementations have been published since the standardization of Rijndael in 2001. In the following, we give a short overview of the most recent implementations of AES on ASICs and FPGAs.* An overview with status quo September 2005 is given in [11].

FPGA implementations mainly focus on high throughput rates. By using techniques like loop-unrolling and pipelining, they are able to report throughput rates up to 21,540 Mbps [12]. Applying such techniques leads to AES hardware implementations that require a huge amount of FPGA resources that are only available for expensive devices and can only be used for high-end applications [12–15]. Considering low-end applications, high throughput rates are not always required (e.g., wireless communications), and high-end FPGAs are too expensive. Only few hardware implementations for low-end FPGAs have been published to date [16–19]. In general, it is difficult to compare the different implementations since they implement different functionalities. For instance, some implementations omit the on-chip key schedule [18] of AES, some only support a fixed key size [13].

Besides FPGA implementations, also numerous ASIC implementations have been published. These implementations achieve throughput rates up to 70,000 Mbps [20]. Alternatively, also compact implementations exist, as shown for instance in [21,22]. In general, high throughput implementations focus on pipelining strategies, whereas compact implementations try to optimize the implementation of (Inv)SubBytes [22–24]. The most compact implementation of AES combines different approaches to minimize the area requirements [25]. Moreover, for ASICs, comparing the different hardware implementations is not straightforward since different target technologies are used.

* A more detailed list on implementations of AES can be found in the AES Lounge at http://www.iaik.tugraz.at/research/krypto/AES/

6.4.3 Compact Hardware Implementation of AES

In this section, we describe a compact AES architecture that is supported by most of the FPGA product families and that can be implemented using inexpensive low-end FPGAs. The design relies on an unconventional but effective hardware architecture that was conceived to map efficiently on reconfigurable hardware like FPGAs from Xilinx [26]. Before we discuss design considerations, we briefly introduce the basic features provided by FPGAs.

6.4.3.1 Basic Features of FPGAs

The basic building blocks of Xilinx FPGAs are configurable logic blocks (CLBs) [26]. CLBs are arranged in a rectangular matrix and are wired by a programable interconnect. A CLB contains four logic cells (LUTs) (eight LUTs for modern devices) that can be programed to have different functionality: combinational logic (an arbitrary Boolean function of four inputs), logic and a register, or synchronous 16×1 bit RAM. Combining two logic blocks allows implementing a 16×1 bit dual-port RAM. Besides CLBs, Xilinx FPGAs offer on-chip block RAM that can store 4096 bits. Block RAM can be configured at ratios between 4096×1 and 256×16, and may have dual-port functionality. Block RAMs are also suitable for implementing synchronous ROMs.

When multiple, fast, and small RAMs are required, distributed (LUT-based) RAMs offer an ideal solution. The benefit is that the RAM cell is adjacent to the logic, and thus, the wiring from the logic to the RAM is negligible. This improves the timing behavior. Multiple distributed RAMs can be merged to either enlarge the address space or the word width. Enlarging the word width is unproblematic (LUTs in parallel), but enlarging the address space can cause performance loss. For instance, a 32×1 bit RAM requires two 4 input LUTs whose outputs need to be multiplexed. This leads to a worse timing behavior and an increased amount of hardware resources. In such cases, it makes sense to use block RAMs instead of distributed RAMs. Using synchronous RAMs and ROMs provides more flexibility for the implementation. Depending on the target technology and available on-chip resources, it can be chosen whether distributed RAM or block RAM should be used for implementing the storage elements.

6.4.3.2 Design Decisions

In the following, we describe the basic design decisions that have been made. We give a brief reasoning and refine it in the detailed description of the architecture. The hardware architecture should implement the complete AES standard—all key lengths should be supported. The implementation of AES should require as few hardware resources as possible, i.e., optimization with respect to area such that it can be implemented on low-end FPGA devices. To achieve this goal, we define a 32-bit architecture. This is also a natural decision since the steps of the round transformation of AES are specified for 32-bit words. Furthermore, we decided in favor of precomputed round keys. This means that the key schedule is able to compute and to store the round keys. Precalculated round keys allow fast encryption or decryption of different data blocks with the same cipher key because no additional key expansion is required. Since the goal is an area-efficient hardware architecture, we decided to implement a nonpipelined approach. Consequently, the same performance for all modes of operations (e.g., ECB and CBC) is reached.

6.4.3.3 Architecture of the AES Implementation

The main components of the AES architecture, as shown in Figure 6.3, are the AMBA APB interface [27], the data unit, the key unit, and the control unit. The key unit calculates the key expansion function. All round keys are precalculated and stored in the key unit. The data unit holds the state and performs all steps of the round transformation: AddRoundKey, (Inv)SubBytes, (Inv)ShiftRows, and (Inv)MixColumns. When encryption or decryption has completed, the ciphertext (plaintext in case of decryption) is stored in the data unit. The control unit receives commands from the AMBA interface and generates control signals for all other modules. In addition to control round key calculation, encryption, and decryption, it also sequences data loading and unloading. In the following, we describe the data unit and the key unit in detail.

6.4.3.3.1 Data Unit

The data unit, schematically depicted in Figure 6.4, stores the state, all intermediate results of the round transformation applied to the state, and the output data when encryption or decryption has completed. An interesting property of the data unit is that the state representation consists of two states. One state contains the actual state values and the other state stores newly calculated values. Figure 6.4 depicts the two states, referred to as StateA and StateB. In each cycle, 32 bits (one row or one column) of either StateA or StateB are altered. Using a second state provides benefits without the need of additional recourses; (Inv)ShiftRows comes for free and no state transposition between column and row operations is required.

Storage elements in FPGAs can be efficiently implemented by using synchronous RAMs because the basic logic elements of FPGAs, called

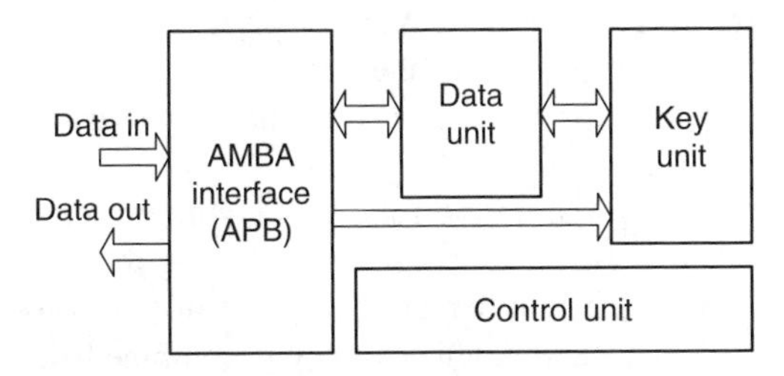

FIGURE 6.3 Architecture of the AES hardware implementation.

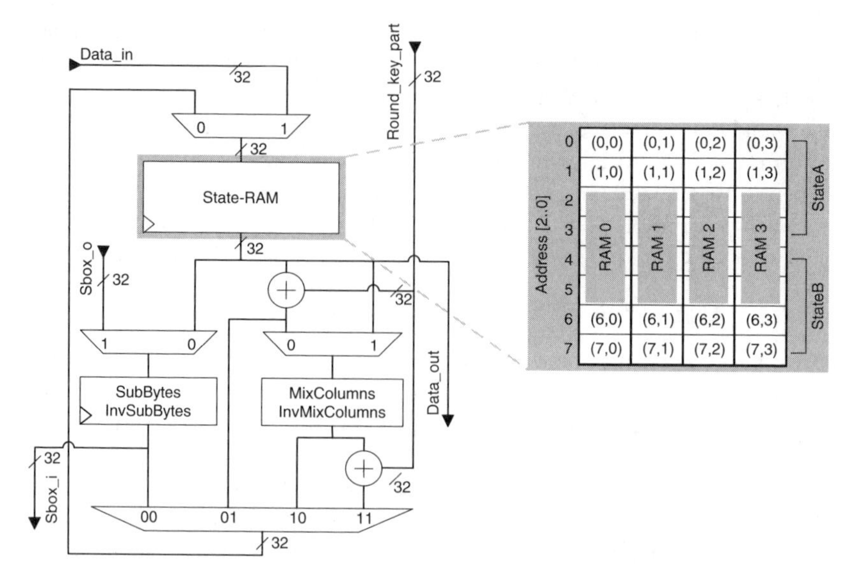

FIGURE 6.4 Architecture of the data unit and the State-RAM.

LUTs, can be configured as 16×1 bit synchronous RAM. Two LUTs provide 16×1 bit synchronous dual-port RAM functionality. Dual-port RAMs allow concurrent reading and writing to the RAM. Due to these technology features, the State-RAM, as depicted in Figure 6.4, is implemented as four slices of 8×8 bit synchronous dual-port RAM to allow addressing the slices individually.

The data unit implements all steps of the round transformation: (Inv)ShiftRows, (Inv)SubBytes, (Inv)MixColumns, and AddRoundKey. The steps AddRoundKey and (Inv)MixColumns are applied to the state column-by-column, whereas the step (Inv)ShiftRows is applied to the state row-by-row. Due to the slice architecture of the State-RAM, it is not possible to read or write from or to the RAM column-by-column. Hence, a transposition of the state is necessary if a row-oriented operation follows a column-oriented operation, or vice versa. Transposition would require a reorganization of the state before further operations can be performed. With two states, transposition can be implemented by accordingly addressing the State-RAM. Furthermore, (Inv)ShiftRows can be combined with transposing the state. As a consequence of this, (Inv)ShiftRows and transposition come for free. In the following, we describe the memory organization and state transposition for encryption. The same approach can be easily modified for decryption.

When a row-oriented operation follows a column-oriented operation (or vice versa), the state must be transposed. Combining row and column transformations minimizes the number of required transpositions: ShiftRows is combined with SubBytes and AddRoundKey is combined with MixColumns

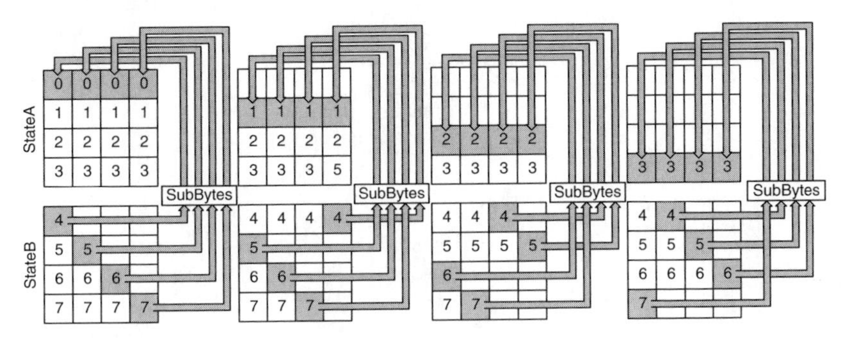

FIGURE 6.5 ShiftRows and SubBytes for encryption.

(see Figure 6.4). This approach requires only one transposition per round. Encryption requires SubBytes followed by ShiftRows. Since ShiftRows does not affect the byte values and SubBytes is applied to each byte of the state individually, the order of both operations does not matter. This fact eases the address generation for the State-RAM.

For explaining the transposition of the state, we consider the state as a 4×4 array: $S = (s_{i,j})$ for $i, j = 0, \ldots, 3$. The ShiftRows transformation described in [10] can then be expressed as follows:

$$S' = \text{ShiftRows}(S) = (s_{i,(j-i) \bmod 4}) \quad \text{for} \quad i, j = 0, \ldots, 3. \tag{6.2}$$

If we replace the state by the transposed state, we obtain

$$S'^{\text{T}} = \text{ShiftRows}(S'^{\text{T}}) = (s_{(i+j) \bmod 4, j}) \quad \text{for} \quad i, j = 0, \ldots, 3. \tag{6.3}$$

Based on Equation 6.3, the addressing of the StateB-RAM can be determined: the indices (i, j) must be substituted with $((i + j) \bmod 4, j)$. Due to the even number of AES rounds for all key lengths, ShiftRows is always applied to StateB only. Thus, the resulting index tuples can be directly mapped to the RAMs. The first part of the tuple index specifies the RAM slice and the second part specifies the RAM address. Since we operate on StateB, we must add an offset of four to the index value to get the correct address. Figure 6.5 shows the transposition of the state, including ShiftRows and SubBytes for encryption.

6.4.3.3.2 Implementation of (Inv)SubBytes

For the implementation of (Inv)SubBytes, we follow the approach presented in [24] since it needs relatively few hardware resources. The implementation is schematically depicted in Figure 6.6. SubBytes is an inversion in GF(2^8) followed by an affine transformation. (Inv)SubBytes is the inverse affine transformation followed by the same byte inversion. To save hardware

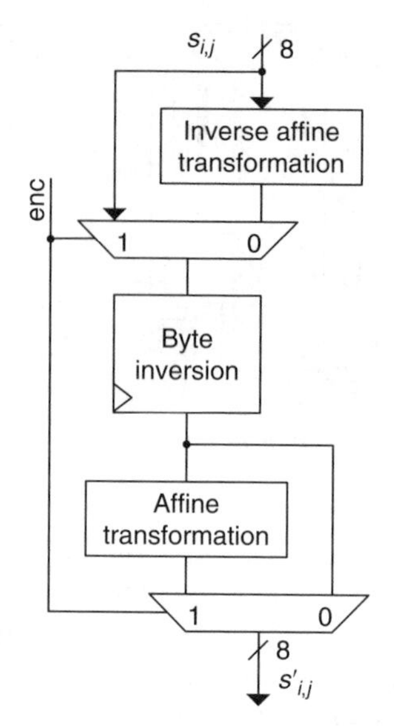

FIGURE 6.6 Implementation of (Inv)SubBytes.

resources, the byte inversion is shared. To implement both SubBytes and (Inv)SubBytes, the byte inversion is bypassed, as shown in Figure 6.6.

The byte inversion and the affine transformation can be described fully combinatorially. However, based on our analysis, it is more efficient to use a synchronous ROM to implement the byte inversion. The look-up table is given in Table 6.5. The affine and the inverse affine transformation can be described by Boolean equations, which are given in Table 6.6.

6.4.3.3.3 Implementation of (Inv)MixColumns

MixColumns is a multiplication of each column of the state with a constant matrix M that is defined as follows:

$$M = \begin{bmatrix} 02_x & 03_x & 01_x & 01_x \\ 01_x & 02_x & 03_x & 01_x \\ 01_x & 01_x & 02_x & 03_x \\ 03_x & 01_x & 01_x & 02_x \end{bmatrix}.$$

For instance, if we compute the first byte of the first column $s'_{0,0}$, we write

$$s'_{0,0} = s_{0,0} \cdot 02_x \oplus s_{1,0} \cdot 03_x \oplus s_{2,0} \cdot 01_x \oplus s_{3,0} \cdot 01_x.$$

So what we need is byte multiplication by 02_x and by $03_x = 02_x \oplus 01_x$. Consequently, we only need to implement multiplication by 02_x. This multiplication can be implemented by using XOR operations and shift operations. To describe it in more detail, we need the irreducible polynomial that defines the finite field for AES

$$m(x) = x^8 + x^4 + x^3 + x + 1.$$

With $m(x)$, we can now define multiplication by 02_x. The values a and b represent bytes, a_i and b_i $(i = 0, \ldots, 7)$ denote the single bits. Multiplication of a by 02_x is then defined as follows:

$$b = a \cdot 02_x = \begin{cases} \text{ShiftLeft}(a,1) & \text{if } a_7 = 0, \\ \text{ShiftLeft}(a,1) \oplus p' & \text{if } a_7 = 1, \end{cases} \tag{6.4}$$

where p' is the 8-bit representation of $x^4 + x^3 + x + 1$, i.e., 00011011_b.

For decryption, the matrix M consists of other entries, namely 09_x, $0B_x$, $0D_x$, and $0E_x$. If we follow the approach presented in [28] to implement MixColumns and (Inv)MixColumns, then we only need multiplication by the following constants: 02_x, 03_x, 08_x, and $0C_x$. The multiplication by these constants can be described based on the multiplication by 02_x:

$$\begin{aligned} b \cdot 03_x &= b \cdot 02_x \oplus b, \\ b \cdot 08_x &= ((b \cdot 02_x) \cdot 02_x) \cdot 02_x, \\ b \cdot 0C_x &= (b \cdot 02_x \oplus b) \cdot 02_x \cdot 02_x. \end{aligned}$$

These multiplications can be implemented fully combinatorially. The Boolean equations are given in Table 6.7 and Table 6.8.

The architecture of the core multiplier that computes 1 byte of MixColumns and (Inv)MixColumns is shown on the left-hand side in Figure 6.7.

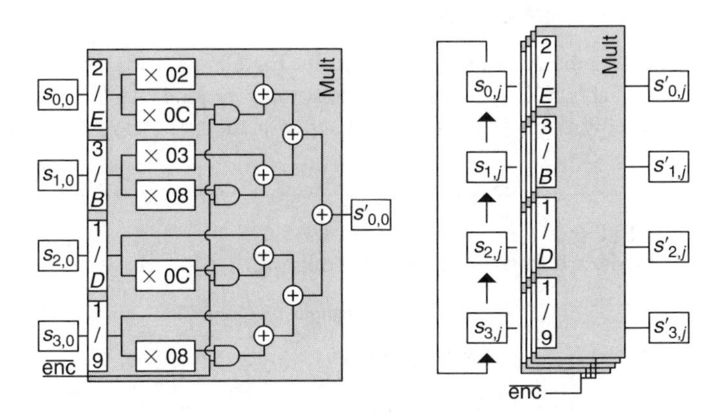

FIGURE 6.7 The core multiplier (left) and (Inv)MixColumns (right).

To process one column of the state we can use four such multipliers, where the inputs of each multiplier are rotated in the same way as the rows of the matrix M. This is shown on the right-hand side in Figure 6.7. It is easy to see that if the signal $enc = 1$, then the multiplication constants are 02_x, 03_x, 01_x, and 01_x, since the constants $0C_x$ and 08_x are masked with the AND gates. If $enc = 0$, then the constants are $0E_x$, $0B_x$, $0D_x$, and 09_x.

6.4.3.3.4 Key Unit

The key unit performs the key expansion function and stores the round keys. For each new cipher key, the round keys are precalculated to allow rapid encryption of subsequent data blocks for the same cipher key—no further key expansion has to be done. Because decryption uses the encryption round keys in the reverse order, the key expansion function must be calculated only once. Hence, the stored round keys are used for both encryption and decryption.

The key expansion function needs the SubBytes functionality. To keep the required hardware resources small, SubBytes is shared between key unit and data unit (multiplexor-input S-box_o in Figure 6.4). This can be done easily because the four SubBytes units are not used by the data unit during the calculation of the round keys.

The memory of the key unit is separated from the memory of the data unit, because the access of a common memory would be a throughput bottleneck. The key store is implemented as a 64×32 bit synchronous single-port RAM.

The key unit supports 128-bit, 192-bit, and 256-bit keys. Compared with supporting only 128-bit keys, only few additional hardware resources are required. Supporting all key lengths increases the needed hardware resources for the key unit by only 7.8%. The size of the key memory for 256-bit keys is the same as that for 128-bit keys. For 128-bit keys, the key expansion function derives 44 round key parts of 32-bit size from the cipher key. This requires a 64×32 bit RAM; 256-bit keys produce 63 round key parts of size 32 bits, fitting the 64×32 bit RAM.

6.4.3.4 Implementation Results

We implemented the AES on a Xilinx Virtex-E XCV1000EBG560-8 device. As shown in Table 6.1, the implementation only requires 1125 CLB-slices and does not require any block RAMs. It supports the complete AES standard

TABLE 6.1
Hardware Resources and Throughput

		Throughput (ECB and CBC) (Mbps)		
# CLB-Slices	# BRAM	AES-128	AES-192	AES-256
1125	0	215	180	156

and supports the CBC mode. Furthermore, it is equipped with a 32-bit AMBA APB interface that eases the integration with processors used in System-on-Chip designs [27]. If we do not consider the CBC mode functionality and the AMBA bus interface, the AES implementation requires about 26% less hardware resources.

The implementation only uses 9.16% of the available logic cells on a Xilinx Virtex-E XCV1000EBG560-8 device. The remaining 90.8% of the logic resources and 100% of the on-chip BRAMs can be used by other applications like a LEON2 or an ARM processor. For a stand-alone application, a low-end FPGA (e.g., Xilinx SpartanII XC2S100-6) is sufficient for implementing the complete AES algorithm. The maximum clock frequency on an XCV1000 FPGA is 161 MHz. At this frequency, a throughput of 215 Mbps for AES-128, 180 Mbps for AES-192, and 156 Mbps for AES-256 is achieved for both ECB mode and CBC mode.

6.5 WHIRLPOOL

6.5.1 Description

The block cipher–based hash function Whirlpool [29] is an iterative hash function. In general, the input message m is divided into i 512-bit message blocks (after applying a padding rule such that the message is a multiple of 512 bits): $m = m_1 m_2 \ldots m_i$. Each of the message blocks is then processed by applying one iteration of the hash function f. The hash value of iteration i is given by $h_i = f(h_{i-1}, m_i)$, where h_{i-1} is the hash value of the previous iteration. For the first message block, an IV is used. For Whirlpool, $h_0 = \mathrm{IV} = 0$. The underlying block cipher, referred to as W, operates in the Miyaguchi–Preneel mode [3], as shown in Figure 6.8.

The block cipher W is strongly based on the structure of the AES. W is a 512-bit block cipher and uses a 512-bit key. The input (plaintext) is the ith message block m_i to be hashed and the cipher key is the intermediate hash value from the previous iteration h_{i-1}. The block cipher W can basically be divided into two parts: the datapath and the key schedule (see also Figure 6.9).

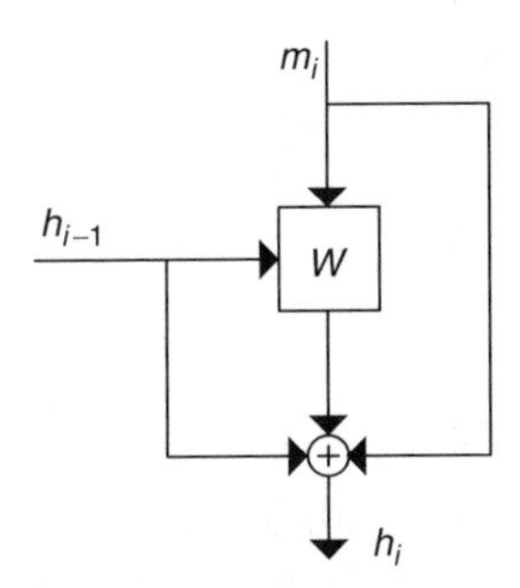

FIGURE 6.8 The Whirlpool hashing function.

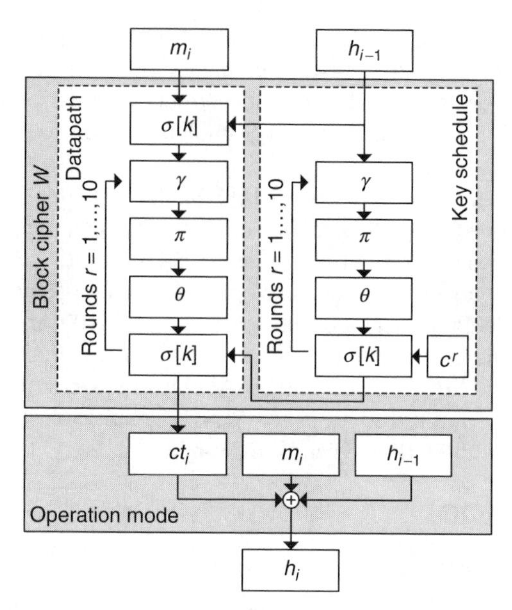

FIGURE 6.9 The Whirlpool dataflow.

The datapath processes the input message m_i by iteratively applying the round transformations for 10 rounds. Each round requires a round key that is derived by the key schedule from the cipher key. The block cipher W uses a 512-bit internal state that is organized as an 8×8 array of bytes. The state stores the input message, the intermediate results for each round, and the ciphertext after 10 rounds. As can be seen in Figure 6.9, both the datapath and the key schedule use the same round transformations. The round transformations are

- Nonlinear layer γ, where a nonlinear S-Box is applied to each byte of the state individually
- Cyclical permutation π, where the bytes of column j are rotated downward by j positions
- Linear diffusion layer θ, where the state is multiplied by a constant matrix
- Key addition $\sigma[k]$, where also round constants c^r are introduced

One round $\rho[k]$ of W is performed as follows:

$$\rho[k] \equiv \sigma[k] \circ \theta \circ \pi \circ \gamma,$$

where the transformations are applied to the state from the right to the left. Figure 6.10 depicts how the round transformations are applied to the state and shows that the state is organized as an 8×8 array of bytes.

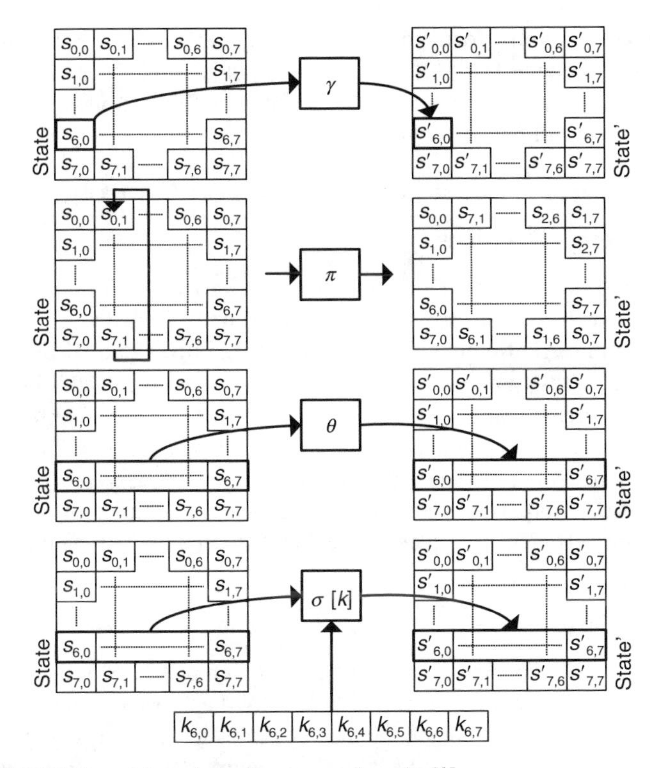

FIGURE 6.10 Round transformations γ, π, θ, and $\sigma[k]$.

A single input message block is processed as shown in Figure 6.9. The cipher key (either h_{i-1} or $h_0 = \text{IV}$) is added to the message block m_i and stored in the state. Then, the round transformation $\rho[k]$ is applied to the state for 10 rounds, with the round key for each round provided by the key schedule. After 10 rounds, the state containing the ciphertext ct_i, the cipher key h_{i-1}, and the input message block m_i are added (Miyaguchi–Preneel operation mode), resulting in the cipher key h_i for the next message block, or the final hash value if the input message has been processed completely.

A closer look at Figure 6.9, confirms the fact that the block cipher W is actually composed of two block ciphers: the datapath with round keys provided by the key schedule and the key schedule with the round constants c^r as round keys. Therefore, we also require a 512-bit internal state for the implementation of the key schedule. As the internal state of the datapath, the state for the key schedule is organized as an 8×8 array of bytes.

6.5.2 Overview of Implementations

Different to AES, only few Whirlpool hardware implementations have been published to date; we are only aware of three implementations on FPGAs.

Two implementations focus on high throughput rates: a throughput of 4900 Mbps is reported in [30] and a throughput of 4480 Mbps is presented in [31,32]. A compact implementation has been published in [33], which is described in the following section.

6.5.3 Compact Hardware Implementation of Whirlpool

6.5.3.1 Design Decisions

Based on the description of Whirlpool, we see a strong similarity to AES. Therefore, we made basically the same design decisions as for the AES implementation. A difference is that we decided in favor of a 64-bit architecture, since the transformations are defined for 64-bit words. In addition, for the Whirlpool implementation, we require that it is area efficient. For Whirlpool, it makes sense to implement on-the-fly round key generation, since in each iteration a new cipher key is used. This avoids any precomputation time that would otherwise decrease the throughput remarkably. Since the aim is an area-efficient architecture, we decided to implement a nonpipelined approach.

6.5.3.2 Whirlpool State

As described in Section 6.5.1, the Whirlpool state is represented as an 8×8 array of bytes. Therefore, we can use a LUT-based RAM approach for the implementation of the state. For implementing the Whirlpool state, we need an 8×64 bit RAM. Since we use LUT-based RAMs, this requires 64 LUTs. With one LUT we can implement a 16×1 bit RAM and therefore eight rows (addresses) of the RAM are not used. This leads to a nice feature: we can implement a second Whirlpool state without additional hardware requirements. By using the LUT-based RAM approach for the implementation of the state, we have the possibility to implement a second state for free. From the second state, we benefit that the round transformation π can be implemented by accordingly addressing the State-RAM in a similar way as we have done for the AES implementation. Note that the π transformation can only be implemented through wiring in the case of a 512-bit datapath. If smaller bit sizes are used, e.g., 64 bits, this transformation requires additional logic and additional cycles. However, by using a second state, this transformation comes for free. To implement the π transformation by accordingly addressing the State-RAM, it is required that we can store single bytes in each row of the State-RAM. Considering this property, we implemented the State-RAM as shown in Figure 6.11: Eight slices of 16×8 bit synchronous dual-port RAM. Dual-port RAM is used to reduce the number of cycles, which in turn increases the throughput. This is due to the fact that dual-port RAM provides concurrent reading and writing. By using a LUT-based RAM approach, the state can be implemented more efficiently than a registers-based approach. For instance, one state requires approximately 512 LUTs if implemented with registers (without counting in additional logic like multiplexors that may be

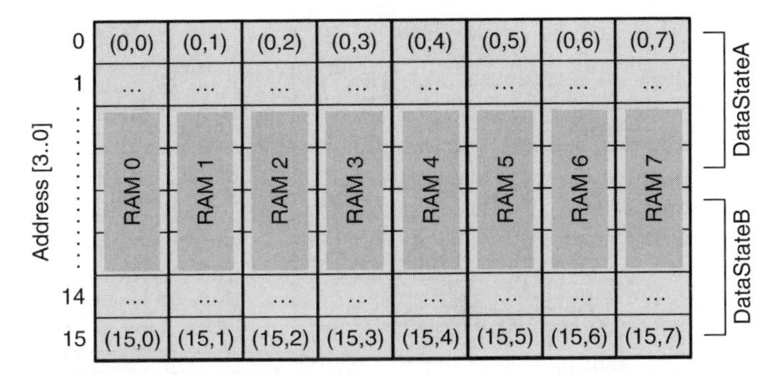

FIGURE 6.11 Eight slices of 16×8 bit synchronous dual-port RAM.

required). This is at least four times more than using a 16×64 bit LUT-based dual-port RAM (128 LUTs), with the additional advantage that the second state comes for free.

As described in Section 6.5.1, we require a Whirlpool state for the datapath and the key schedule. For both parts, we use the same representation of the state. For the remainder of this section, we refer to the state of the datapath as DataState and the state of the key schedule as KeyState. Since we have two states, we refer to the first state as DataStateA and the second as DataStateB, respectively, KeyStateA and KeyStateB. The labeling for the DataState is shown in Figure 6.11.

6.5.3.3 Fully Interleaved Hash Computation

In this section, we present the overall architecture of the proposed Whirlpool implementation. The architecture is schematically depicted in Figure 6.12. As can be seen in Figure 6.12, the transformations γ and θ are reused by the datapath and the key schedule. Note that the transformation π does not appear in Figure 6.12, since it can be implemented by addressing the State-RAM. This is described later in this section. In the following, we show how a 512-bit input message is processed and why the sharing of γ and θ works without the need of additional cycles. This is summarized in Table 6.2. For the description of the architecture's dataflow we use the following phases: data loading, round part A, round part B, and operation mode. The pseudo-code is given in Table 6.3, where π(data) means that data are stored according to the π transformation:

- Phase: data loading. This mode represents the initial state of one Whirlpool iteration. Eight 64-bit words of the 512-bit input message block are loaded and the initial key value (h_{i-1}) is added. The result is then stored in DataStateA and in ModeRAM. The ModeRAM stores the result of the addition of h_{i-1} and m_i, which is required for the

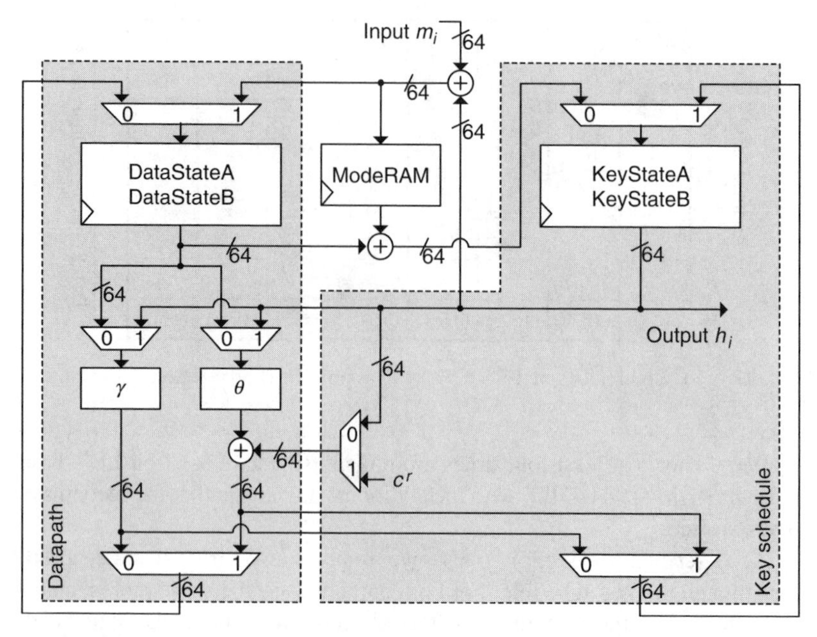

FIGURE 6.12 Architecture of the Whirlpool implementation.

Miyaguchi–Preneel operation mode. Since we have to read the values of the KeyStateA for the initial key addition, we can compute the γ transformation of the key schedule concurrently. After data loading the ModeRAM contains the values for the operation mode, and the key schedule (KeyStateB) holds the result of the γ transformation.

- Phase: round part A. In this mode, the γ transformation is applied to DataStateA and at the same time we compute the θ transformation for KeyStateB. Moreover, the round constants c^r are added. After eight cycles, KeyStateA holds the next round key and DataStateB stores the result of the γ transformation.

TABLE 6.2
Sharing of γ and θ between Datapath and Key Schedule

Phase	γ	θ
Data loading	KeyState	—
Round part A	DataState	KeyState
Round part B	KeyState	DataState

TABLE 6.3
Interleaved Hash Computation

Phase data loading (8 cycles)			
	DSA	$\Leftarrow$	$m_i \oplus$ KSA
	MR	$\Leftarrow$	$m_i \oplus$ KSA
	KSB	$\Leftarrow$	$\pi(\gamma(\text{KSA}))$
Phase round part A (8 cycles/iteration)			
$r = 1, \ldots, 10$:	DSB	$\Leftarrow$	$\pi(\gamma(\text{DSA}))$
	KSA	$\Leftarrow$	$\theta(\text{KSA}) \oplus c^r$
Phase round part B (8 cycles/iteration)			
$r = 1, \ldots, 9$:	DSA	$\Leftarrow$	$\theta(\text{DSB}) \oplus$ KSA
	KSB	$\Leftarrow$	$\pi(\gamma(\text{KSA}))$
$r = 10$:	DSA	$\Leftarrow$	$\theta(\text{DSB}) \oplus$ KSA
Phase operation mode (8 cycles)			
	KSA	$\Leftarrow$	DSA $\oplus$ MR

Notation:
DSA, DSB . . . DataStateA, DataStateB
KSA, KSB . . . KeyStateA, KeyStateB
MR . . . ModeRAM

- Phase: round part B. Now the θ transformation is applied to DataStateB and the round keys from KeyStateA are added. Since we read the round keys from KeyStateA for the key addition, we compute again the γ transformation for KeyStateA concurrently. After eight cycles, we start again with round part A.
- Phase: operation mode. After 10 iterations of round part A and round part B, DataStateA holds the ciphertext. For the mode of operation, we read the data of DataStateA and the data of ModeRAM. These two values are added and stored in KeyStateA, which holds the hash value of the current iteration after eight cycles. One iteration is now completed—for the next input message block we start again with the mode data loading. If the input message has been processed completely, the eight 64-bit words of KeyStateA are unloaded resulting in the 512-bit hash value.

Based on the description of the different phases, the total number of cycles required to process one message block (without unloading) is $8 + 10 \cdot (8 + 8) + 8 = 176$. The 176 cycles result from 8 cycles required in each phase and 10 iterations of phases round part A and round part B.

So far, we did not discuss how the ModeRAM has been implemented. The ModeRAM has to store eight 64-bit words. Therefore, we need an 8×64 bit array, as for the DataState and the KeyState. We decided again to use a LUT-based RAM approach since we can save hardware resources compared with a register implementation. The savings are based on the same reasoning as for the DataState and the KeyState. However, since the ModeRAM acts only as a register, we do not need a second state or the dual-port functionality. Therefore, we implemented the ModeRAM as an 8×64 bit synchronous single-port RAM. The 10 round constants c^r required for the key schedule are defined as the first 10 entries of the substitution box γ. For the implementation, we used a 10×64 bit synchronous LUT-based ROM.

6.5.3.4 Implementation of the Round Transformations

In this section, we show the implementation of the round transformations π, γ, and θ. Note that the round key addition $\sigma[k]$ is a simple XOR operation of the state and the round key, as described in Section 6.5.1.

6.5.3.4.1 Implementation of π

Now we show how we can implement the transformation π by accordingly addressing the State-RAM (DataState and KeyState). An example of the state addressing is given in Figure 6.13 for row 0, row 4, and row 7 of DataStateA. The addressing can be expressed by the formula

$$\text{DataStateB}((i + j \bmod 8) + 8, j) = \text{DataStateA}(i, j), \qquad (6.5)$$

where $i, j = 0, \ldots, 7$. For instance, the byte $s_{4,5}$ of DataStateA is stored in $s_{9,5}$ of DataStateB. To implement π, the data are read row-by-row from DataStateA and stored in the DataStateB according to the index substitution given in Equation 6.5. After all rows have been processed, the values are read row-by-row

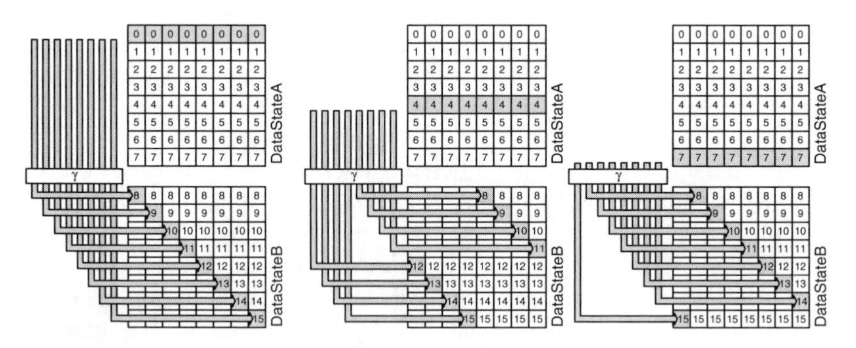

FIGURE 6.13 State addressing to implement π for row 0, row 4, and row 7 of DataStateA.

from DataStateB and the result of the θ and $\sigma[k]$ transformation is stored in DataStateA. The same procedure applies for the KeyState.

6.5.3.4.2 *Transformation* γ

For the implementation of the nonlinear layer γ, we chose the proposed representation given in [29] (see Figure 6.14). The byte substitution in the finite field GF(2^8) is mapped to the composite field GF($(2^4)^2$). In the smaller field GF(2^4), so-called mini-boxes are used. Each mini-box has a 4-bit input and a 4-bit output. A block diagram of the γ transformation is shown in Figure 6.14. The functionality of the mini-boxes can be described by Boolean equations, i.e., γ can be implemented fully combinatorially. However, our analysis has shown that it is more efficient in terms of area requirements and combinatorial delay to implement the five mini-boxes with LUT-based ROMs. The look-up tables for each mini-box are given in Table 6.10. Since γ is defined for bytes, it can also be implemented by using one look-up table with $2^8 = 256$ entries. This requires a storage of 2048 bits. Using only one look-up table requires more hardware resources than the mini-boxes approach, where each mini-box is implemented using a LUT-based ROM. This is due to the fact that each mini-box consists of 16 entries of 4-bit size, resulting in $5 \times 16 \times 4 = 320$ bits, compared with 2048 bits. Even if the mini-box approach requires 12 additional XOR gates (see Figure 6.14), it still requires remarkably less resources than the approach with only one look-up table. To implement γ for the proposed 64-bit architecture we need 8 byte substitutions, each consisting of 5 mini-boxes and 12 XOR gates.

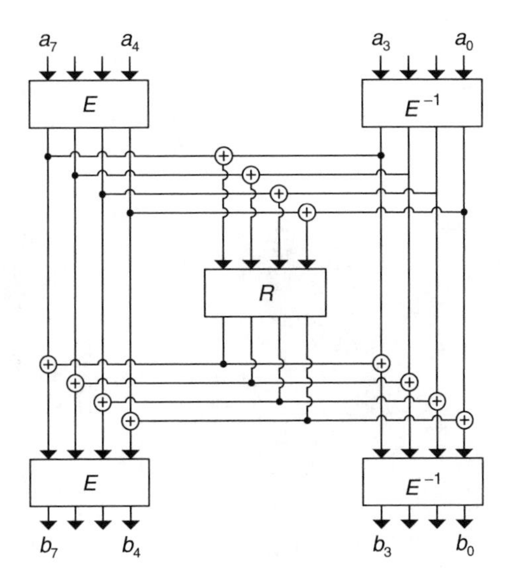

FIGURE 6.14 Implementation of γ.

6.5.3.4.3 Transformation θ

The linear transformation θ is defined as the multiplication of the state with a constant matrix C. The matrix C is defined as follows:

$$C = \begin{bmatrix} 01_x & 01_x & 04_x & 01_x & 08_x & 05_x & 02_x & 09_x \\ 09_x & 01_x & 01_x & 04_x & 01_x & 08_x & 05_x & 02_x \\ 02_x & 09_x & 01_x & 01_x & 04_x & 01_x & 08_x & 05_x \\ 05_x & 02_x & 09_x & 01_x & 01_x & 04_x & 01_x & 08_x \\ 08_x & 05_x & 02_x & 09_x & 01_x & 01_x & 04_x & 01_x \\ 01_x & 08_x & 05_x & 02_x & 09_x & 01_x & 01_x & 04_x \\ 04_x & 01_x & 08_x & 05_x & 02_x & 09_x & 01_x & 01_x \\ 01_x & 04_x & 01_x & 08_x & 05_x & 02_x & 09_x & 01_x \end{bmatrix}.$$

Since we use a 64-bit approach, we have to process one row at a time. For instance, the 8-bit output $s'_{0,0}$ is computed as follows:

$$\begin{aligned} s'_{0,0} = {} & s_{0,0} \cdot 01_x \oplus s_{0,1} \cdot 09_x \oplus s_{0,2} \cdot 02_x \oplus s_{0,3} \cdot 05_x \oplus \\ & s_{0,4} \cdot 08_x \oplus s_{0,5} \cdot 01_x \oplus s_{0,6} \cdot 04_x \oplus s_{0,7} \cdot 01_x. \end{aligned}$$

Therefore, to process one row (64 bits), 5 multiplications for each output byte are required resulting in 40 multiplications. For the computation of one output byte we use a multiplier that takes as input one row of the state (see Figure 6.15). To process one 64-bit row, we need eight such multipliers. Since the matrix C is circular, i.e., the coefficients of each column are rotated downward, we can use the same multiplier eight times. For each multiplier, the single input bytes are rotated (by accordingly wiring) in the same way as the coefficients of the matrix.

The number of required multiplications for one 64-bit row can still be reduced. This is due to the fact that we can reuse the multiplication with 04_x and 08_x for the multiplication with $09_x = 08_x \oplus 01_x$ and $05_x = 04_x \oplus 01_x$.

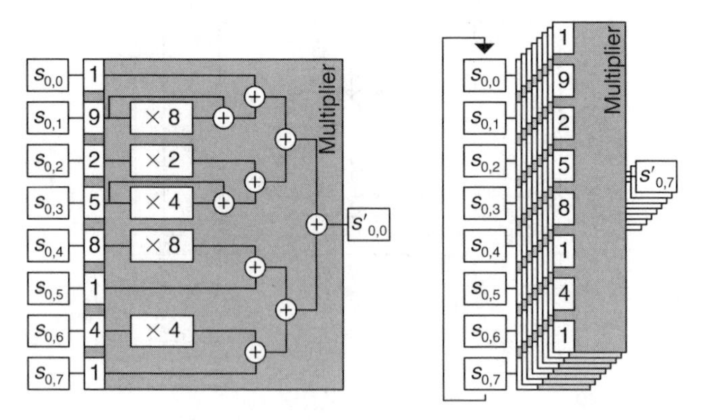

FIGURE 6.15 The core multiplier (left) and θ (right).

This reduces the required multipliers from 40 to 24. This has already been considered by the designers of Whirlpool when choosing the matrix C.

The multiplication of the state bytes with the constants 02_x, 04_x, and 08_x can be described fully combinatorially in the same way as for AES (see Section 6.4). The only difference is that Whirlpool uses a different irreducible polynomial

$$p(x) = x^8 + x^4 + x^3 + x^2 + 1. \tag{6.6}$$

Therefore, the multiplication by 02_x is defined as

$$b = a \cdot 02_x = \begin{cases} \text{ShiftLeft}(a,1) & \text{if } a_7 = 0, \\ \text{ShiftLeft}(a,1) \oplus p' & \text{if } a_7 = 1, \end{cases} \tag{6.7}$$

where p' is the 8-bit representation of $x^4+x^3+x^2+1$, i.e., 00011101_b. Multiplication by 04_x is performed by computing twice the multiplication with 02_x, i.e., $b=a \cdot 04_x=(a \cdot 02_x) \cdot 02_x$. The same holds for $b=a \cdot 08_x=(a \cdot 04_x) \cdot 02_x$. These multiplications can be described fully combinatorially. The Boolean equations are given in Table 6.9.

6.5.3.5 Implementation Results

Table 6.4 lists the required hardware resources for the proposed Whirlpool implementation. The datapath requires most resources since it includes the two transformations γ and θ. If these transformations would not be reused, the datapath and the key schedule would require approximately the same hardware resources. This emphasizes the importance of reusing these expensive transformations for a compact implementation. Note that reusing the transformations does not need any additional cycles. This works because the datapath uses γ, whereas the key schedule uses θ and vice versa.

For the implementation of Whirlpool we used a Xilinx Virtex 2P xc2vp40-7fg676 device. On this device, we achieve a throughput of 382 Mbps at a frequency of 131 MHz requiring 1456 CLB-slices and no

TABLE 6.4
Required Hardware Resources for the Whirlpool Implementation

Module	CLB-Slices	%	BRAMs
Datapath	679	46.6	0
Key schedule	367	25.2	0
Operation mode	128	8.8	0
Control unit	102	7.0	0
64-bit AMBA interface	180	12.4	0
Total	1456	100	0

BRAMs. If we do not consider the AMBA bus interface, the Whirlpool implementation requires about 12% less hardware resources.

On the given device, the implementation uses only 6% of the available hardware resources and, most notable, 100% of the available BRAMs are free for use by other applications such as a LEON2 or ARM processor. Nevertheless, the proposed architecture can also be implemented as a stand-alone application for low-end devices such as SPARTAN2.

6.6 APPENDICES

6.6.1 Further Implementation Details of AES

In this section, we give further details of the implementation of (Inv)SubBytes and (Inv)MixColumns.

6.6.1.1 Byte Inversion in GF(2^8)

As described in Section 6.4.3, the byte inversion is implemented by using a synchronous ROM. The look-up table is given in Table 6.5.

6.6.1.2 Affine and Inverse Affine Transformation

The affine and inverse affine transformation of (Inv)SubBytes can be described by the Boolean equations given in Table 6.6, where a_i and b_i $(i = 0, \ldots, 7)$ represent bytes values.

TABLE 6.5
Byte Inversion $b = b_l\, b_r$ in GF(2^8) in Hexadecimal Notation

		b_r															
		0	1	2	3	4	5	6	7	8	9	A	B	C	D	E	F
b_l	0	00	01	8D	F6	CB	52	7B	D1	E8	4F	29	C0	B0	E1	E5	C7
	1	74	B4	AA	4B	99	2B	60	5F	58	3F	FD	CC	FF	40	EE	B2
	2	3A	6E	5A	F1	55	4D	A8	C9	C1	0A	98	15	30	44	A2	C2
	3	2C	45	92	6C	F3	39	66	42	F2	35	20	6F	77	BB	59	19
	4	1D	FE	37	67	2D	31	F5	69	A7	64	AB	13	54	25	E9	09
	5	ED	5C	05	CA	4C	24	87	BF	18	3E	22	F0	51	EC	61	17
	6	16	5E	AF	D3	49	A6	36	43	F4	47	91	DF	33	93	21	3B
	7	79	B7	97	85	10	B5	BA	3C	B6	70	D0	06	A1	FA	81	82
	8	83	7E	7F	80	96	73	BE	56	9B	9E	95	D9	F7	02	B9	A4
	9	DE	6A	32	6D	D8	8A	84	72	2A	14	9F	88	F9	DC	89	9A
	A	FB	7C	2E	C3	8F	B8	65	48	26	C8	12	4A	CE	E7	D2	62
	B	0C	E0	1F	EF	11	75	78	71	A5	8E	76	3D	BD	BC	86	57
	C	0B	28	2F	A3	DA	D4	E4	0F	A9	27	53	04	1B	FC	AC	E6
	D	7A	07	AE	63	C5	DB	E2	EA	94	8B	C4	D5	9D	F8	90	6B
	E	B1	0D	D6	EB	C6	0E	CF	AD	08	4E	D7	E3	5D	50	1E	B3
	F	5B	23	38	34	68	46	03	8C	DD	9C	7D	A0	CD	1A	41	1C

TABLE 6.6
Affine and Inverse Affine Transformation of (Inv)SubBytes

$b = \text{Affine}(a)$	$b = \text{Affine}^{-1}(a)$
$b_0 = \overline{a_0} \oplus a_4 \oplus a_5 \oplus a_6 \oplus a_7$	$b_0 = a_2 \oplus \overline{a_5} \oplus a_7$
$b_1 = a_0 \oplus a_1 \oplus \overline{a_5} \oplus a_6 \oplus a_7$	$b_1 = a_0 \oplus a_3 \oplus a_6$
$b_2 = a_0 \oplus a_1 \oplus a_2 \oplus a_6 \oplus a_7$	$b_2 = a_1 \oplus a_4 \oplus \overline{a_7}$
$b_3 = a_0 \oplus a_1 \oplus a_2 \oplus a_3 \oplus a_7$	$b_3 = a_0 \oplus a_2 \oplus a_5$
$b_4 = a_0 \oplus a_1 \oplus a_2 \oplus a_3 \oplus a_4$	$b_4 = a_1 \oplus a_3 \oplus a_6$
$b_5 = \overline{a_1} \oplus a_2 \oplus a_3 \oplus a_4 \oplus a_5$	$b_5 = a_2 \oplus a_4 \oplus a_7$
$b_6 = a_2 \oplus a_3 \oplus a_4 \oplus a_5 \oplus \overline{a_6}$	$b_6 = a_0 \oplus a_3 \oplus a_5$
$b_7 = a_3 \oplus a_4 \oplus a_5 \oplus a_6 \oplus a_7$	$b_7 = a_1 \oplus a_4 \oplus a_6$

TABLE 6.7
Byte Multiplication with 02_x and 03_x for AES

$b = a \cdot 02_x$	$b = a \cdot 03_x$
$b_0 = a_7$	$b_0 = a_0 \oplus a_7$
$b_1 = a_0 \oplus a_7$	$b_1 = a_0 \oplus a_1 \oplus a_7$
$b_2 = a_1$	$b_2 = a_1 \oplus a_2$
$b_3 = a_2 \oplus a_7$	$b_3 = a_2 \oplus a_3 \oplus a_7$
$b_4 = a_3 \oplus a_7$	$b_4 = a_3 \oplus a_4 \oplus a_7$
$b_5 = a_4$	$b_5 = a_4 \oplus a_5$
$b_6 = a_5$	$b_6 = a_5 \oplus a_6$
$b_7 = a_6$	$b_7 = a_6 \oplus a_7$

TABLE 6.8
Byte Multiplication with 08_x and $0C_x$ for AES

$b = a \cdot 08_x$	$b = a \cdot 0C_x$
$b_0 = a_5$	$b_0 = a_5 \oplus a_6$
$b_1 = a_5 \oplus a_6$	$b_1 = a_5 \oplus a_7$
$b_2 = a_6 \oplus a_7$	$b_2 = a_0 \oplus a_6$
$b_3 = a_0 \oplus a_5 \oplus a_7$	$b_3 = a_0 \oplus a_1 \oplus a_5 \oplus a_6 \oplus a_7$
$b_4 = a_1 \oplus a_5 \oplus a_6$	$b_4 = a_1 \oplus a_2 \oplus a_5 \oplus a_7$
$b_5 = a_2 \oplus a_6 \oplus a_7$	$b_5 = a_2 \oplus a_3 \oplus a_6$
$b_6 = a_3 \oplus a_7$	$b_6 = a_3 \oplus a_4 \oplus a_7$
$b_7 = a_4$	$b_7 = a_4 \oplus a_5$

6.6.1.3 Byte Multiplication by Constants in (Inv)MixColumns

For the implementation of (Inv)MixColumns we need to implement byte multiplication with the constants 02_x, 03_x, 08_x, and $0C_x$. These multiplications are described by the Boolean equations listed in Table 6.7 and Table 6.8.

TABLE 6.9
Byte Multiplication with 02_x, 04_x, and 08_x for Whirlpool

$b = a \cdot 02_x$	$b = a \cdot 04_x$	$b = a \cdot 08_x$
$b_0 = a_7$	$b_0 = a_6$	$b_0 = a_5$
$b_1 = a_0$	$b_1 = a_7$	$b_1 = a_6$
$b_2 = a_1 \oplus a_7$	$b_2 = a_0 \oplus a_6$	$b_2 = a_5 \oplus a_7$
$b_3 = a_2 \oplus a_7$	$b_3 = a_1 \oplus a_6 \oplus a_7$	$b_3 = a_0 \oplus a_5 \oplus a_6$
$b_4 = a_3 \oplus a_7$	$b_4 = a_2 \oplus a_6 \oplus a_7$	$b_4 = a_1 \oplus a_5 \oplus a_6 \oplus a_7$
$b_5 = a_4$	$b_5 = a_3 \oplus a_7$	$b_5 = a_2 \oplus a_6 \oplus a_7$
$b_6 = a_5$	$b_6 = a_4$	$b_6 = a_3 \oplus a_7$
$b_7 = a_6$	$b_7 = a_5$	$b_7 = a_4$

TABLE 6.10
Look-Up Tables for Mini-Boxes E, E^{-1}, and R

in_x	$E(in_x)$	$E^{-1}(in_x)$	$R(in_x)$
0	1	F	7
1	B	0	C
2	9	D	B
3	C	7	D
4	D	B	E
5	6	E	4
6	F	5	9
7	3	A	F
8	E	9	6
9	8	2	3
A	7	C	8
B	4	1	A
C	A	3	2
D	2	4	5
E	5	8	1
F	0	6	0

6.6.2 Further Implementation Details for Whirlpool

6.6.2.1 Byte Multiplication by Constants

As described in Section 6.5.3, we need byte multiplication by the constants 02_x, 04_x, and 08_x for the implementation of the transformation θ. These multiplications can be implemented fully combinatorially by the Boolean equations given in Table 6.9.

6.6.2.2 Look-Up Tables for γ

As described in Section 6.5.3, the mini-boxes E, E^{-1}, and R for the γ transformation are implemented using look-up tables, which are given in Table 6.10.

REFERENCES

1. A.W. Dent and C.J. Mitchell, *User's Guide to Cryptography and Standards*, Artech House, 2005.
2. National Institute of Standards and Technology (NIST), Recommendation for Block Cipher Modes of Operation—Methods and Techniques, NIST Special Publication SP 800–38a, Dec. 2001, available online at http://csrc.nist.gov/publications/nistpubs
3. A.J. Menezes, P.C. van Oorschot, and S.A. Vanstone, *Handbook of Applied Cryptography*, CRC Press, 1997.
4. C.M. Campbell Jr., Design and specification of cryptographic capabilities, *NBS Special Publication 500–27: Computer Security and the Data Encryption Standard*, U.S. Department of Commerce, National Bureau of Standards, 1977, pp. 54–66.
5. K. Brincat and C.J. Mitchell, New CBC-MAC forgery attacks, *Information Security and Privacy—ACISP 2001*, LNCS 2119, Springer-Verlag, 2001, pp. 3–14.
6. N. Borisov, I. Goldberg, and D. Wagner, Intercepting mobile communications: the insecurity of 802.11, *Seventh Annual International Conference on Mobile Computing and Networking*, ACM, 2001, pp. 180–189.
7. eSTREAM—the ECRYPT stream cipher project, http://www.ecrypt.eu.org/stream.
8. M. Bellare, R. Canetti, and H. Krawczyk, Keyed hash functions and message authentication, *Advances in Cryptology—CRYPTO '96*, LNCS 1109, Springer-Verlag, 1996, pp. 1–15.
9. M. Feldhofer and C. Rechberger, A case against currently used hash functions in RFID protocols, *IAIK Technical Report 2006/005*, 2006.
10. National Institute of Standards and Technology (NIST), Advanced Encryption Standard (AES), Federal Information Processing Standards Publication 197 (FIPS PUB 197), November. 2001.
11. M. Feldhofer, K. Lemke, E. Oswald, F.-X. Standaert, T. Wollinger, and J. Wolkerstorfer, State of the Art in Hardware Architectures, ECRYPT Deliverable No. D.VAM2, September 2005, available online at http://www.iaik.tugraz.at/research/krypto/AES/VAM2-IAIK-17-D.VAM2–1_0.pdf.

12. A. Hodjat and I. Verbauwhede, A 21.54 Gbits/s fully pipelined AES processor on FPGA, *Twelfth IEEE Symposium on Field-Programmable Custom Computing Machines—FCCM 2004*, IEEE Computer Society, 2004, pp. 308–309.
13. P. Chodowiec, P. Khuon, and K. Gaj, Fast implementations of secret-key block ciphers using mixed inner- and outer-round pipelining, *Ninth ACM/SIGDA International Symposium on Field Programmable Gate Arrays—FPGA 2001*, ACM Press, 2001, pp. 94–102.
14. M. McLoone and J. McCanny, High performance single chip FPGA Rijndael algorithm implementations, *Workshop on Cryptographic Hardware and Embedded Systems—CHES 2001*, LNCS 2162, Springer-Verlag, 2001, pp. 65–76.
15. N.A. Saqib, F. Rodríguez-Henríquez, and A. Díaz-Pérez, Two approaches for a single-chip FPGA implementation of an encryptor/decryptor AES core, *Thirteenth International Conference on Field Programmable Logic and Application—FPL 2003*, LNCS 2778, Springer-Verlag, 2003, pp. 303–312.
16. P. Chodowiec and K. Gaj, Very compact FPGA implementation of the AES algorithm, *Workshop on Cryptographic Hardware and Embedded Systems—CHES 2003*, LNCS 2779, Springer-Verlag, 2003, pp. 319–333.
17. A. Dandalis, V. Prasanna, and J. Rolim, A comparative study of performance of AES final candidates using FGPAs, *The Third Advanced Encryption Standard (AES) Candidate Conference*, 2000, available online at http://csrc.nist.gov/CryptoToolkit/aes/round2/conf3/aes3agenda.html.
18. V. Fischer and M. Drutarovsky, Two methods of Rijndael implementation in reconfigurable hardware, *Workshop on Cryptographic Hardware and Embedded Systems—CHES 2001*, LNCS 2162, Springer-Verlag, 2001, pp. 77–92.
19. N. Pramstaller and J. Wolkerstorfer, A universal and efficient AES coprocessor for field programmable logic arrays, *Fourteenth International Conference on Field Programmable Logic and Application—FPL 2004*, LNCS 3203, Springer-Verlag, 2004, pp. 565–574.
20. A. Hodjat and I. Verbauwhede, Area-throughput trade-offs for fully pipelined 30 to 70 Gbits/s AES processors, *IEEE Transactions on Computers*, 55(4):366–372, 2006.
21. S. Mangard, M. Aigner, and S. Dominikus, A highly regular and scalable AES hardware architecture, *IEEE Transactions on Computers*, 52(4):483–491, April 2003.
22. A. Satoh, S. Morioka, K. Takano, and S. Munetoh, A compact Rijndael hardware architecture with S-box optimization, *Advances in Cryptology—ASIACRYPT 2001*, LNCS 2248, Springer-Verlag, 2001, pp. 239–254.
23. D. Canright, A very compact S-box for AES, *Workshop on Cryptographic Hardware and Embedded Systems—CHES 2005*, LNCS 3659, Springer-Verlag, 2005, pp. 441–455.
24. J. Wolkerstorfer, E. Oswald, and M. Lamberger, An ASIC implementation of the AES SBoxes, *Cryptographer's Track at the RSA Conference 2002*, LNCS 2271, Springer-Verlag, 2002, pp. 67–78.
25. M. Feldhofer, J. Wolkerstorfer, and V. Rijmen, AES implementation on a grain of sand, *IEE Proceedings of Information Security*, 152(1):13–20, 2005.
26. Xilinx Incorporated, Silicon Solutions—Virtex Series FPGAs, http://www.xilinx.com/products.

27. ARM Limited, AMBA 2.0 Specification, available online at http://www.arm.com.
28. J. Wolkerstorfer, An ASIC implementation of the AES-MixColumn operation, *Austrochip 2001*, Austria, 2001, pp. 129–132.
29. P.S.L.M. Baretto and V. Rijmen, The Whirlpool Hashing Function, 2000, revised in May 2003, available online at http://paginas.terra.com.br/informatica/paulobarreto/WhirlpoolPage.html.
30. M. McLoone, C. McIvor, and A. Savage, High-speed hardware architecture of the whirlpool hash function, *IEEE International Conference on Field-Programmable Technology—FPT 2005*, IEEE, 2005, pp. 147–162.
31. P. Kitsos and O. Koufopavlou, Efficient architecture and hardware implementation of the whirlpool hash function, *IEEE Transactions on Consumer Electronics*, 50(1):208–213, 2004.
32. P. Kitsos and O. Koufopavlou, Whirlpool hash function: Architecture and VLSI implementation, *IEEE International Symposium on Circuits Systems—ISCAS'04*, pp. II—893–896, Vol. 2.
33. N. Pramstaller, C. Rechberger, and V. Rijmen, A compact FPGA implementation of the hash function whirlpool, *Fourteenth ACM/SIGDA International Symposium on Field Programmable Gate Arrays—FPGA 2006*, ACM Press, 2006, pp. 159–166.

7 Side-Channel Analysis Attacks on Hardware Implementations of Cryptographic Algorithms

Siddika Berna Örs, Bart Preneel, and Ingrid Verbauwhede

CONTENTS

7.1 INTRODUCTION

Traditionally, the main task of cryptographic hardware is the acceleration of operations frequently used in cryptosystems or the acceleration of a complete cryptographic algorithm. In applications, hardware devices are also required to store secret or private keys securely. Hence, a cryptographic device must prevent the extraction of any sensitive information. A side-channel analysis (SCA) attack takes advantage of implementation-specific characteristics to recover the secret parameters involved in the computation. It is therefore less general, but often more powerful than classical cryptanalysis.

SCA attacks can be divided into two groups as active and passive attacks, according to the ability of the attacker. Active attacks targeting the keys in cryptographic devices are commonly referred to as tamper attacks; they have a long history in the field of cryptography [1]. In these attacks, the attacker has to reach the internal circuitry of the cryptographic device. There are two kinds of attacks: probing attack [2] and fault induction attack [3,4]. A probing attack consists of inserting sensors into the device, to directly examine the content of memory zones or the data circulating on a bus. A fault induction attack works by disturbing the behavior of the device to induce errors in the computation.

Passive attacks were recognized in the cryptographic community as a major threat in 1996, when the first article about timing attacks (TAs) [5] was published. In a passive attack, the adversary uses the standard functionality of the cryptographic device. The physical and the electrical effects of the functionality of the device are then used for the attack. There are many different types of effects. If these effects unintentionally deliver information about the key that is used inside the device, then they deliver side-channel information and are called side channels.

Passive attacks are divided into four groups according to the side-channel information that they exploit. Timing attacks (TA) exploit the timing information on the cryptographic hardware. Power attacks (PA) use the dynamic power consumption of the cryptographic hardware during the execution of the cryptographic algorithm. Electromagnetic attacks (EMA) use the electromagnetic (EM) radiation of the cryptographic hardware during the execution of the cryptographic algorithm. Acoustic (sound) analysis attacks exploit the sound coming out of the cryptographic hardware during the execution of the cryptographic algorithm. All the groups of the passive attacks have two types: simple and differential analysis attacks.

In this chapter, we introduce the passive attacks that we have conducted on the hardware implementations of an elliptic curve cryptosystem (ECC) over GF(p), the advanced encryption standard (AES), and the data encryption standard (DES). We also summarize the previous work on these side-channel attacks.

7.2 SIMPLE ATTACKS

In a simple analysis attack, an attacker uses the side-channel information from one measurement directly to determine parts of the secret key. A simple analysis attack exploits the relationship between the executed operations and the side-channel information.

7.3 DIFFERENTIAL ATTACKS

In differential analysis attack, many measurements are used to filter out noise. A differential analysis attack exploits the relationship between the processed data and the side-channel information.

In differential analysis attacks, an attacker uses a so-called hypothetical model of the attacked device. The quality of this model is dependent on the knowledge of the attacker. The model is used to predict several values for the side-channel information of a device.

These predictions are compared with the real, measured side-channel information of the device. Comparisons are performed by applying statistical methods on the data. We use the distance-of-mean test and the correlation analysis in our attacks shown in the following sections.

7.3.1 Distance-of-Mean Test

A distance-of-mean test begins by running the cryptographic algorithm for N random values of input. For each of the N inputs, I_i, a discrete time side-channel signal, $S_i[j]$, is collected and the corresponding output, O_i, may also be collected. The side-channel signal $S_i[j]$ is a sampled version of the side-channel output of the device during the portion of the algorithm that is attacked. The index i corresponds to the I_i that produced the signal and the index j corresponds to the time of the sample. The $S_i[j]$ is split into two sets using a partitioning function, $D(\cdot)$: $S_0 = \{S_i[j] | D(\cdot) = 0\}$, $S_1 = \{S_i[j] | D(\cdot) = 1\}$.

The next step is to compute the average side-channel signal for each set:

$$A_0[j] = \frac{1}{|S_0|} \sum_{S_i[j] \in S_0} S_i[j], A_1[j] = \frac{1}{|S_1|} \sum_{S_i[j] \in S_1} S_i[j],$$

where $|S_0| + |S_1| = N$. By subtracting the two averages, a discrete time differential side-channel bias signal, $T[j]$, is obtained $T[j] = A_0[j] - A_1[j]$. Selecting an appropriate D function results in a differential side-channel bias signal that can be used to verify guessed portions of the secret key.

7.3.2 Correlation Analysis

For the correlation analysis, the model predicts the amount of side-channel information for a certain moment of the execution. These predictions are correlated to the real side-channel information. This correlation can be measured with the Pearson correlation coefficient [6]. Let t_i denote the ith measurement data and T the set of measurements. Let p_i denote the prediction of the model for the ith measurement and P the set of such predictions. Then we calculate

$$C(T,P) = \frac{E(T \cdot P) - E(T) \cdot E(P)}{\sqrt{\mathrm{Var}(T) \cdot \mathrm{Var}(P)}} \quad -1 \leq C(T,P) \leq 1. \tag{7.1}$$

In Equation 8.1, $E(T)$ denotes the expected (average) measurement data of the set of measurements T and $\mathrm{Var}(T)$ denotes the variance of the set of measurements T. T and P are said to be uncorrelated, if $C(T, P)$ equals zero. Otherwise, they are said to be correlated. If their correlation is high, that is, if $C(T, P)$ is close to $+1$ or -1, it is usually assumed that the prediction of the model, and thus the key hypothesis, is correct.

7.4 TIMING ATTACKS

The differences in the processing time of a hardware or a software system may vary with the code sequence and the processed data sets. Checking time may, in unsecured systems, retrieve secret information [5,7].

An unsecured hardware or software system shows data dependencies because of differences in timing according to different operations executed. Addition and multiplication may be distinguished. Assume that we want to calculate the following operations $z = x + y$ and $z = x \times y$, with x and y which are m-bit binary numbers. The execution time of one of the implementations of the addition operation takes $T_A = m$ clock cycles. If we use this addition implementation as the basis for a multiplication implementation, then its execution time is $T_M = (3(m-1)m)/2$. Hence, for the same bit-length operands, the one with shorter execution time is an addition operation.

As the timing depends on the bit length of the operands, by just using the timing information of one operation, even the big values with higher bit length are distinguished from the smaller ones with smaller bit length. The same problem arises if the test of specific values and a following dependent branch in the program code is not secured.

7.4.1 Simple Timing Attack on FPGA Implementation of ECC

In this section, we conduct a simple timing attack (STA) against an field programmable gate array (FPGA) implementation of an ECC over GF(p) [8–10]. The basic operation for ECC algorithms is point or scalar multiplication, denoted as $Q = [k]P$, k is an integer, and P and Q are elliptic curve (EC) points. This operation can be calculated by using the double-and-add algorithm, as shown in Algorithm 1a.

Step 5 of Algorithm 1a is an EC point addition, and Step 3 of Algorithm 1a is an EC point doubling which can be realized by Algorithm 2 and Algorithm 3, respectively.

For EC point addition and also for EC point doubling, 14 states are needed. Because completing one $\pm$ operation takes a shorter time than one $*$ operation, the latency of one state is the same as for one $*$. Hence, the total execution time of an EC point addition is $14T_*$, with T_* the latency of one $*$. The total execution time of an EC point doubling is $8T_* + 6T_\pm$, with $T_\pm$ the latency of one $\pm$.

If we use Algorithm 1a for a 160-bit EC point multiplication with an ℓ-bit key with the most significant bit (MSB) of the key equals to 1, the latency of one point multiplication is

$$T_{\text{PMUL}} = (\ell - 1)T_{\text{PDB}} + (w - 1)T_{\text{PAD}} = (8\ell + 14w - 22)T_* + 6(\ell - 1)T_\pm,$$

where w is the Hamming weight of the binary representation of the key [11], and T_{PDB} and T_{PAD} are the latency of EC point doubling and addition, respectively. It means that somebody who knows the execution time of one $*$ and $\pm$ and can measure the execution time of one 160-bit EC point multiplication can learn the Hamming weight of the key by using the earlier expression. Hence, Algorithm 1a is vulnerable to simple TA attack due to the conditional branch at Step 4.

Algorithm 1. Elliptic Curve Point Multiplication: (a) Double-and-Add (b) Always Double-and-Add

(a)

Require: EC point $P = (x, y)$, integer k, $0 < k < M$, $k = (k_{\ell-1}, k_{\ell-2}, \ldots, k_0)_2, k_{\ell-1} = 1$ and M

Ensure: $Q = [k]P = (x', y')$

1: $Q \leftarrow P$
2: **for** i from $\ell - 2$ down to 0 **do**
3: $\quad Q \leftarrow 2Q$
4: $\quad$ **if** $k_i = 1$ **then**
5: $\quad\quad Q \leftarrow Q + P$
6: $\quad$ **end if**
7: **end for**

(b)

Require: EC point $P = (x, y)$, integer k, $0 < k < M$, $k = (k_{\ell-1}, k_{\ell-2}, \ldots, k_0)_2, k_{\ell-1} = 1$ and M

Ensure: $Q = [k]P = (x', y')$

1: $Q \leftarrow P$
2: **for** i from $\ell - 2$ down to 0 **do**
3: $\quad Q_1 \leftarrow 2Q$
4: $\quad Q_2 \leftarrow Q_1 + P$
5: $\quad$ **if** $k_i = 0$ **then**
6: $\quad\quad Q \leftarrow Q_1$
7: $\quad$ **else**
8: $\quad\quad Q \leftarrow Q_2$
9: $\quad$ **end if**
10: **end for**

Algorithm 2. Elliptic Curve Point Addition over GF(p)

Require: $P_1 = (x, y, 1, a)$, $P_2 = (X_2, Y_2, Z_2, V_2)$
Ensure: $P_1 + P_2 = P_3 = (X_3, Y_3, Z_3, V_3)$

1. $T_1 \leftarrow Z_2 * Z_2$
2. $T_2 \leftarrow x * T_1$
3. $T_1 \leftarrow T_1 * Z_2$ $T_3 \leftarrow X_2 - T_2$
4. $T_1 \leftarrow y * T_1$
5. $T_4 \leftarrow T_3 * T_3$ $T_5 \leftarrow Y_2 - T_1$
6. $T_2 \leftarrow T_2 * T_4$
7. $T_4 \leftarrow T_4 * T_3$ $T_6 \leftarrow T_2 + T_2$
8. $Z_3 \leftarrow Z_2 * T_3$ $T_6 \leftarrow T_4 + T_6$
9. $T_3 \leftarrow T_5 * T_5$
10. $T_1 \leftarrow T_1 * T_4$ $X_3 \leftarrow T_3 - T_6$
11. $V_3 \leftarrow Z_3 * Z_3$ $T_2 \leftarrow T_2 - X_3$
12. $T_3 \leftarrow T_5 * T_2$
13. $V_3 \leftarrow V_3 * V_3$ $Y_3 \leftarrow T_3 - T_1$
14. $V_3 \leftarrow a * V_3$

Algorithm 3. Elliptic Curve Point Doubling over GF(p)

Require: $P_1 = (X_1, Y_1, Z_1, V_1)$
Ensure: $2P_1 = P_3 = (X_3, Y_3, Z_3, V_3)$

1.	$T_1 \leftarrow Y_1 * Y_1$	$T_2 \leftarrow X_1 + X_1$
2.	$T_3 \leftarrow T_1 * T_1$	$T_2 \leftarrow T_2 + T_2$
3.	$T_1 \leftarrow T_2 * T_1$	$T_3 \leftarrow T_3 + T_3$
4.	$T_2 \leftarrow X_1 * X_1$	$T_3 \leftarrow T_3 + T_3$
5.	$T_4 \leftarrow Y_1 * Z_1$	$T_3 \leftarrow T_3 + T_3$
6.	$T_5 \leftarrow T_3 * V_1$	$T_6 \leftarrow T_2 + T_2$
7.		$T_2 \leftarrow T_6 + T_2$
8.		$T_2 \leftarrow T_2 + V_1$
9.	$T_6 \leftarrow T_2 * T_2$	$Z_3 \leftarrow T_4 + T_4$
10.		$T_4 \leftarrow T_1 + T_1$
11.		$X_3 \leftarrow T_6 - T_4$
12.		$T_1 \leftarrow T_1 - X_3$
13.	$T_2 \leftarrow T_2 * T_1$	$V_3 \leftarrow T_5 + T_5$
14.		$Y_3 \leftarrow T_2 - T_3$

To get rid of this weakness we use the algorithm proposed by Coron [12]: we execute always a point doubling and a point addition, independent of the value of the current key bit. After finishing both point operations, we select the needed result according to the value of the current key bit, as shown in Algorithm 1b.

If we use Algorithm 1b for a 160-bit EC point multiplication with an ℓ-bit key with the MSB of the key equals to 1, then the latency of one point multiplication is $T_{\text{PMUL}} = (\ell - 1)(T_{\text{PDB}} + T_{\text{PAD}}) = (\ell - 1) \quad (22T_* + 6T_\pm)$. This latency depends only on the key bit length ℓ, but not on the Hamming weight of the key.

7.4.2 Differential Timing Attack on Implementation of AES

In this section, we conduct a differential timing attack on a hardware implementation of AES [13]. One of the operations in AES is the S-Box() transformation in which the output byte is calculated by Algorithm 4. MultInv() operation in Algorithm 4 is the multiplicative inverse in the finite field GF(2^8) and AffTrans() operation is an affine transformation over GF(2), described in [13].

Algorithm 4. S-Box Operation in AES

Require: $in = (in_1 \; in_0)_8$
Ensure: $out = \text{S} - \text{Box}(in) = (out_1 \; out_0)_8$
1: **if** $in = (00)_8$ **then**
2: $\quad out = (00)_8$
3: **else**
4: $\quad out = \text{MultInv}(in)$
5: **end if**
6: $out = \text{AffTrans}(out)$

There are 16 S-Boxes in AES, and each takes 1 byte of the state as an input. The input of the first S-Box operation in the first round is the first byte of output of the AddRoundKey(Plaintext,Key) = Plaintext $\oplus$ Key. Step 2 is executed in shorter time than Step 4 in Algorithm 4. Hence, the attacker's steps are as follows in a differential timing attack:

1. Feed the hardware with N plaintexts.
2. Measure the time taken for encrypting each of them and form an $N \times 1$ matrix M_1 with these timing data.
3. Calculate $Plaintext_{15} \oplus Key_{15}$ for N plaintexts for each possible 256 values of the first byte of the key and for each plaintext.
4. Form an $N \times 256$ matrix M_2 with the expected time of S-box $(Plaintext_{15} \oplus Key_{15})$ operation.

Now the attacker should choose a statistical analysis method described in Section 8.3 for finding the first byte of the key. If he chooses the correlation analysis, then he should find the correlation between M_1 and each column of M_2. The highest correlation gives the right first byte of the key.

7.4.3 Previous Attacks

TA attacks have been important significantly in the last few years. In June 1998, a TA attack could be performed on a smart card, compromising a software test code for the Rivest, Shamir, Adelman (RSA) [14] public key cryptosystem. After analyzing 300,000 timing tests, a 512-bit RSA key could be determined. The overall time for this attack has been specified to be only a few minutes [15]. In this study, the individual bits of the RSA key were tested sequentially.

Schindler [16] demonstrated a further evolution of TA attack, breaking the barriers of the RSA Chinese remainder theorem (CRT) [17] applications. The attack can only be effectively performed if the so-called Montgomery Algorithm [18] is used for calculation of the RSA, and if CRT is used. This attack is improved by using an error-correction strategy in [19,20].

Handschuh and Heys [21] showed that the implementations of Rivest's RC5 [22] that take time for a rotation that is linear in the number of left shifts, are vulnerable to a TA attack. The attack recovers the extended secret key table with only 2^{20} ciphertexts from the sole knowledge of the total amount of rotations carried out during the encryption.

Hevia and Kiwi [23] studied the vulnerability of two implementations of the DES cryptosystem under a TA attack. They showed that a TA attack yields the Hamming weight of the key used by the DES implementations and that all the design characteristics of the target system could be inferred from timing measurements.

Koeune and Quisquater [24] explained how to perform a TA attack on Rijndael. They used the fact that MixColumn operation can be implemented very efficiently and the execution time can depend on the data processed.

The international data encryption algorithm (IDEA) is a product block cipher designed by Lai et al. [25]. IDEA can cryptanalyzed with a piece of side-channel information: whether one of the inputs into one of the multiplications is zero. Since the multiplication is done modulo $2^{16} + 1$, a zero operand is treated as a special case. Some implementations bypass the multiplication completely and simply patch in the correct value. Kelsey et al. [26] used this information and the ciphertexts for attacking the IDEA block cipher.

7.4.4 Countermeasures

The obvious countermeasure for a timing attack is executing the operations in constant time independent of the processed data. All the previous works in the literature try to solve this problem.

Most timing attacks exploit the modular reduction occurring in a Montgomery multiplication [18]. Therefore, Dhem [27], Walter [28,29], and Hachez and Quisquater [30] propose several countermeasures that typically consist of removing the time variation in this multiplication.

Kocher [5] suggests a countermeasure consisting of randomizing the exponent in RSA by adding a random multiple of $\varphi(n)$, a modification that does not effect the final result.

Using square-and-multiply-always algorithm during the exponentiation allows to hide the Hamming weight of the keys. Using double-and-add-always algorithm proposed by Coron [12] during the EC point multiplication allows to hide the Hamming weight of the keys. This countermeasure increases the computation time by about 30%.

As a second countermeasure against timing attack on ECC, Izu and Takagi [31,32] propose the binary right to left point multiplication algorithm by executing point addition and doubling in parallel.

7.5 POWER ATTACKS

Nowadays, complementary metal oxide semiconductor (CMOS) is by far the most commonly used technology to implement digital integrated circuits. The dominating factor for the power consumption of a CMOS gate is the dynamic power consumption [33]. The current absorbed by one inverter from V_{CC} shown in Figure 7.1 is used to charge the load capacitor C_L at the output of the inverter. The voltage on the load capacitor is the output level of the inverter either logic 0 (V_{CC} V) or 1 (0 V). The current–voltage relation of a capacitor is defined as $i_C(t) = C\mathrm{d}/\mathrm{d}t\ v(t)$. Hence, if the capacitor voltage does not change by the time then the current flow on the capacitor is zero, otherwise different from zero. The following transition situations can occur at the output:

$$\begin{aligned}
0 \rightarrow 0 \quad & i_{C_L}(t) = 0, \\
0 \rightarrow 1 \quad & i_{C_L}(t) = C_L V_{CC}, \\
1 \rightarrow 0 \quad & i_{C_L}(t) = -C_L V_{CC}, \\
1 \rightarrow 1 \quad & i_{C_L}(t) = 0.
\end{aligned}$$

If we measure the current absorbed from the source by an ampermeter connected between V_{CC} and p-channel metal oxide semiconductor (PMOS) transistor in Figure 7.1a, then we observe a current only during the $0 \rightarrow 1$ transition at the output of the inverter. This transition depends on the input of the inverter, so the processed data in the gate. The power attack uses this simple fact that by just observing the current consumption of a gate we can learn some information about the processed data, and if this data has some relation with the secret information then we gain some information about the secret by power analysis of the circuit.

7.5.1 SIMPLE POWER ATTACK ON FPGA IMPLEMENTATION OF ECC

In this section, we conduct a simple power attack (SPA) against an FPGA implementation of ECC over GF(p) [8–10]. The power consumption trace of

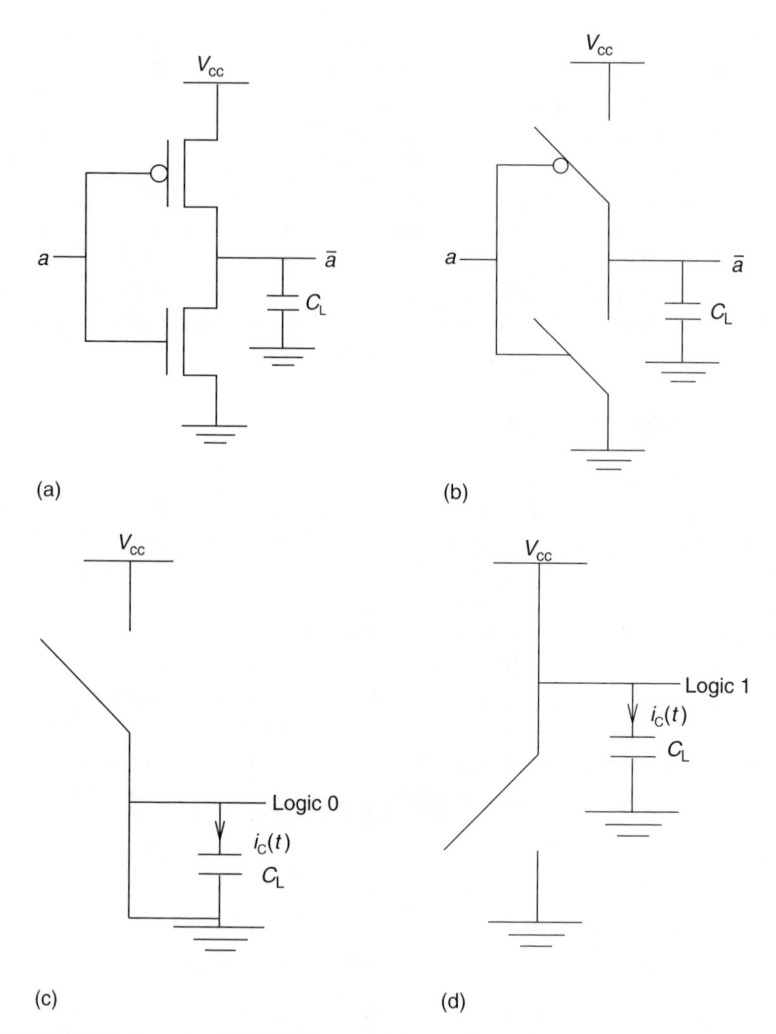

FIGURE 7.1 A static CMOS inverter: (a) transistor structure, (b) switch model, (c) input = 1, output = logic 0, and (d) input = 1, output = logic 1.

a 160-bit EC point multiplication is shown in Figure 7.2. The EC point multiplication is implemented with Algorithm 1a. It can be easily seen from Figure 7.2 that the key used during this measurement is 1001100.

We have changed our design to work with Algorithm 1b as a countermeasure for the attack given earlier [8]. The current consumption trace of one EC point multiplication is shown in Figure 7.2b. It follows from Figure 7.2b that it is no longer possible to attack this circuit by SPA.

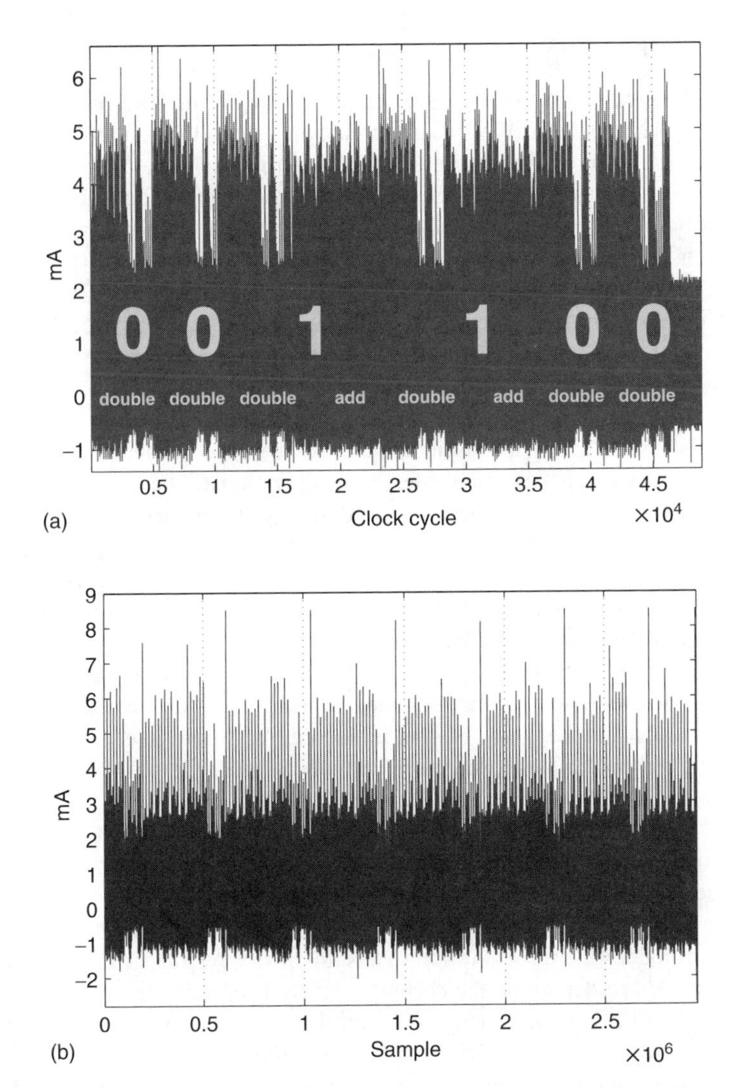

FIGURE 7.2 Power consumption trace of a 160-bit elliptic curve point multiplication over GF(p) with (a) Algorithm 1a and (b) Algorithm 1b.

7.5.2 Differential Power Attack on ASIC Implementation of AES

In this section, we present our differential power attack (DPA) on the application specific integrated circuit (ASIC) implementation of the AES [8,34,35]. The target for our DPA was the 8 MSBs of the state after the initial key addition operation.

7.5.2.1 DPA Using Simulated Data

We have tested our attack with simulated data before making real measurements. This approach enabled us to estimate the difficulty of a real attack, that is, an attack using real measurements. To predict the dynamic power consumption of the state, behavioral hardware description language (HDL) simulations of the ASIC implementation were used. An advantage of this approach is that it allows to simulate attacks in an early stage of the design flow. Another reason for using a HDL simulation was that we did not reset the chip after each AES execution. At the beginning of an AES execution, the state still contained some value related to the previous AES execution. Hence, without the HDL simulation, we could have only predicted the Hamming weight of the state, but not the dynamic power consumption.

In the first step of this simulated attack, we have produced a so-called simulated power consumption file. For this purpose, we have chosen N random plaintexts and one fixed, but random key. After each first encryption round (clock cycle), the simulator has written the total number of bit changes between the previous and the current values of the state to this file. Hence, the simulator has produced a file that contains an $N \times 1$ matrix ($N = 10{,}000$), M_1, with values between 0 and 128.

Then we calculated an $N \times 2^L$ matrix M_2. Each column of the matrix M_2 contains the prediction for the bit changes in the state for a particular guess of the L attacked key bits of the initial key addition. We calculate the correlation coefficients between the predictions of all the possible keys and M_1 as $c_i = C(M_1, M_2(1{:}N, i))$, $i = 0, \ldots, 2^L - 1$.

We expect that only one value, corresponding to the correct L-key bits, leads to a high correlation coefficient. Figure 7.3a shows that this is indeed the case.

We have already demonstrated that our attack setup works well together with our model. The only question that remains is how many measurements, N, are needed to determine the correct key. To determine this minimum, we have calculated the correlation coefficient between M_1 and M_2 for different values of $N{:}c_{i,j} = C(M_1, M_2(1{:}i, j))$, $i = 1, \ldots, 10{,}000$, $j = 0, \ldots, 2^L - 1$.

As shown in Figure 7.3b, after approximately 400 plaintexts, the correct L MSBs can be distinguished from the wrong L MSBs. Hence, for the simulated attack, 400 measurements are sufficient to find the correct L MSBs of the key.

7.5.2.2 DPA Using Measured Data

In this section, we present the results of our DPA on the ASIC implementation of AES using real, measured data. We have encrypted the same N plaintexts with the same key as that used in the first step of Section 7.5.2. The initial key addition operation occurs during the first clock cycle. The result of this operation is written into the state at the rising edge of the second clock

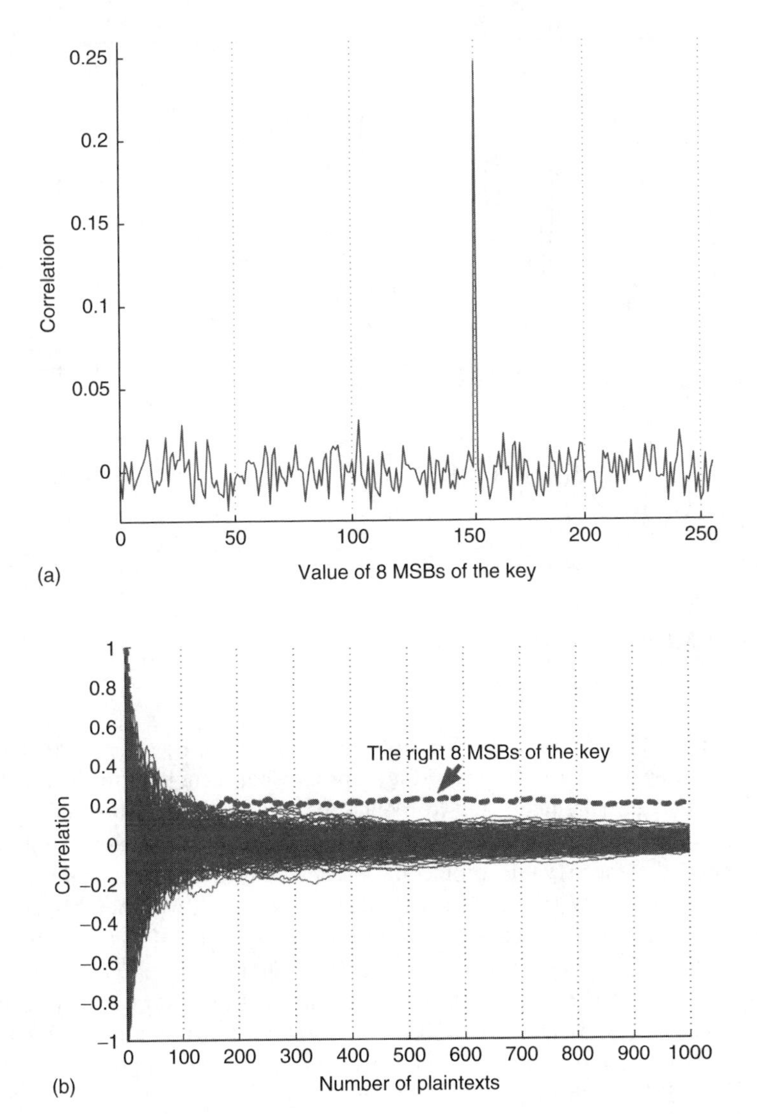

FIGURE 7.3 Correlation between M_1 and all the columns of M_2: (a) with 10,000 plaintexts (b) as a function of the number of measurements.

cycle. Hence, we have measured the current consumption during the first two clock cycles of the encryption operation. With these measurements, we have produced an $N \times 1000$ matrix, M_3. The power trace of one of these measurements is shown in Figure 7.4.

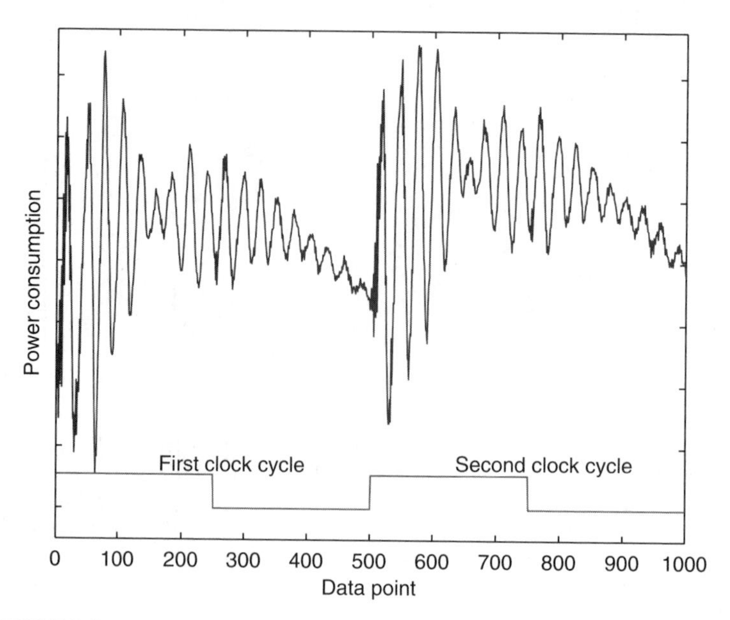

FIGURE 7.4 Power consumption trace of a measurement.

To identify the correct L MSBs of the key we have used the correlation coefficient again. We have applied a preprocessing technique to reduce the noise in our measurements and to reduce the amount of measurement data. The preprocessing technique essentially consists of averaging. We have calculated the mean value of the measurement data in the second clock cycle as follows: $M_4(i) = E(M_3(i, D+1{:}2D))$, $i = 1, \ldots, N$, where D is the number of data points measured during one clock cycle. $M_3(i, D+1{:}2D)$ is the vector that consists of the ith row and the columns between $D+1$ and $2D$ of M_3. We used these preprocessed measurements as input for our correlation analysis $c_i = C(M_4, M_2(1{:}N, i))$, $i = 0, \ldots, 2^L - 1$. As shown in Figure 7.5a, the highest correlation occurs at $i = 153$. This value corresponds to 0×99, which are the 8 MSBs of the key.

As in Section 7.5.2, N was taken as 10,000. However, we are interested in the smallest number of measurements that allow for a successful attack. To find the minimal number of measurements, we have calculated the following correlation coefficients: $c_{i,j} = C(M_4(1{:}i), M_2(1{:}i, j))$, $i = 1, \ldots, N$, $j = 0, \ldots, 2^L - 1$. It is shown in Figure 7.5b that after approximately 4000 measurements the correct and the wrong 8 MSBs of the key can be distinguished.

The attack with simulated data in Section 7.5.2 needs about 400 measurements to deduce the correct key. Taking into account that the 4000 measurements are the averages of 64,000 real measurements, we conclude that we need 160 times more data to deduce the correct key.

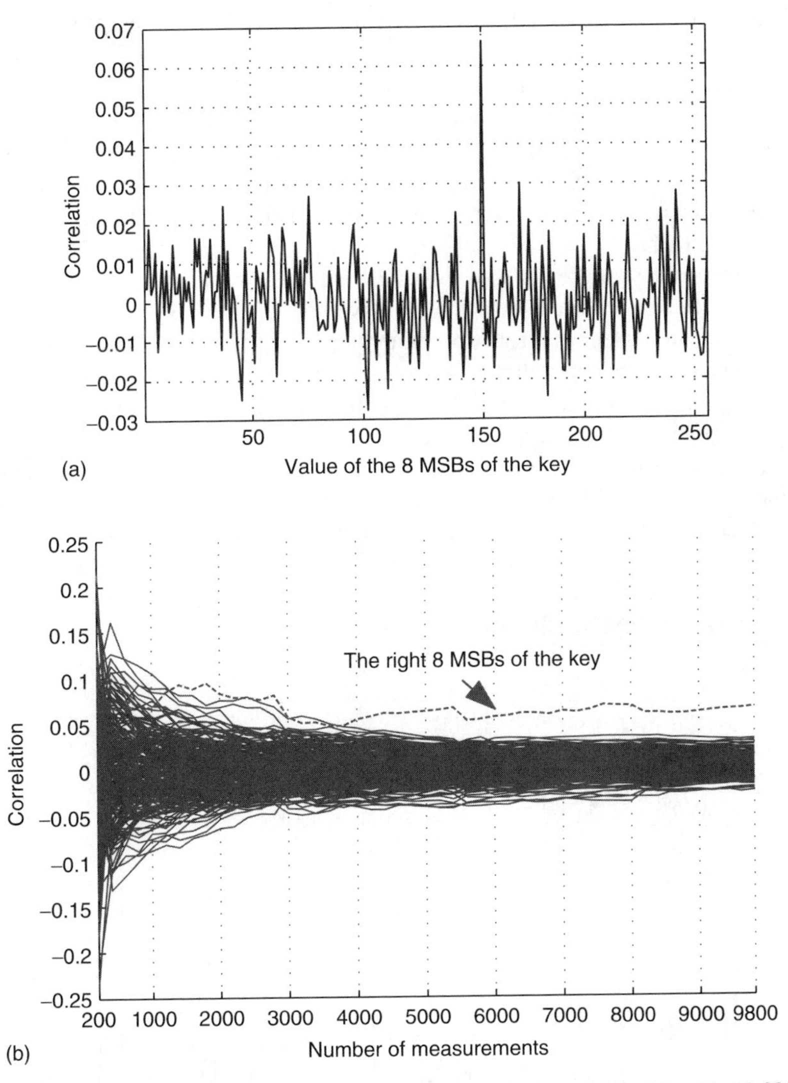

FIGURE 7.5 Correlation between all the columns of M_2 and M_4: (a) with 10,000 measurements (b) as a function of the number of measurements.

7.5.3 DPA on FPGA Implementation of DES

We show an example for DPA against the sequential DES [36] implementation of Rouvroy et al. [37] that takes one clock cycle to perform one round. We use correlation analysis to implement a DPA on the FPGA implementation of DES [8,38]. This time our target is the 4 MSBs of the register L that are affected by the 6 MSBs of the round key 16 [38]. It corresponds to the output bits of S-box $S0$.

The number N of measurements for this experiment was 4096. For each of the N encrypted plaintexts, we predict the number of bit transitions inside our target register bits between rounds 15 and 16 for the 2^6 key guesses. The result of the prediction is an $N \times 2^6$ matrix M_1 containing integers between 0 and 4.

Since the same key is used for all the measurements, the power consumption of the key schedule is fixed and may be considered as a DC component that we can neglect as a first approximation.

Then we measure the power consumption of the FPGA during each encryption for 16 clock cycles and we store the maximum value of each encryption cycle to an $N \times 16$ matrix M_2 for 16 rounds (clock cycles) of DES.

In the correlation phase, we compute the correlation coefficient between the column 16 of M_2 and all the columns of M_1. If the attack is successful, we expect that only one value, corresponding to the correct key guess, leads to a high correlation coefficient. As it is shown in Figure 7.6, the highest correlation occurs when the key guess is $1E_{hex} = 30_{dec}$. This value corresponds to the correct 6 MSBs of the round key 16.

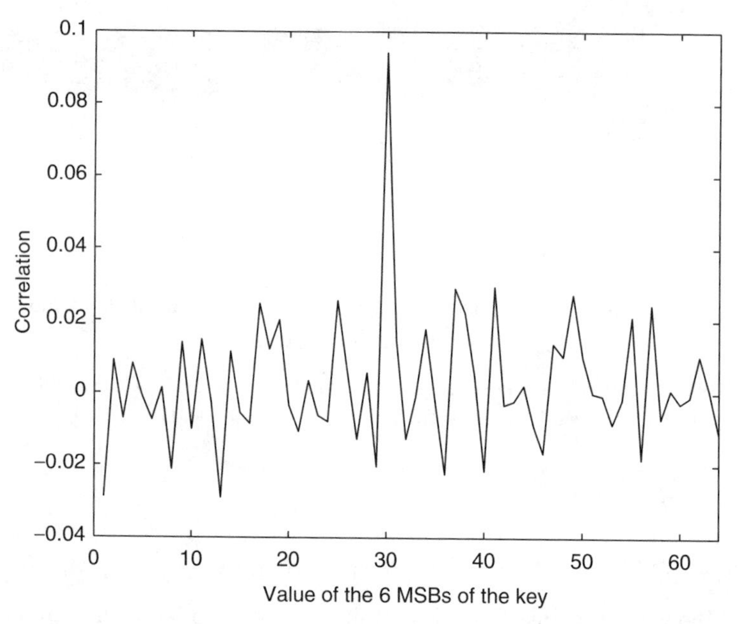

FIGURE 7.6 Correlation coefficient of all the 2^6 key guesses for the practical attack on the FPGA implementation of the DES ($N = 4096$).

7.5.4 Previous Attacks

The first practical implementation of a power analysis attack on the DES was reported by Kocher et al. [39]. Since then, several companies and universities have developed the skills to conduct these measurements in practice; these skills include knowledge about statistics, the properties of the attacked cryptographic algorithm, and the measurement setup. Koeune and Standaert [40] present the state of the art side-channel attacks.

Schindler et al. [41] present an approach to optimize the efficiency of differential side-channel cryptanalysis against block ciphers by advanced stochastic methods. They demonstrate that the adaptation of probability densities is clearly advantageous regarding the correlation method, especially, if multiple leakage signals at different times can be jointly evaluated.

Tiri and Verbauwhede [42] point out that an actual DPA must be performed with the correct accuracy on the power simulation model, as the quality of the resistance assessment of a countermeasure is only as good as the simulation model.

Mangard [43] describes how to determine the complete secret key in Rijndael by using Hamming weight information from a few subkeys. Novak [44] shows a side-channel attack on a substitution block of Rijndael, which is usually implemented as a table lookup operation. The attack is based on identifying equal intermediate results from power measurements, while the actual values of these intermediates remain unknown.

Biham and Shamir [45] present an attack which can determine the secret key of the DES uniquely by attacking several subkeys. Messerges et al. [46] review and analyze the power analysis techniques used to attack DES.

Megarajan [47] proposes an attack based on the comparison of the repeated parts of an algorithm. Joye et al. [48] provide an analysis of second-order DPA and compute what one expects from second-order attacking any randomized algorithm.

Mangard et al. [49] show that glitches occurring in circuits of masked gates make these circuits susceptible to classical first-order DPA. They provide a thorough theoretical analysis of the DPA resistance of masked gates in the presence of glitches and simulation results that confirm the theoretical elaborations. Mangard et al. [50] have mounted attacks on the output of logic gates. Based on simulations and physical measurements, they show that the unmasked and masked implementations leak side-channel information because of glitches at the output of logic gates. It turns out that masking the AES S-Boxes does not prevent DPA, if glitches occur in the circuit.

Peeters et al. [51] describe an improvement of the previously introduced higher-order techniques allowing to defeat masked implementations. The proposed technique is based on the efficient use of the statistical distributions of the power consumption in an actual design.

7.5.5 Countermeasures

The goals of power analysis countermeasures are reducing the correlation between the power consumption data and the secret data and obscuring the power consumption measurements. There are two different types of countermeasures as software and hardware. Surveys about the countermeasures are given in [52,53].

7.5.5.1 Software Countermeasures

Time randomization was discussed in [12,32,54–63]. In this type of countermeasures, the operations occur during random intervals in an execution. This is done by using no-operations (NOPs), using dummy variables and instructions, and data balancing (representation of the data is done to make the Hamming weight constant). Permuting the execution (rearranged instructions) is proposed by Goubin and Patari [64].

Masking techniques are studied in [64–75]. Gomulkiewicz and Kutylowski [76] show that masking is not always useful. The authors present an attack against an addition implementation, based on the observation of the Hamming weight of the sequence of carry that occurs during the bitwise addition. Apart from the efficiency of this attack, of more interest is the fact that this attack is not hindered by masking; in fact, the authors note that this could even make the attack easier.

To obtain DPA-resistant applications, it cannot be tolerated that the software or hardware performs many cryptographic operations on known inputs with the same secret information. In addition, not too many cryptographic operations should occur on the inputs that vary according to a known scheme with keys that vary according to a known scheme. Borst and Bosselaers et al. [53,77] demonstrated how to take countermeasures at the protocol level. They proposed to use more key levels in a typical smart card application.

7.5.5.2 Hardware Countermeasures

Increasing the measurement noise was the idea of Kocher et al. [39] by a hardware noise generator as a random number generator (RNG). The design of this approach may be relatively simple, and it is an effective way to resist power analysis attacks. But it is expensive to implement and might be easy to disable through tampering and it is not energy efficient.

Shamir [78] and Coron and Goubin [66] proposed power signal filtering to obscure the measurements. The design of this approach may be relatively simple and it is an effective way to resist attacks, but it requires a change to the hardware and might be easy to disable through tampering. Two types of filters were proposed; a passive filter in which physical limitations restrict the size of an on-chip capacitor and an active filter in which compensation techniques are likely to lag behind power supply changes. This countermeasure does not hide the EM radiation information of the device. The source of the EM radiation of the device is the internal current flow on the wires of the device, and this countermeasure does not change the current flow that depends on the processed data.

There are also novel circuit designs. Shamir [78] proposed detachable power supplies. Securing algorithm at the logic level was the idea of Tiri and Verbauwhede [79]. The method employs logic gates with a power consumption, which is independent of the data signals and therefore the technique removes the foundation for DPA. Fischer and Gammel [80] refine the model for the power consumption of CMOS gates, taking into account the side channel of glitches. They propose a family of masked gates, which is theoretically secure in the presence of glitches if certain practically controllable implementation constraints are imposed. Mace et al. [81] present principles and concepts for the secured design of cryptographic integrated circuits (ICs). To achieve a secure implementation of those structures, they propose to use a binary decision diagrams (BDDs) approach to design and determine the secured structures in dynamic current mode logic. Popp and Mangard [82] describe a novel SCA-resistant logic style called masked dual-rail precharged logic (MDPL). It is a masked and dual-rail precharge logic style and can be implemented using common CMOS standard cell libraries.

Asynchronous circuits are used as a countermeasure [83,84]. The power consumption and EM radiation are reduced, but the execution time depends on the data processed, so they are vulnerable to timing attacks. Golic [85] used reversible logic to reverse computation, which returns the consumed energy during the computation back to the circuit.

Mangard [86] has identified the hardware countermeasures that influence the number of samples needed in DPA. Based on these properties, he proposed formulas that allow the calculation of lower bounds for the number of samples needed in DPA.

7.6 ELECTROMAGNETIC ATTACKS

The sudden current pulse that occurs during the transition of the output of a CMOS gate, mentioned in Section 7.5, causes a sudden variation of the EM field surrounding the chip, which can be monitored by inductive probes that are particularly sensitive to the related impulse. The electromotive force across the sensor (Lentz law) relates to the variation of magnetic flux as follows [87]:

$$V = -\frac{\mathrm{d}\phi}{\mathrm{d}t} \quad \text{and} \quad \phi = \int\!\!\int \vec{B} \cdot \mathrm{d}\vec{A},$$

where V is the output voltage of the probe, ϕ is the magnetic flux sensed by probe, t is the time, $\vec{B}$ is the magnetic field, and $\vec{A}$ is the perpendicular area that it penetrates.

The Biot–Savart law relates magnetic fields to their current sources. Finding the magnetic field resulting from a current distribution involves the vector product, and is inherently a calculus problem when the distance from the current to the field point is continuously changing:

$$d\vec{B} = \frac{\mu_0 I d\vec{L} \times \vec{r}}{4\pi r^2},$$

where $d\vec{L}$ is length of conductor carrying electric current I and $\vec{r}$ is unit vector to specify the direction of the vector distance r from the current to the field point.

EM radiation itself consists of two components, the electrical and the magnetic field vectors [88]. In theory, both components can be measured individually or in their interaction. Capacitive sensors mainly capture the electrical field components, while antennas and coils are able to acquire both electrical and magnetic components, and Hall sensors and so-called SQUIDs (super conducting quantum interference devices) mainly detect the pure magnetic field components.

7.6.1 Simple Electromagnetic Attack on FPGA Implementation of ECC

In this section, we conduct a simple electromagnetic analysis (SEMA) attack on an FPGA implementation of an EC processor over GF(p) [8–10]. In our measurement, we connect an antenna directly to an oscilloscope, as shown in Figure 7.7 [8,89–91].

The EM radiation trace of a 160-bit EC point multiplication is shown in Figure 7.8 [8,89,90]. The SEMA attack is implemented on an EC processor over GF(p) [8–10], which uses Algorithm 1a for EC point multiplication. It can be easily seen from Figure 7.8 that the key used during this measurement is 11001100, because there is a clear difference between the traces of EC point addition and doubling. The SEMA attack was successful because of the conditional branch in Step 4 of Algorithm 1a.

As a countermeasure to this attack, we have implemented the EC point multiplication with Algorithm 1b. One EM measurement of this architecture is shown in Figure 7.9a.

7.6.2 Differential Electromagnetic Attack on FPGA Implementation of ECC

In this section, we conduct a differential electromagnetic attack (DEMA) on an FPGA implementation of an EC processor over GF(p) [10]. The EM radiation trace of one EC point multiplication is shown in Figure 7.9a.

FIGURE 7.7 The measurement setup. The loop antenna is placed vertically on the FPGA.

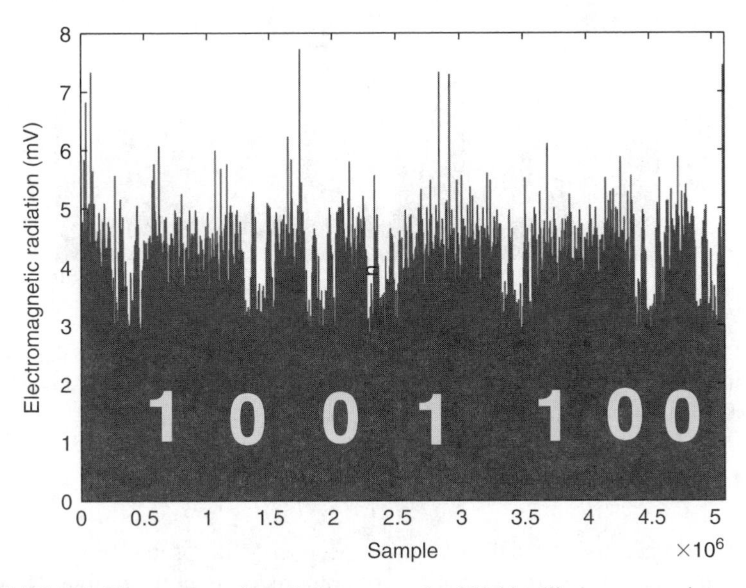

FIGURE 7.8 Electromagnetic radiation trace of a 160-bit elliptic curve point multiplication over GF(p) with Algorithm 1a.

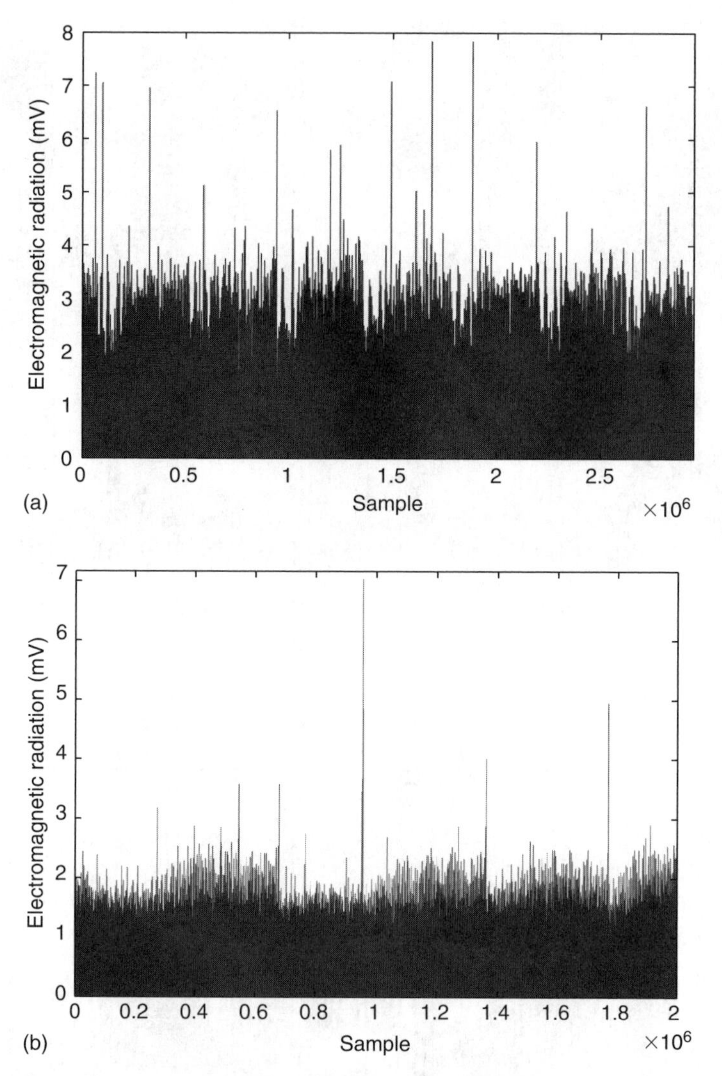

FIGURE 7.9 Electromagnetic radiation trace of a 160-bit elliptic curve point multiplication (ECPM) over GF(p) with Algorithm 1b: (a) complete and (b) around the attack point.

The target for our DEMA is the second MSB of the key, k_{l-2}, in Algorithm 1b. There are two temporary point registers in the design, Q_1 and Q_2. These temporary points and the output point Q are updated in the following order: $Q = P$, $Q_1 = 2P$, $Q_2 = 3P$,

$$Q = \begin{cases} 2P & \text{if } k_{l-2} = 0, \\ 3P & \text{if } k_{l-2} = 1, \end{cases} \qquad Q_1 = \begin{cases} 4P & \text{if } k_{l-2} = 0. \\ 6P & \text{if } k_{l-2} = 1. \end{cases}$$

Our choice for the measurement point is the fifth spike shown in Figure 7.9b. This spike corresponds to the second update of Q_1 after the second EC point doubling.

We have produced an EM radiation file. For this purpose, we have chosen N random points on the EC and one fixed, but random key, k. The FPGA executes N point multiplications such that $Q_i = [k]P_i$, for $i = 1, 2, \ldots, N$. We have measured the EM radiation of the FPGA during 2400 clock cycles around the second update of Q_1. The clock frequency applied to the chip was around 300 kHz, and the sampling frequency of the oscilloscope was 250 MHz. With these measurements, we have produced M_1, in which $M_1(i)$ is the ith measurement. The EM radiation trace of one of these measurements is shown in Figure 7.9b.

We have applied a preprocessing technique to reduce the amount of measurement data in every clock cycle. We have found the maximum value of the measurement data in each clock cycle and taken the data in 20 clock cycles around the clock cycles that correspond to the five spikes in Figure 7.9b. Thus, M_2 has 100 columns and N rows. We used the discrete Fourier transform to find the exact clock frequency and the number of samples per clock cycle.

We have implemented the EC point multiplication with Algorithm 1b in the C programming language. During the execution of the EC point multiplications, the C program computes the number of bits that change from 0 to 1 in some registers at the step corresponding to the fifth spike shown in Figure 7.9b. The number of transitions is used as the EM radiation prediction.

We have produced two EM radiation prediction matrices, M_3 and M_4, for the $k_{l-2} = 0$ and $k_{l-2} = 1$ guesses, respectively. M_3 and M_4 have one column for the fifth spike and N rows for the N EC points. We use the prediction matrices M_3 (for $k_{l-2} = 0$ guess) and M_4 (for $k_{l-2} = 1$ guess) to split the measurements in M_2 into sets. For each guess, we divide the N measurements into two sets. First, we calculate the mean value of the prediction matrix M_3, $E(M_3)$. Measurement by measurement, we check if the predicted value is lower than the average value. If so, we put the measurement in set $S_{1,1}$, otherwise in set $S_{1,2}$. Then we calculate the mean value for each of the two sets and calculate the bias signal as $T_1 = E(S_{1,2}) - E(S_{1,1})$. We do the same for the prediction matrix M_4, the sets are now called $S_{2,1}$ and $S_{2,2}$ and the bias signal is T_2.

The current consumption bias signals for $k_{l-2} = 0$ and $k_{l-2} = 1$ guesses are shown in Figure 7.10. The figure shows a high peak on the expected spot on the trace for the $k_{l-2} = 1$ guess. Hence, the decision for the right key bit is equal to 1.

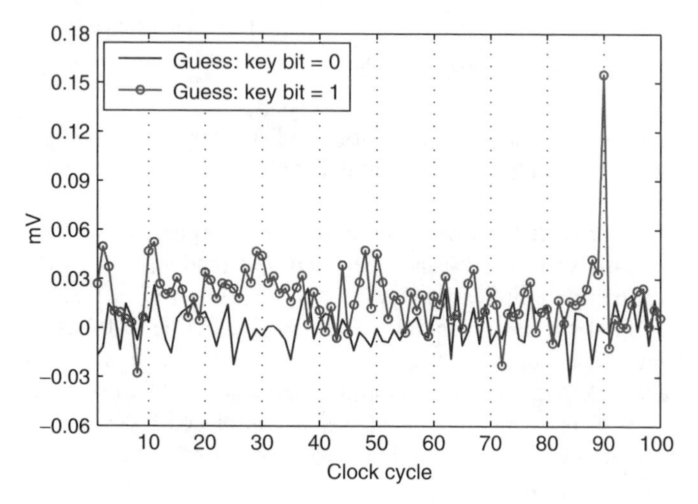

FIGURE 7.10 Electromagnetic radiation bias signals for the $k_{l-2}=0$ and $k_{l-2}=1$ guesses.

Figure 7.11 shows the change in the amplitude of all the clock cycles of the current consumption bias signals for the $k_{l-2}=1$ guess. The number of measurements on these traces is the number of measurements in the sets $S_{2,1}$, $S_{2,2}$ described earlier. The number of measurements in these sets is nearly the same. Hence, we should multiply the number of measurements seen in

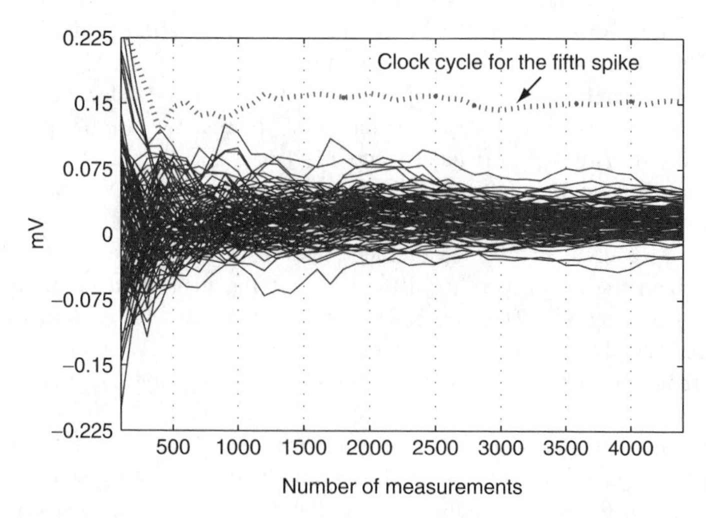

FIGURE 7.11 Change in the amplitude of the electromagnetic radiation bias signal for the $k_{l-2}=1$ guess and all clock cycles.

Figure 7.11 by two to find the needed number of measurements. As it is shown in Figure 7.11, 2000 measurements are needed to distinguish the right clock cycle from the wrong ones.

7.6.3 Previous Attacks

It is well known that the U.S. government has been aware of EM leakage since the 1950s. The resulting standards are called TEMPEST; partially declassified documents can be found in [92]. The first published papers are work of Quisquater and Samyde [93] and the Gemplus team [94]. Quisquater and Samyde showed that it is possible to measure the EM radiation from a smart card. Their measurement setup consisted of a sensor which was a simple flat coil, a spectrum analyzer or an oscilloscope, and a Faraday cage. Quisquater and Samyde also introduced the terms SEMA and DEMA. The work of Gemplus deals with experiments on three algorithms: DES, RSA, and COMP128. They observed the feasibility of EMAs and compared them with PA in favor of the first. Namely, EM emanation can also exploit local information and, although more noisy, the measurements can be performed from a distance. This fact broadens the spectrum of targets to which SCA attacks can be applied. They are not limited to smart cards and similar tokens but also include secure sockets layer (SSL) accelerators and many other cryptographic devices.

According to Agrawal et al., there are two types of emanations: intentional and unintentional [95,96]. The first type results from direct current flows. The second type is caused by various couplings, modulations (amplitude modulation (AM) and frequency modulation (FM)), and so on. The two papers mentioned earlier deal exclusively with intentional emanations. On the contrary, the real advantage over other SCA attacks lies in exploring unintentional emanations [95,96]. More precisely, EM leakage consists of multiple channels. Therefore, compromising information can be available even for DPA-resistant devices that can be detached from the measurement equipment.

Mangard [97] showed that near-field EM attacks can be conducted even with a simple handmade coil. In addition, he showed that measuring the far-field emissions of a smart card connected to a power supply unit also suffices to determine the secret key used in the smart card.

Carlier et al. [98] showed that EM side channels from an FPGA implementation of AES can be effectively used by an attacker to retrieve some secret information. They worked close to the FPGA and by this way were able to get rid of the effects of other computations made at the same time. They also introduced a new Square EM Attack.

Until now, most papers on EMA applied similar techniques as power analysis while apparently much more information is available to be explored. It is likely that future work also deals with combinations of EMA with other side-channel attacks.

7.6.4 Countermeasures

Very few articles describe countermeasures against an EMA analysis. A complete shielding of smart card controllers, known from devices used in electronic data processing, is possible, but an attacker could simply remove the shield before analysis, making this countermeasure worthless [88].

With these presumptions in mind, EMA countermeasures have to reach much further than the commonly known PA defense systems, because of the fact that EMAs may provide information about small chip areas, whereas the PA measurement only yields data concerning the supply current of the complete chip.

7.7 ACOUSTIC ATTACKS

Recently, Shamir and Tromer [99] present their results using the sound of a central processing unit (CPU) as a side-channel information. The oldest eavesdropping channel, namely acoustic emanations, has received little attention. Preliminary analysis of Shamir and Tromer of acoustic emanations from personal computers shows them to be a surprisingly rich source of information on CPU activity.

Several desktop and laptop computers have been tested and in all cases it was possible to distinguish an idle CPU from a busy CPU. For some computers, it was also possible to distinguish various patterns of CPU operations and memory access. This can be observed for artificial cases (e.g., loops of various CPU instructions) and also for real-life cases (e.g., RSA decryption).

A low-frequency (kHz) acoustic source can yield information on a much faster (GHz) CPU in two ways. First, when the CPU is carrying out a long operation, it may create a characteristic acoustic spectral signature. Second, temporal information about the length of each operation is learnt and this can be used to mount TA, especially when the attacker can affect the input to the operation.

7.7.1 Countermeasures

One obvious countermeasure is to use sound dampening equipment, such as sound-proof boxes, that are designed to sufficiently attenuate all relevant frequencies. Conversely, a sufficiently strong wide-band noise source can mask the informative signals, though ergonomic concerns may render this unattractive. Careful circuit design and high-quality electronic components can probably reduce the emanations. Alternatively, one can employ known algorithmic techniques to reduce the usefulness of the emanations to the attacker. These techniques ensure the rough-scale behavior of the algorithm is independent of the inputs it receives; they usually carry some performance penalty, but are often already used to thwart other side-channel attacks.

REFERENCES

1. R. Anderson and M. Kuhn. Tamper resistance—a cautionary note. In D. Tygar, editor, *Proceedings of the 2nd USENIX Workshop on Electronic Commerce*, pp. 1–11, Oakland, CA, November 18–21, 1996.
2. O. Kommerling and M.G. Kuhn. Design principles for tamper resistant smartcard processors. In *Proceedings of the USENIX Workshop on Smartcard Technology*, Chicago, IL, May 10–11, 1999.
3. D. Boneh, R.A. DeMillo, and R.J. Lipton. On the importance of checking cryptographic protocols for faults (extended abstract). In W. Fumy, editor, *Advances in Cryptology: Proceedings of EUROCRYPT'97*, volume 1233 of *Lecture Notes in Computer Science*, pp. 37–51, Konstanz, Germany, May 11–15, 1997. Springer-Verlag.
4. M. Joye, A.K. Lenstra, and J.-J. Quisquater. Chinese remaindering based cryptosystem in the presence of faults. *Journal of Cryptology*, 4(12):241–245, 1999.
5. P. Kocher. Timing attacks on implementations of Diffie-Hellman, RSA, DSS and other systems. In N. Koblitz, editor, *Advances in Cryptology: Proceedings of CRYPTO'96*, volume 1109 of *Lecture Notes in Computer Science*, pp. 104–113, Santa Barbara, CA, August 18–22, 1996. Springer-Verlag.
6. G.M. Clarke and D. Cooke. *A Basic Course in Statistics*. Arnold London, 4th ed., 1998.
7. M. Janke and P. Laackmann. Power and timing analysis attacks against security controllers. Auinfineon Technologies Publication, Secure, 5/2002, pp. 40–44.
8. S.B. Ors. Hardware Design of Elliptic Curve Cryptosystems and Side-Channel Attacks. PhD thesis, Katholieke Universiteit Leuven, Faculteit Toegepaste Wetenschappen, Departement Elektrotechniek, Kasteelpark Arenberg 10, 3001 Leuven (Heverlee), Belgium, February 2005.
9. S.B. Ors, L. Batina, and B. Preneel. Hardware implementation of elliptic curve processor over *GF(p)*. New European Schemes for Signatures, Integrity and Encryption (NESSIE) working paper NES/DOC/KUL/WP5/023, Department of Electrical Engineering, ESAT/COSIC, Katholieke Universiteit Leuven, October 10, 2002.
10. S.B. Ors, L. Batina, B. Preneel, and J. Vandewalle. Hardware implementation of an elliptic curve processor over *GF(p)*. In *IEEE 14th International Conference on Application-specific Systems, Architectures and Processors (ASAP)*, pp. 433–443, The Hague, The Netherlands, June 24–26, 2003.
11. N. Mentens, P. Rommens, and M. Verhelst. Timing and Power Analysis Attacks on the Hardware Implementation of Elliptic Curve Cryptosystems Over GF(p) and GF (2^m). Master's thesis, Katholieke Universiteit Leuven, Departement Elektrotechniek—ESAT, Kasteelpark Arenberg 10, B 3001 Heverlee, Belgium, May 2003.
12. J.-S. Coron. Resistance against differential power analysis for elliptic curve cryptosystems. In Ç.K. Koç and C. Paar, editors, *Proceedings of the 1st International Workshop on Cryptographic Hardware and Embedded Systems (CHES)*, volume 1717 of *Lecture Notes in Computer Science*, pp. 292–302, Worcester, MA, August 12–13, 1999. Springer-Verlag.
13. National Institute of Standards and Technology. FIPS 197: Advanced Encryption Standard, November 2001.

14. R.L. Rivest, A. Shamir, and L. Adleman. A method for obtaining digital signatures and public-key cryptosystems. *Communications of the ACM*, 21(2):120–126, 1978.
15. J.F. Dhem, F. Koeune, P.A. Leroux, P. Mestre, J.-J. Quisquater, and J.L. Willems. A practical implementation of the timing attack. Technical Report CG-1998/1, UCL Crypto Group, Universite Catholique de Louvain, Belgium, 1998.
16. A. Menezes, P. van Oorschot, and S. Vanstone. *Handbook of Applied Cryptography*. CRC Press, 1997.
17. W. Schindler. A timing attack against RSA with the Chinese remainder theorem. In C. Paar and Çetin Koç, editors, *Proceedings of the 2nd International Workshop on Cryptographic Hardware and Embedded Systems (CHES)*, volume 1965 of *Lecture Notes in Computer Science*, pp. 109–124, Worcester, MA, August 17–18, 2000. Springer-Verlag.
18. P. Montgomery. Modular multiplication without trial division. *Mathematics of Computation*, 44:519–521, 1985.
19. G. Hachez, F. Koeune, and J.-J. Quisquater. Timing attack: what can be achieved by a powerful adversary? In A. Barbé, E.C. van der Meulen, and P. Vanroose, editors, *Proceedings of the 20th Symposium on Information Theory in the Benelux*, pp. 63–70, May 1999.
20. W. Schindler, F. Koeune, and J.-J. Quisquater. Unleashing the full power of timing attack. Technical Report CG-2001/3, UCL Crypto Group, 2001.
21. H. Handschuh and H.M. Heys. A timing attack on RC5. In S.E. Tavares and H. Meijer, editors, *Proceedings of Selected Areas in Cryptography (SAC)*, volume 1556 of *Lecture Notes in Computer Science*, pp. 306–318, Kingston, Ontario, Canada, August 17–18, 1998. Springer-Verlag.
22. R.L. Rivest. The RC5 encryption algorithm. In Bart Preneel, editor, *Proceedings of the 2nd International Workshop on Fast Software Encryption (FSE)*, volume 1008 of *Lecture Notes in Computer Science*, pp. 86–96, Leuven, Belgium, December 14–16, 1994. Springer-Verlag.
23. A. Hevia and M.A. Kiwi. Strength of two data encryption standard implementations under timing attacks. In C.L. Lucchesi and A.V. Moura, editors, *Proceedings of the 3rd Latin American Symposium on Theoretical Informatics*, volume 1380 of *Lecture Notes in Computer Science*, pp. 192–205, Campinas, Brazil, April 20–24, 1998. Springer-Verlag.
24. F. Koeune and J.-J. Quisquater. A timing attack against Rijndael. Technical Report CG-1999/1, UCL Crypto Group, Louvain-la-Neuve, 1999.
25. X. Lai, J.L. Massey, and S. Murphy. Markov ciphers and differential cryptanalysis. In D.W. Davies, editor, *Advances in Cryptology: Proceedings of EUROCRYPT'91*, volume 547 of *Lecture Notes in Computer Science*, p. 1738, Brighton, UK, April 1991. Springer-Verlag.
26. J. Kelsey, B. Schneier, D. Wagner, and C. Hall. Side channel cryptanalysis of product ciphers. *Journal of Computer Security*, 8(2/3), pp. 141–158, 2000.
27. J.F. Dhem. Design of an Efficient Public-Key Cryptographic Library for RISC-Based Smart Cards. PhD thesis, Universite Catholiqué de Louvain, UCL Crypto Group, Laboratoire de microelectronique (DICE), May 1998.
28. C.D. Walter. Montgomery exponentiation needs no final subtraction. *Electronic Letters*, 35(21):1831–1832, October 1999.
29. C.D. Walter. MIST: An efficient, randomized exponentiation algorithm for resisting power analysis. In B. Preneel, editor, *Proceedings of RSA 2002*

Cryptographers' Track, volume 2271 of *Lecture Notes in Computer Science*, pp. 53–66, San Jose, CA, February 18–22, 2002. Springer-Verlag.
30. G. Hachez and J.-J. Quisquater. Montgomery exponentiation with no final subtractions: Improved results. In Ç.K. Koç and C. Paar, editors, *Proceedings of the 2nd International Workshop on Cryptographic Hardware and Embedded Systems (CHES)*, volume 1965 of *Lecture Notes in Computer Science*, pp. 293–301, Worcester, MA, August 17–18, 2000.
31. T. Izu and T. Takagi. A fast parallel elliptic curve multiplication resistant against side channel attacks. Technical Report CORR 2002–03, the Centre for Applied Cryptographic Research (CACR), University of Waterloo, 2002.
32. T. Izu and T. Takagi. A fast parallel elliptic curve multiplication resistant against side channel attacks. In D. Naccache and P. Paillier, editors, *Proceedings of the 5th International Workshop on Practice and Theory in Public Key Cryptosystems (PKC)*, volume 2274 of *Lecture Notes in Computer Science*, pp. 280–296, Paris, France, February 12–14, 2002. Springer-Verlag.
33. S.-M. Kang and Y. Leblebici. *CMOS Digital Integrated Circuits: Analysis and Design*. McGraw Hill, 2002.
34. S.B. Ors, F.K. Gürkaynak, E. Oswald, and B. Preneel. Power-analysis attack on an ASIC AES implementation. *Embedded Cryptographic Hardware: Design and Security*. Nova Science, NY, 2004.
35. S.B. Ors, F.K. Gürkaynak, E. Oswald, and B. Preneel. Power-analysis attack on an ASIC AES implementation. In *Proceedings of the International Conference on Information Technology (ITCC)*, pp. 546–552, Las Vegas, NV, April 5–7, 2004.
36. National Institute of Standards and Technology. FIPS 46–3: Data Encryption Standard, October reaffirmed 1999.
37. G. Rouvroy, F.-X. Standaert, J.-J. Quisquater, and J.D. Legat. Design strategies and modified descriptions to optimize cipher FPGA implementations: Fast and compact results for DES and triple-DES. In *Proceedings of the 9th International Workshop on Field-Programmable Logic and Applications (FPL)*, volume 2778 of *Lecture Notes in Computer Science*, pp. 181–193, Lisbon, Portugal, September 1–3, 2003. Springer-Verlag.
38. F.-X. Standaert, S.B. Örs, B. Preneel, and J.-J. Quisquater. Power analysis attacks against FPGA implementations of DES. In *Proceedings of International Conference on Field-Programmable Logic and Its Applications (FPL), Lecture Notes in Computer Science*, Antwerp, Belgium, August 30 to September 1, 2004. Springer-Verlag.
39. P. Kocher, J. Jaffe, and B. Jun. Differential power analysis. In M. Wiener, editor, *Advances in Cryptology: Proceedings of CRYPTO'99*, volume 1666 of *Lecture Notes in Computer Science*, pp. 388–397, Santa Barbara, CA, August 15–19, 1999. Springer-Verlag.
40. F. Koeune and F.-X. Standaert. A tutorial on physical security and side-channel attacks. In A. Aldini, R. Gorrieri, and F. Martinelli, editors, *Proceedings of the Foundations of Security Analysis and Design III: FOSAD*, volume 3655 of *Lecture Notes in Computer Science*, pp. 78–108, September 2005. Springer-Verlag.
41. W. Schindler, K. Lemke, and C. Paar. A stochastic model for differential side channel cryptanalysis. In J.R. Rao and B. Sunar, editors, *Proceedings of the 7th International Workshop on Cryptographic Hardware and Embedded Systems (CHES)*, volume 3659 of *Lecture Notes in Computer Science*, pp. 30–46, Edinburgh, UK, August 29 to September 1, 2005. Springer-Verlag.

42. K. Tiri and I. Verbauwhede. Simulation models for side-channel information leaks. In *Proceedings of the 42nd Design Automation Conference*, pp. 228–233, Anaheim, CA, June 13–17, 2005. ACM.
43. S. Mangard. A simple power-analysis (SPA) attack on implementations of the AES key expansion. In P.J. Lee and C.H. Lim, editors, *Proceedings of the 5th International Conference on Information Security and Crytography (ICISC)*, volume 2587 of *Lecture Notes in Computer Science*, pp. 343–358, Seoul, Korea, November 2002. Springer-Verlag.
44. R. Novak. Side-channel attack on substitution blocks. In J. Zhou, M. Yung, and Y. Han, editors, *Proceedings of the 1st International Conference on Applied Cryptography and Network Security (ACNS)*, volume 2846 of *Lecture Notes in Computer Science*, pp. 307–318, Kunming, China, October 16–19, 2003. Springer-Verlag.
45. E. Biham and A. Shamir. Power analysis of the key scheduling of the AES candidates. In *Proceedings of the 2nd Advanced Encryption Standard (AES) Candidate Conference*, Rome, Italy, 1999.
46. T.S. Messerges, E.A. Dabbish, and R.H. Sloan. Investigations of power analysis attacks on smartcards. In *Proceedings of the USENIX Workshop on Smartcard Technology*, Chicago, IL, May 10–11, 1999.
47. B. Megarajan. Combinational power analysis on smart cards. Technical report, Department of Electrical and Computer Engineering, Oregon State University, Corvallis, Oregon, 2002.
48. Joye, P. Paillier, and B. Schoenmakers. On second-order differential power analysis. In J.R. Rao and B. Sunar, editors, *Proceedings of the 7th International Workshop on Cryptographic Hardware and Embedded Systems (CHES)*, volume 3659 of *Lecture Notes in Computer Science*, pp. 293–308, Edinburgh, UK, August 29 to September 1, 2005. Springer-Verlag.
49. S. Mangard, T. Popp, and B.M. Gammel. Side-channel leakage of masked CMOS gates. In A. Menezes, editor, *Proceedings of the RSA Conference—Cryptographers' Track (CT-RSA)*, volume 3376 of *Lecture Notes in Computer Science*, pp. 351–365, San Francisco, CA, February 14–18, 2005. Springer-Verlag.
50. S. Mangard, N. Pramstaller, and E. Oswald. Successfully attacking masked AES hardware implementations. In *Proceedings of the 7th International Workshop on Cryptographic Hardware and Embedded Systems (CHES)*, volume 3659 of *Lecture Notes in Computer Science*, pp. 157–171, Edinburgh, UK, August 29 to September 1, 2005. Springer-Verlag.
51. E. Peeters, F.-X. Standaert, N. Donckers, and J.-J. Quisquater. Improved higher-order side-channel attacks with FPGA experiments. In J.R. Rao and B. Sunar, editors, *Proceedings of the 7th International Workshop on Cryptographic Hardware and Embedded Systems (CHES)*, volume 3659 of *Lecture Notes in Computer Science*, pp. 309–323, Edinburgh, UK, August 29 to September 1, 2005. Springer-Verlag.
52. T.S. Messerges. Power Analysis Attacks and Countermeasures on Cryptographic Algorithms. PhD thesis, University of Illinois, 2002.
53. J. Borst. Block Ciphers: Design, Analysis and Side-Channel Analysis. PhD thesis, K.U. Leuven, September 2001.
54. S. Chari, C.S. Jutla, J.R. Rao, and P. Rohatgi. Towards sound approaches to counteract power-analysis attacks. In M. Wiener, editor, *Advances in Cryptology: Proceedings of CRYPTO'99*, volume 1666 of *Lecture Notes in Computer Science*, pp. 398–412, Santa Barbara, CA, August 15–19, 1999. Springer-Verlag.

55. J. Daemen and V. Rijmen. Resistance against implementation attacks: A comparative study of the AES proposals. In *Proceedings of the 2nd Advanced Encryption Standard (AES) Candidate Conference*, March 1999. http://csrc.nist.gov/CryptoToolkit/aes/round1/conf2/papers/daemen.pdf.
56. M.A. Hasan. Power analysis attacks and algorithmic approaches to their countermeasures for Koblitz curve cryptosystems. In Ç.K. Koç and C. Paar, editors, *Proceedings of the 2nd International Workshop on Cryptograpic Hardware and Embedded Systems (CHES)*, volume 1965 of *Lecture Notes in Computer Science*, pp. 93–108, Worcester, MA, August 17–18, 2000. Springer-Verlag.
57. M.A. Hasan. Power analysis attacks and algorithmic approaches to their countermeasures for Koblitz curve cryptosystems. *IEEE Transactions on Computers*, 50(10):1071–1083, 2001.
58. M. Joye and C. Tymen. Protections against differential analysis for elliptic curve cryptography. In Ç.K. Koç, D. Naccache, and C. Paar, editors, *Proceedings of the 3rd International Workshop on Cryptographic Hardware and Embedded Systems (CHES)*, volume 2162 of *Lecture Notes in Computer Science*, pp. 377–390, Paris, France, May 13–16, 2001. Springer-Verlag.
59. P.Y. Liardet and N.P. Smart. Preventing SPA/DPA in ECC systems using the Jacobi form. In Ç.K. Koç, D. Naccache, and C. Paar, editors, *Proceedings of the 3rd International Workshop on Cryptographic Hardware and Embedded Systems (CHES)*, volume 2162 of *Lecture Notes in Computer Science*, pp. 391–401, Paris, France, May 13–16, 2001. Springer-Verlag.
60. D. May, H. Muller, and N. Smart. Random register renaming to foil DPA. In Ç.K. Koç, D. Naccache, and C. Paar, editors, *Proceedings of the 3rd International Workshop on Cryptographic Hardware and Embedded Systems (CHES)*, volume 2162 of *Lecture Notes in Computer Science*, pp. 28–38, Paris, France, May 13–16, 2001. Springer-Verlag.
61. E. Oswald and M. Aigner. Randomized addition–subtraction chains as a countermeasure against power attacks. In Ç.K. Koç, D. Naccache, and C. Paar, editors, *Proceedings of the 3rd International Workshop on Cryptograpic Hardware and Embedded Systems (CHES)*, volume 2162 of *Lecture Notes in Computer Science*, pp. 39–50, Paris, France, May 14–16, 2001. Springer-Verlag.
62. J.C. Ha and S.J. Moon. Randomized signed-scalar multiplication of ECC to resist power attacks. In B.S. Kaliski Jr., Ç.K. Koç, and C. Paar, editors, *Proceedings of the 4th International Workshop on Cryptographic Hardware and Embedded Systems (CHES)*, volume 2523 of *Lecture Notes in Computer Science*, pp. 129–143, Redwood Shores, CA, August 13–15, 2002. Springer-Verlag.
63. L. Benini, A. Macii, E. Macii, E. Omerbegovic, M. Poncino, and F. Pro. Energy-aware design techniques for differential power analysis protection. In *Proceedings of the 40th Design Automation Conference (DAC)*, Anaheim, CA, June 2–6, 2003.
64. L. Goubin and J. Patari. DES and differential power analysis the "duplication" method. In Ç.K. Koç and C. Paar, editors, *Proceedings of the 1st International Workshop on Cryptographic Hardware and Embedded Systems (CHES)*, volume 1717 of *Lecture Notes in Computer Science*, pp. 158–172, Worcester, MA, August 12–13, 1999. Springer-Verlag.
65. T.S. Messerges. Securing the AES finalists against power analysis attacks. In B. Schneier, editor, *Proceedings of the 7th International Workshop on Fast*

Software Encryption Workshop (FSE), volume 1978 of *Lecture Notes in Computer Science*, New York, April 2000. Springer-Verlag.

66. J.-S. Coron and L. Goubin. On Boolean and arithmetic masking against differential power analysis. In Ç.K. Koç and C. Paar, editors, *Proceedings of the 2nd International Workshop on Cryptographic Hardware and Embedded Systems (CHES)*, volume 1965 of *Lecture Notes in Computer Science*, pp. 231–237, Worcester, MA, August 17–18, 2000. Springer-Verlag.
67. M.-L. Akkar and C. Giraud. An implementation of DES and AES, secure against some attacks. In Ç.K. Koç, D. Naccache, and C. Paar, editors, *Proceedings of the 3rd International Workshop on Cryptographic Hardware and Embedded Systems (CHES)*, volume 2162 of *Lecture Notes in Computer Science*, pp. 309–318, Paris, France, May 13–16, 2001. Springer-Verlag.
68. L. Goubin. A sound method for switching between Boolean and arithmetic masking. In Ç.K. Koç, D. Naccache, and C. Paar, editors, *Proceedings of the 3rd International Workshop on Cryptographic Hardware and Embedded Systems (CHES)*, volume 2162 of *Lecture Notes in Computer Science*, pp. 3–15, Paris, France, May 13–16, 2001. Springer-Verlag.
69. J.D. Golic and C. Tymen. Multiplicative masking and power analysis of AES. In B.S. Kaliski Jr., Ç.K. Koç, and C. Paar, editors, *Proceedings of the 4th International Workshop on Cryptographic Hardware and Embedded Systems (CHES)*, volume 2535 of *Lecture Notes in Computer Science*, pp. 198–212, Redwood Shores, CA, August 13–15, 2002. Springer-Verlag.
70. E. Trichina, D. De Seta, and L. Germani. Simplified adaptive multiplicative masking for AES. In B.S. Kaliski Jr., Ç.K. Koç, and C. Paar, editors, *Proceedings of the 4th International Workshop on Cryptographic Hardware and Embedded Systems (CHES)*, volume 2535 of *Lecture Notes in Computer Science*, pp. 187–197, Redwood Shores, CA, August 13–15, 2002. Springer-Verlag.
71. E. Trichina. Combinational logic design for AES subbyte transformation on masked data. Cryptology ePrint Archive: Report 2003/236, 2003.
72. N. Pramstaller, E. Oswald, S. Mangard, F.K. Gürkaynak, and S. Hane. A masked AES ASIC implementation. In *Proceedings of Austrochip*, pp. 77–81, Villach, Austria, October 8, 2004.
73. N. Pramstaller, F.K. Gürkaynak, S. Haene, H. Kaeslin, N. Felber, and W. Fichtner. Towards an AES crypto-chip resistant to differential power analysis. In *Proceedings of the 30th European Solid-State Circuits Conference (ESSCIRC)*, Leuven, Belgium, September 21–23, 2004. IEEE.
74. E. Oswald, S. Mangard, N. Pramstaller, and V. Rijmen. A side-channel analysis resistant description of the AES s-box. In H. Gilbert and H. Handschuh, editors, *Proceedings of the 12th International Workshop on Fast Software Encryption (FSE)*, volume 3557 of *Lecture Notes in Computer Science*, pp. 413–423, Paris, France, February 21–23, 2005. Springer-Verlag.
75. E. Oswald and K. Schramm. An efficient masking scheme for AES software implementations. In J. Song, T. Kwon, and M. Yung, editors, *Proceedings of the 6th International Workshop on Information Security Applications (WISA)*, volume 3786 of *Lecture Notes in Computer Science*, pp. 292–305, Jeju Island, Korea, August 22–24, 2005. Springer-Verlag.

76. M. Gomulkiewicz and M. Kutylowski. Hamming weight attacks on cryptographic hardware—breaking masking defenses. In D. Gollmann, G. Karjoth, and M. Waidner, editors, *Proceedings of the 7th European Symposium on Research in Computer Security* (ESORICS), volume 2502 of *Lecture Notes in Computer Science*, pp. 90–103, Zurich, Switzerland, October 14–16, 2002. Springer-Verlag.
77. A. Bosselaers, R. Govaerts, and J. Vandewalle. Comparison of three modular reduction functions. In E.F. Brickell, editor, *Advances in Cryptology: Proceedings of CRYPTO'92*, volume 740 of *Lecture Notes in Computer Science*, pp. 175–186, Santa Barbara, CA, August 1993. Springer-Verlag.
78. A. Shamir. Protecting smart cards from passive power analysis with detached power supplies. In C. Paar and Ç.K. Koç, editors, *Proceedings of the 2nd International Workshop on Cryptographic Hardware and Embedded Systems (CHES)*, volume 1965 of *Lecture Notes in Computer Science*, pp. 71–77, Worcester, MA, August 17–18, 2000. Springer-Verlag.
79. K. Tiri and I. Verbauwhede. Securing encryption algorithms against DPA at the logic level: Next generation smart card technology. In C. Walter, Ç.K. Koç, and C. Paar, editors, *Proceedings of the 5th International Workshop on Cryptographic Hardware and Embedded Systems (CHES)*, volume 2779 of *Lecture Notes in Computer Science*, pp. 125–136, Cologne, Germany, September 7–10, 2003. Springer-Verlag.
80. W. Fischer and B.M. Gammel. Masking at gate level in the presence of glitches. In J.R. Rao and B. Sunar, editors, *Proceedings of the 3rd International Workshop on Cryptographic Hardware and Embedded Systems (CHES)*, volume 3659 of *Lecture Notes in Computer Science*, pp. 187–200, Edinburgh, UK, August 29 to September 1, 2005. Springer-Verlag.
81. F. Mace, F.-X. Standaert, J.-J. Quisquater, and J.-D. Legat. A design methodology for secured ICs using dynamic current mode logic. In V. Paliouras, J. Vounckx, and D. Verkest, editors, *Proceedings of the 15th International Workshop on Integrated Circuit and System Design (PATMOS)*, volume 3728 of *Lecture Notes in Computer Science*, pp. 550–560, Leuven, Belgium, September 20–23, 2005. Springer-Verlag.
82. T. Popp and S. Mangard. Masked dual-rail pre-charge logic: DPA-resistance without routing constraints. In J.R. Rao and B. Sunar, editors, *Proceedings of the 7th International Workshop on Cryptographic Hardware and Embedded Systems (CHES)*, volume 3659 of *Lecture Notes in Computer Science*, pp. 172–186, Edinburg, UK, August 29 to September 1, 2005. Springer-Verlag.
83. J.J.A. Fournier, S. Moore, H. Li, R. Mullins, and G. Taylor. Security evaluation of asynchronous circuits. In C. Walter, Ç.K. Koç, and C. Paar, editors, *Proceedings of the 5th International Workshop on Cryptographic Hardware and Embedded Systems (CHES)*, volume 2779 of *Lecture Notes in Computer Science*, pp. 137–151, Cologne, Germany, September 7–10, 2003. Springer-Verlag.
84. S. Moore, R. Anderson, R. Mullins, G. Taylor, and J. Fournier. Balanced self-checking asynchronous logic for smart card applications. *Journal of Microprocessors and Microsystems*, 27(9), pp. 421–430, October 2003.
85. J.D. Golic. DeKaRT: A new paradigm for key-dependent reversible circuits. In C. Walter, Ç.K. Koç, and C. Paar, editors, *Proceedings of the 5th International Workshop on Cryptographic Hardware and Embedded Systems (CHES)*, volume

2779 of *Lecture Notes in Computer Science*, pp. 98–112, Cologne, Germany, September 7–10, 2003. Springer-Verlag.

86. S. Mangard. Hardware countermeasures against DPA—a statistical analysis of their effectiveness. In T. Okamoto, editor, *Topics in Cryptology—CT-RSA—The Cryptographers' Track at the RSA Conference*, volume 2964 of *Lecture Notes in Computer Science*, pp. 222–235, San Francisco, CA, February 23–27, 2004. Springer-Verlag.
87. R.A. Serway. *Physics for Scientists and Engineers.* Saunders Golden Sunburst Series. Saunders College Publishing, 1996.
88. P. Hofreiter and P. Laackmann. Electromagnetic espionage from smart cards attacks and countermeasures. Infineon Technologies Publication, Secure, 02/2002, pp. 40–43.
89. P. Buysschaert and E. De Mulder. Elektromagnetische analyse (EMA) van een FPGA implementatie van een elliptische krommen cryptosysteem. Master's thesis, Katholieke Universiteit Leuven, Departement Elektrotechniek—ESAT, Kasteelpark Arenberg 10, B 3001 Heverlee, Belgium, May 2004.
90. E. De Mulder, P. Buysschaert, S.B. Ors, P. Delmotte, B. Preneel, G. Vandenbosch, and I. Verbauwhede. Electromagnetic analysis attack on a FPGA implementation of an elliptic curve cryptosystem. In *Proceedings of the International Conference on "Computer as a tool (EUROCON),"* Sava Center, Belgrade, Serbia and Montenegro, November 21–24, 2005. IEEE.
91. E. De Mulder, S.B. Ors, B. Preneel, and I. Verbauwhede. Differential electromagnetic attack on an FPGA implementation of elliptic curve cryptosystems. In *Proceedings of the World Automation Congress (WAC) 2006, the 5th International Forum on Multimedia and Image Processing (IFMIP)*, page in print, Budapest, Hungary, July 24–27, 2006.
92. NSA. NSA TEMPEST Documents. http://www.cryptome.org/nsa-tempest.htm.
93. J.-J. Quisquater and D. Samyde. Electromagnetic analysis (EMA): Measures and counter-measures for smart cards. In I. Attali and T. Jensen, editors, *Proceedings of the International Conference on Research in Smart Cards: Smart Card Programming and Security (E-smart)*, volume 2140 of *Lecture Notes in Computer Science*, pp. 200–210, Cannes, France, September 19–21, 2001. Springer-Verlag.
94. K. Gandolfi, C. Mourtel, and F. Olivier. Electromagnetic analysis: Concrete results. In Ç.K. Koç, D. Naccache, and C. Paar, editors, *Proceedings of the 3rd International Workshop on Cryptographic Hardware and Embedded Systems (CHES)*, volume 2162 of *Lecture Notes in Computer Science*, pp. 255–265, Paris, France, May 13–16, 2001. Springer-Verlag.
95. D. Agrawal, B. Archambeault, J.R. Rao, and P. Rohatgi. The EM side-channel(s): Attacks and assessment methodologies. In B.S. Kaliski Jr., Ç.K. Koç, and C. Paar, editors, *Proceedings of the 4th International Workshop on Cryptographic Hardware and Embedded Systems (CHES)*, volume 2523 of *Lecture Notes in Computer Science*, pp. 29–45, Redwood Shores, CA, August 13–15, 2002. Springer-Verlag.
96. D. Agrawal, B. Archambeault, S. Chari, J.R. Rao, and P. Rohatgi. Advances in side-channel cryptanalysis. *RSA Laboratories Cryptobytes*, 6(1):20–32, Spring 2003.

97. S. Mangard. Exploiting radiated emissions—EM attacks on cryptographic ICs. In *Proceedings of Austrochip*, Linz, Austria, October 3, 2003.
98. V. Carlier, H. Chabanne, E. Dottax, and H. Pelletier. Electromagnetic side channels of an FPGA implementation of AES. Cryptology ePrint Archive-2004/145, 2004. http://eprint.iacr.org/.
99. A. Shamir and E. Tromer. Acoustic cryptanalysis. Preliminary proof-of-concept presentation, 2004. http://www.wisdom.weizmann.ac.il/tromer/acoustic/.

8 Security Enhancement Layer for Bluetooth

Panu Hämäläinen, Marko Hännikäinen, and Timo D. Hämäläinen

CONTENTS

8.1 INTRODUCTION

During the recent years, wireless networking technologies have achieved a significant role as telecommunications media because of their flexibility and convenience in numerous usage scenarios. Of the wireless technologies, Bluetooth (Bluetooth Special Interest Group, Bellevue, Washington, USA) [1] has become the default choice for low-cost, low-power, short-range, and personal area communications. It is specified by Bluetooth Special Interest Group (SIG) [2], an industry consortium established for developing and advancing the wireless technology. Originally, Bluetooth was only intended as a simple serial cable replacement for electronic devices. However, presently the technology supports various more advanced functionalities, such as ad hoc networking and access point operation for Internet connections. The ongoing development extends Bluetooth with new features, including support for quality of service, higher data rates, multicasting, and lower power consumption. Currently, Bluetooth can be found in mobile phones, personal digital assistants (PDAs), laptops, printers, digital cameras, headsets, portable payment terminals (e.g., for facilitating credit card payments in restaurants), cars, and medical equipments. The application area expands as new products with the Bluetooth capability are constantly introduced [3].

In addition to low-cost and robust operation, Bluetooth applications often require protected communications. For example, confidential data transfers between personal devices and transactions with payment terminals must be protected from outsiders. Due to the wireless link, Bluetooth transmissions are available and devices discoverable to anyone within the radio coverage. Therefore, the Bluetooth specification [1] defines security services for authentication and confidentiality. Unfortunately, researchers have identified several vulnerabilities in the security design. The vulnerabilities originate from the usage of a personal identification number (PIN) in key generation, improper key management and authentication, and the possibility of tracking Bluetooth devices. The security level of the Bluetooth encryption algorithm has turn out to be significantly lower than the key sized permits to expect. A major weakness is that transmitted data are only protected with noncryptographic checksums instead of proper message authentication codes (MACs) [4].

This chapter proposes a novel enhanced security layer (ESL) for improving the security of Bluetooth technology. The security level is increased by replacing the original Bluetooth encryption scheme with a design based on advanced encryption standard (AES) [5]. Except for Bluetooth, AES is currently employed in all significant wireless short-range technologies, that is, IEEE 802.11 [6] (Institute of Electrical and Electronics Engineers Inc., Piscataway, NJ, USA), IEEE 802.15.3 [7], IEEE 802.15.4 [8], and ZigBee [9] (ZigBee Alliance, http://www.zigbee.org). Furthermore, ESL adds MACs to the transmitted data for cryptographic integrity protection and data origin authentication. ESL includes two additional authentication and key exchange protocols, one using public keys and the other secret keys. The protocols can be used for agreeing on ESL keys as well as standard Bluetooth PINs. The proposed security layer is placed on top of the standard Bluetooth controller interface, which allows integrating it as an additional module into any standard Bluetooth chip or as a software layer into a host device.

A prototype implementation of ESL is also presented in this chapter. AES and the supported modes of operation are implemented in hardware for high performance and energy efficiency [10]. In addition to the improved security, the ESL prototype offers an easy-to-use application programming interface (API) for Bluetooth devices by hiding low-level management commands from applications.

The rest of the chapter is organized as follows. Section 8.2 gives an overview of the Bluetooth technology, its security design, and known security problems. The design of ESL is presented in Section 8.3. Section 8.4 describes the ESL prototype implementation as well as compares its performance and resource consumption with the standard Bluetooth security design. Finally, Section 8.5 concludes the chapter.

8.2 BLUETOOTH OVERVIEW

The Bluetooth technology consists of several protocol layers ranging from the physical radio and link layer (baseband) to object exchange and service discovery protocols. In addition, Bluetooth SIG has specified a number of profiles [2], which define a selection of messages, procedures, and protocols required for supporting a specific service. The portion of the protocol stack considered in this work is depicted in Figure 8.1. Host controller interface (HCI) separates the stack into two parts, Bluetooth host and Bluetooth controller. It provides the host with a low-level, uniform interface to the hardware capabilities of the controller. The host is connected to the HCI firmware through a physical bus, such as universal asynchronous

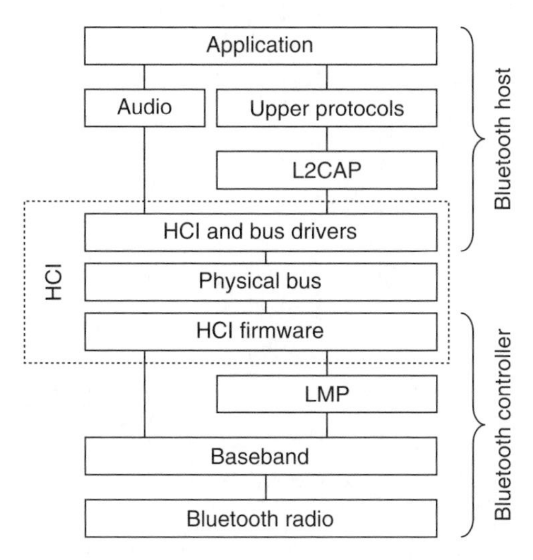

FIGURE 8.1 Bluetooth protocol stack. The stack is separated into a Bluetooth host and a Bluetooth controller by HCI. Typically the host side stack is implemented as software and the controller side as firmware and hardware.

receiver transmitter (UART) or universal serial bus (USB). IEEE has also adopted the lowest Bluetooth layers and standardized them in its IEEE 802.15.1 standard [11].

Originally, the transmission rate of the Bluetooth radio was 1 Mbit/s. The newest specification extends the rate to 3 Mbit/s [1]. There are two types of Bluetooth links: asynchronous connectionless (ACL) and synchronous connection oriented (SCO). ACL is used for data transfers and SCO for audio. Both the links have several network packet types of different lengths. The host transmits and receives data in HCI data packets. The HCI packets are fragmented to and assembled from the ACL and SCO network packets by the Bluetooth controller. The highest data payload rate is 723.2 kbit/s with the 1 Mbit/s radio and 2178.1 kbit/s with the 3 Mbit/s radio over an asymmetric ACL link with the largest ACL packets [1].

Communications between Bluetooth devices can be point-to-point or point-to-multipoint. A Bluetooth network is called a "piconet." It consists of a master and up to seven active slave devices. Piconets can be linked together to form a larger network, "scatternet", as illustrated in Figure 8.2. Link manager protocol (LMP), residing below HCI, manages piconets using the services of the baseband. The protocol layers between HCI and standard upper protocols are related to multiplexing, segmentation, service discovery, and serial port emulation. Audio transmissions can bypass the higher protocols and use baseband services directly through HCI.

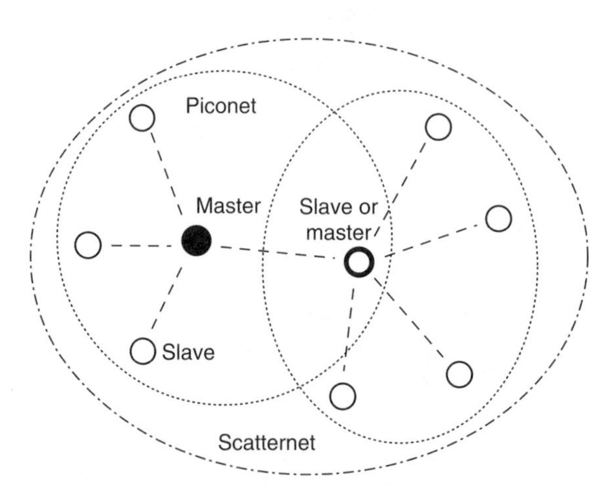

FIGURE 8.2 Bluetooth network topology. A single network consisting of a master and slaves is called a piconet. Several piconets can be connected together to form a scatternet.

8.2.1 Standard Bluetooth Security

A Bluetooth device can operate in three security-related modes [1]. In the nonsecure mode, the device does not initiate any security procedures. A device operating in the service level–enforced security mode does not initiate security procedures before the channel establishment at the logical link control and adaptation protocol (L2CAP) layer. In the link level–enforced security mode, security procedures are initiated before the Bluetooth link has been established. As the service level–enforced security mode supports different security policies for parallel applications, the link level–enforced security mode enforces the same link layer security level for all connections [12].

In addition to the operating modes, Bluetooth specifies security levels for devices and services [1,12]. A device is trusted if it has been previously authenticated and marked as trusted. A trusted device has unrestricted access to services. Unknown devices and devices that have previously been authenticated but not explicitly marked as trusted are untrusted. They have only restricted access to services. In the service level–enforced security mode, services can choose to require authorization, authentication, and encryption. When a service requires authorization, access is automatically granted only to trusted devices.

When security procedures are applied, Bluetooth implements key management, entity authentication, and confidentiality. The security processing is carried out at the baseband and controlled by link manager (LM) according to the requirements of the higher protocol layers. Due to the large number of adjustable parameters, Bluetooth SIG has published additional recommendations for configuring the security services in different profiles [13]. Furthermore, the ambiguities of the Bluetooth specification related to the encryption of piconet broadcasts have been clarified [14].

8.2.1.1 Entity Authentication and Key Management in Bluetooth

The Bluetooth security is based on three types of link keys: initialization keys, combination keys, and master keys. The earlier versions of the Bluetooth specification included a fourth type of a link key called unit key but its usage is deprecated in the newest specification because of severe security problems. The 128-bit link key is used in entity authentication. Depending on its type and the desired level of protection, the link key is also used for generating an encryption key. The Bluetooth key hierarchy is illustrated in Figure 8.3.

An initialization key is typically used only when two Bluetooth devices establish a connection for the first time. The key is generated from a PIN code and is supplied to both the devices. A combination key is generated from information shared between two Bluetooth devices. The sharing of the generation information is protected with the effective link key. A master key is a temporary key distributed by the piconet master and used for protecting broadcast transmissions.

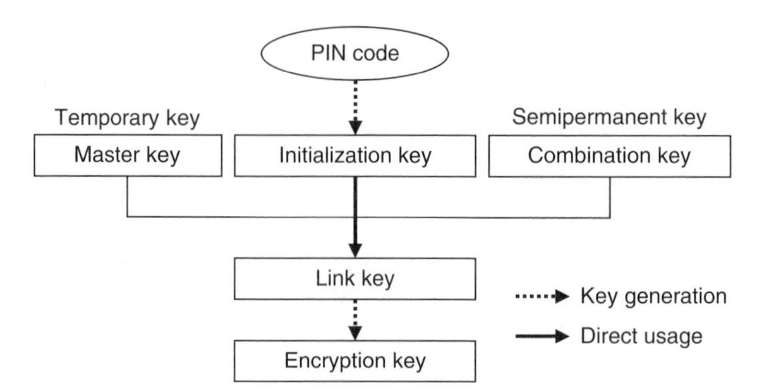

FIGURE 8.3 Bluetooth key hierarchy. The initialization key is derived from the PIN code and used as a link key for protecting the exchange of another type of key. The exchanged link key is used for authentication and optionally for generating an encryption key.

After exchanging a new link key, the devices verify its correctness and each other's identities by subsequently running a challenge–response authentication protocol. The devices are identified by unique 48-bit Bluetooth addresses. The procedure involving the creation of an initialization key, using it to protect the exchange of a new link key, and running the authentication protocol with the new link key is called "pairing." Key generations and the authentication protocol use algorithms referred to as E_1, E_{21}, E_{22}, and E_3. They are all based on the SAFER+ block cipher [15].

8.2.1.2 Confidentiality and Integrity in Bluetooth

Bluetooth provides confidentiality through optionally encrypting network packet payloads. Encryption is performed with a proprietary stream cipher called E_0, which is based on four parallel linear feedback shift registers (LFSR). It generates a key stream, which is XORed with plaintext to produce ciphertext and vice versa. Before proceeding with encryption, devices agree on the size of the encryption key, which can vary between 8 and 128 bits. The encryption key is derived from the current link key and parameters are provided by the piconet master. E_0 is initialized with the encryption key and the real-time clock of the master. The clock ensures that each key stream produced with the same encryption key is different and thwarts the initialization vector attacks of the wired equivalent privacy (WEP) protocol of IEEE 802.11 [16]. For integrity verification, a keyless cyclic redundancy check (CRC) checksum is computed and appended to the payload before encryption in the same way as in WEP.

8.2.1.3 Bluetooth Security Vulnerabilities

The Bluetooth security design has been found vulnerable to a number of attacks. By exploiting the vulnerabilities, an attacker can, for example, obtain confidential data from Bluetooth-enabled mobile phones, inject viruses, make unauthorized phone calls, and send short messages [17–19]. Not all the exploits are directly caused by the security design but rather by its implementation and configuration in end products. Despite that the Bluetooth technology has been designed for short-range communications, attackers can easily expand their range and attempt attacking devices from the distance [19,20].

The security is completely based on the PIN code [21,22]. Therefore, the PIN should always be long and randomly chosen. However, the Bluetooth specification permits fixed PINs and PINs of only 8-bit long, and also defines a default value for the code (zero). Since a method for automatically exchanging PINs has not been defined, users tend to choose short (typically four digits) and easily memorable values [22]. When the PIN code is poorly chosen, an attacker can perform off-line search for the code after eavesdropping on pairing or after masquerading as the initiator of the pairing procedure [21]. If the attackers have not been present during pairing, they can also claim the link key lost and make the victim devices rerun pairing [21,23]. Due to the weaknesses, Bluetooth SIG recommends performing pairing in a private location [13]. Due to more robust and IEEE 802.11 WLAN-compatible solutions, researchers have proposed the 802.1X framework [24] and Diffie–Hellman-based key exchange mechanisms to be used for link key establishments [25]. However, these kinds of solutions have not yet been specified by Bluetooth SIG.

In addition to the PIN code vulnerabilities, a weakness in the entity authentication of Bluetooth is that only devices are authenticated. For example, it has been reported that switching the owner and the subscriber identity module (SIM) of a mobile phone does not always require reauthenticating the Bluetooth connections of the phone [26]. An attacker is also able to fake two Bluetooth devices to believe that they are directly communicating with each other by simply relaying traffic between them [27]. To work and be beneficial, the relaying attack requires that the two devices do not hear each other and that they do not invoke the optional encryption after authentication. An adequate cryptographic integrity protection mechanism can thwart the attack as well. More advanced man-in-the-middle attacks, which are applicable because of the missing integrity protection, are described [28]. The attacker is able to compromise encrypted connections also by exploiting the frequency-hopping mechanism of Bluetooth [1]. The clocks of the victim devices are unsynchronized, which causes their hopping sequences to have different offsets, preventing them from hearing each other.

The noncryptographic integrity protection based on the encrypted CRC checksum can be attacked in the same way as that of WEP [16,28], allowing an attacker to manipulate Bluetooth transmissions without detection. Particularly, the usage of the stream cipher makes this applicable. If the transmitted plaintext is known, the attackers can change the packet contents into whatever they wish by flipping bits. For example, it is possible to convey higher-protocol data into a different destination by altering the higher-layer addressing fields, which often reside at known locations in Bluetooth packets. To prevent all the attacks described in [28], it is stated that adding MACs into Bluetooth transmissions is inevitable. The protection should cover management packets also to prevent denial-of-service (DoS) attacks performed by tampering with them.

The negotiable key size is a threat to the Bluetooth encryption scheme since a malicious party can purposely make a device use a short key and thus alleviate attacking against encryption [28,29]. Furthermore, the applications and users of Bluetooth devices are not aware of the negotiation procedure or agreed encryption key sizes. In addition to the protocol attacks, weaknesses in the Bluetooth stream cipher E_0 have been exposed [21,30–33]. The cipher has appeared to provide a significantly lower level of security than a 128-bit-key algorithm should provide. For example, if the cipher is used outside the Bluetooth technology and allowed to produce long key streams, it is far too weak [32]. Within the constraints of Bluetooth, a practical attack for recovering the encryption key can be performed after discovering the first 24 key stream bits of about 2^{24} packets [33]. Even though the E_0 attacks have not yet been exploited in practice, these are alerting results as they correspond to the ones that led to the complete insecurity of WEP [34]. Generally, instead of destining to proprietary solutions, such as E_0, it is more secure to use solutions that have been developed through a public process and those that are widely used and trusted, such as AES.

When a device has been implemented according to the older versions of the Bluetooth specification, it may use its unit key as the link key. This exposes all the traffic protected with the key in the past and in the future to other devices with which the key has been shared [21–23]. It allows impersonation as well.

Bluetooth-enabled devices can be tracked as they constantly advertise their unique addresses [21,22,26]. This introduces a threat to a person's location privacy as Bluetooth is typically used in personal devices that people carry with them. An anonymity scheme for thwarting the threat has been proposed [25]. Bluetooth SIG has also discussed about addressing this threat [35] but so far support for location privacy or anonymity has not been specified [1].

8.3 ENHANCED SECURITY LAYER (ESL) FOR BLUETOOTH

To address the weaknesses of the Bluetooth security design, ESL for protecting Bluetooth links is proposed in this work. ESL specifically aims at fixing the shortcomings of the Bluetooth encryption algorithm and the lack of cryptographic integrity protection as well as improving the entity authentication. For protecting data transfers, ESL supports four well-scrutinized operation modes from which the application can choose the preferred one according to its security requirements. In addition, two enhanced entity authentication and key agreement protocols are included.

8.3.1 Placement of ESL in Bluetooth Protocol Stack

The ESL architecture is presented in Figure 8.4. As shown, ESL is placed above HCI. Generally, Bluetooth technology is provided as fixed chips, which implement the Bluetooth functionalities below HCI. Application developers have only access to the Bluetooth controller through the standard HCI. Therefore, to improve the security, the most universally applicable method is to add the enhancements above HCI. This way ESL can be integrated as an additional module into any standard Bluetooth controller or host. Another advantage of placing ESL on the top of HCI and not into the baseband is that the method results in lower packet overhead. Added protocol fields, such as MACs, are transmitted in the HCI data packets instead of every Bluetooth network packet. Despite that tampering with the packets is still possible at the baseband layer, the tampering attempts are detected at the ESL layer. The drawback is that only the messages that originate from above HCI can be protected.

8.3.2 Confidentiality and Integrity in ESL

ESL replaces the E_0 cipher with AES [5]. AES is a symmetric cipher that encrypts data in 128-bit blocks, supporting key sizes of 128, 192, and 256 bits. As several other block ciphers, AES consists of successive, similar iteration rounds. Depending on the chosen key size, the number of the rounds is 10, 12, or 14. Each round mixes the data with a round key, which is generated from the encryption key. As in the other significant short-range wireless technologies, the 128-bit key version is used in ESL. AES decryption requires inverting the iterations resulting in a different datapath. However, the operation modes of ESL only require the forward cipher and thus save resources.

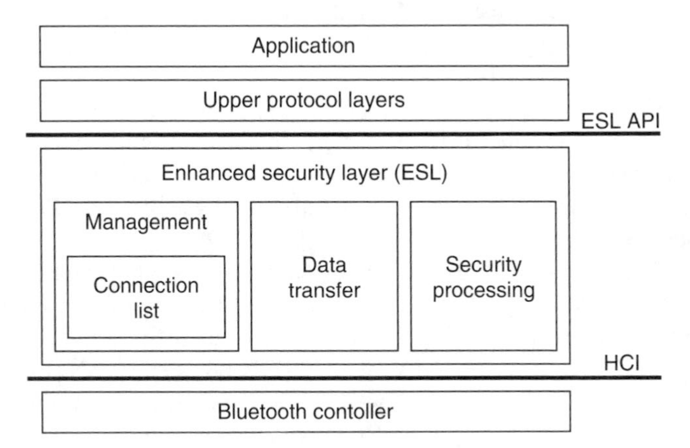

FIGURE 8.4 Architecture of Bluetooth ESL. ESL is placed on top of the standard HCI and accessed through the ESL API.

Applying an encryption algorithm alone without a proper encryption mode is not secure. The counter (CTR) mode [36] is generally regarded as a good choice and it is also used in ESL. CTR has a proven security bound [37] and it provides most performance trade-offs for implementations [36]. In CTR, a block cipher produces a key stream from a secret key and a counter. The key stream is generated a block at a time by encrypting counter values until the stream length matches the data length. After each algorithm pass the counter is incremented. The stream is XORed with the plaintext to get the ciphertext and vice versa. If the data length is not a multiple of the block length, only the required bits of the last key stream block are used. It is important that the same counter value is used only once during the lifetime of a key. Another security requirement is that at maximum 2^{64} counter values are used per key.

As the CTR mode can only provide confidentiality and not integrity, it is required that the encrypted data are accompanied with MAC. Otherwise, the bit manipulation attacks of the standard Bluetooth encryption still apply. In ESL, MACs are computed using the cipher block chaining MAC (CBC-MAC) technique [38]. This allows using the same algorithm for both encryption and integrity protection. CBC [38] is a feedback encryption mode in which the previous ciphertext block is XORed with the plaintext block before encryption. CBC-MAC operates in the same way, except that only the result of the last encryption is output as MAC. The security of CBC-MAC has been proven for fixed-sized messages [39]. As a solution for protecting variable-sized messages, it is proposed that messages are prefixed with their lengths before the CBC-MAC computations [39]. The solution is used in ESL.

ESL supports plain CBC-MAC (MAC mode) and two combinations of the CTR encryption and CBC-MAC computation. In the combined modes, MAC can be computed over the plaintext (MAC-then-encrypt mode) or over the ciphertext (encrypt-then-MAC mode). The phases use separate keys. In the MAC-then-encrypt mode, MAC is encrypted. The MAC mode can be used for decreasing processing requirements in applications that only require authenticity. It has been shown that the encrypt-then-MAC mode is generally secure [40]. However, it has also been suggested that it should be the plaintext that is authenticated [41]. Adding the other mode to an implementation that supports one of the modes requires only little additional resources. Thus, both the modes are supported by ESL.

ESL also supports the special MAC-and-encrypt mode called CTR with CBC-MAC (CCM) [42], which uses the same key for encryption and integrity protection. Originally, CCM was proposed to the IEEE 802.11i working group for improving the security of the IEEE 802.11 WLAN and it was also adopted. CCM has been adopted in the other significant short-range wireless technologies as well. The CCM components, CTR and CBC-MAC, have been well known for decades but CCM is a new definition for their combined usage. The security of the CCM mode has been proven [43].

While the CTR encryption can process arbitrary length data, CBC-MAC requires that the input is padded to match a block boundary. In ESL, the last input data block is padded with zeroes if required in all the modes. The padding is ignored in CTR. The MAC size can be chosen to be 64 or 128 bits, which allows trade-offs between the protection level and the communication overhead. By default, the MAC size is 64 bits, in which case the 64 least significant bits of the MAC computation output are discarded. MAC is appended to the HCI payload data.

To prevent repeated counter values, the CTR is composed of concatenated nonce and block counter in all the combined modes. The nonce is a constant value for a transmission, and the block counter is incremented for each data block within the transmission. The nonce is provided and managed by the application. For example, recommendations for choosing the nonce in the CCM mode are presented [42]. Good practice is to include at least the sender's Bluetooth address and the transmission's sequence number in it. The nonce size was chosen to be 96 bits to allow sufficient amount of information. Thus, the block counter size is 32 bits. In addition to a nonce, in the CTR input of CCM, there are fixed flags, which are regarded as a part of the nonce. When the nonce space has been exhausted or 2^{64} blocks encrypted, new keys must be agreed on. In addition, it is not allowed to use the same key across the ESL operation modes.

Since ESL is located above the standard HCI, only the HCI data packet payloads can be encrypted. However, the application can still protect the known lower-layer header fields (e.g., Bluetooth addresses) with MAC, even if they were not placed into the Bluetooth packets by ESL or the application (e.g., only a connection handle is used for addressing in the HCI data packets). The application must ensure that the nonce can be generated at the receiving device. It can be predefined or transmitted in the HCI data packet. The nonce does not have to be kept secret as long as it is protected with MAC. When the protected nonce includes the sequence number of the transmission, it can also be used for providing the freshness of transmissions.

The format of an ESL packet, placed in the HCI data payload, is presented in Figure 8.5. The maximum size of the application data per ESL packet depends on the chosen MAC size. It is assumed that the complete nonce is transmitted in an ESL packet. If a combined mode is chosen, the ESL payload field is encrypted. The HCI payload is transmitted in a Bluetooth ACL network packet. If the HCI payload does not fit into a single network packet, the Bluetooth controller fragments the payload across several network packets. In the figure, the Bluetooth ACL packet is the ACL data packet with the largest payload size [1].

8.3.3 Entity Authentication and Key Agreement in ESL

To support different usage scenarios and processing requirements, ESL provides two entity authentication and key agreement methods for link

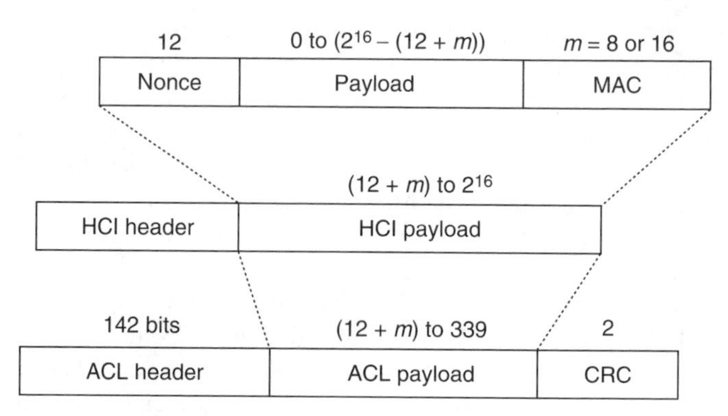

FIGURE 8.5 ESL, HCI ACL, and Bluetooth ACL packet formats. An ESL packet is placed in the payload of a HCI data packet, which is then transmitted in one or more Bluetooth network packets. The figure presents the largest HCI data packet. The field sizes without units are in bytes.

establishments, one based on public keys and the other on secret keys. The handshake portion of the widely employed and trusted transport layer security (TLS) protocol [44] was chosen as the public-key protocol and the authenticated key exchange protocol 2 (AKEP2) [4] as the secret-key protocol. AKEP2 has a security proof [45]. While AKEP2 requires that devices have preshared a secret key, the TLS handshake can be used for authenticating devices that are previously unknown to each other, using public-key certificates. The advantage of AKEP2 is that it has lower processing requirements and can be implemented with the AES-based confidentiality and integrity protection procedures of ESL. Both the protocols result in shared temporary secrets that are used as keying material for generating ESL keys as well as PIN codes. The authentication message exchanges are treated as regular HCI data transfers by the lower protocol layers.

8.3.4 Restrictions to Standard Bluetooth Security

The support for the standard Bluetooth security is included in ESL for the interoperability with devices that do not contain the new security features. However, to improve the security of the standard procedures, the usage is restricted to a subset of the supported parameter combinations. According to the results of [23], it can be estimated that discovering a PIN code of 12 decimal digits requires about 80 d of processing for a state-of-the-art PC. This can currently be seen as the limit for most feasible attacks. Hence, ESL requires the PIN size to be at least 12 digits. A longer PIN should be used if the PIN is not changed for a long period of time or if the exchanged data are valuable enough for 80 d of processing. Authentication without invoking encryption and protecting connections with a unit key is not allowed. Either of the ESL key exchange protocols can be used for automatic PIN exchange to prevent poorly chosen values and to make the link setup more convenient for users. The encryption key size negotiation cannot be improved by ESL as this requires modifications to the standard HCI and Bluetooth controller implementations.

8.3.5 ESL Components

As depicted in Figure 8.4, the tasks of ESL can be divided into three components: security processing, data transfer, and management. ESL is accessed through the ESL API. The security-processing entity performs the computations related to the new security features. It appends MACs to the transmitted application data and verifies the received MACs. Packets with failing MACs are dropped and the higher protocol layer is notified of the failures. The security-processing entity is not used if only the standard Bluetooth security without the enhanced authentication protocols is applied.

The data transfer entity conveys application data between the ESL API and the Bluetooth controller by constructing and decoding HCI data packets. If only the standard Bluetooth security is used, in transmission the entity places the application data into a HCI data packet payload field and forwards the HCI packet to the Bluetooth controller. On the reception of a HCI data packet, the entity decodes the application payload from the packet and gives it to the ESL API. If the AES-based protection or the ESL authentication protocols are used, the payload to be transmitted in a HCI data packet is received from the security-processing entity. Similarly, the payload of the received HCI data packet is first processed by the security-processing entity.

The management entity controls the other entities and Bluetooth links. It initiates the Bluetooth controller, establishes and closes connections, and provides keys to the security-processing entity. The entity constructs HCI commands and receives information from the Bluetooth controller in HCI events. The controller initiation prepares the device to function as a slave or a master and, if the standard security is used, provides the controller with the standard security parameters. The management entity runs the two ESL authentication protocols using the services of the security-processing entity as well as forces the ESL restrictions for the standard security parameters. It maintains a connection list that contains handles to the established Bluetooth links and their ESL parameters, including ESL keys, operation mode, and the chosen MAC length. If the new security features are not used for a link, the management entity sets ESL to bypass the enhanced security processing.

The ESL processing and HCI are hidden behind the ESL API. It provides high-level procedures for device initiation, connection establishment, sending and receiving application data, disconnecting, and handling failures. After a connection has been established, the sent and the received application data are transparently processed by ESL. The application only needs to provide nonces. The API procedures are described in more detail in the following section.

8.4 PROTOTYPE IMPLEMENTATION OF BLUETOOTH ESL

Altera Excalibur EPXA10 DDR Development Kit [46] (Altera Corporation, San Jose, California, USA), presented in Figure 8.6, has been used as the ESL prototype implementation platform. The main component is the EPXA10-F1020-C2 programmable chip, which consists of an integrated 32-bit ARM922T (ARM Ltd., Cambridge, UK) processor core and an Altera APEX20KE-like programmable logic device (PLD). The PLD consists of a large number of programmable logic elements (LEs) and embedded system blocks (ESB) for implementing a variety of memory functions. ARM9 and the PLD are connected through two advanced microcontroller bus architecture (AMBA) high-performance bus (AHB) bridges, a shared dual-port RAM (DPRAM), and interrupt lines. In the implementation, a 256 megabyte SDRAM was used as an external memory.

The daughter card of Ericsson (Telefonaktiebolaget LM Ericsson, Stockholm, Sweden) Bluetooth Starter Kit [47] was used as the Bluetooth controller. It provides the host with HCI via UART or USB. The radio transmits at 1 Mbit/s. s. The card was connected to an expansion header of the development kit and accessed via UART. The communications between Bluetooth devices use the ACL link. The used ACL packet types can be defined with a HCI command.

The architecture of the ESL prototype is presented in Figure 8.7. The components implemented in the hardware (PLD) are direct memory access (DMA), UART and its control, security processing and control, and processor

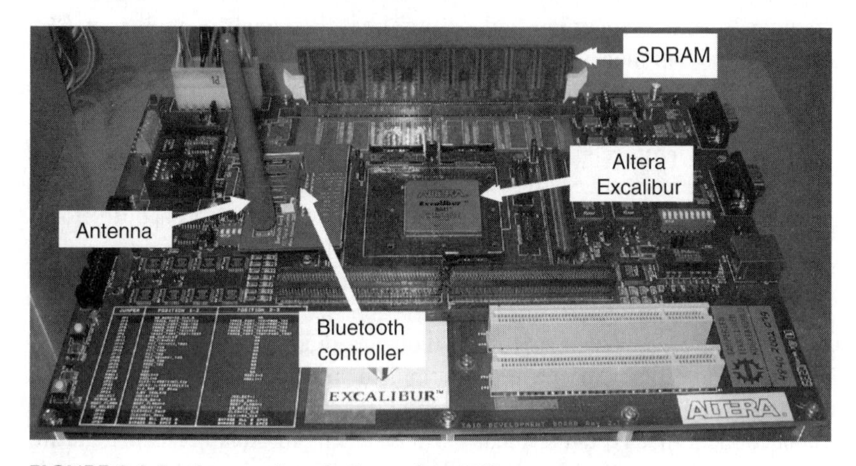

FIGURE 8.6 Implementation platform of the ESL prototype.

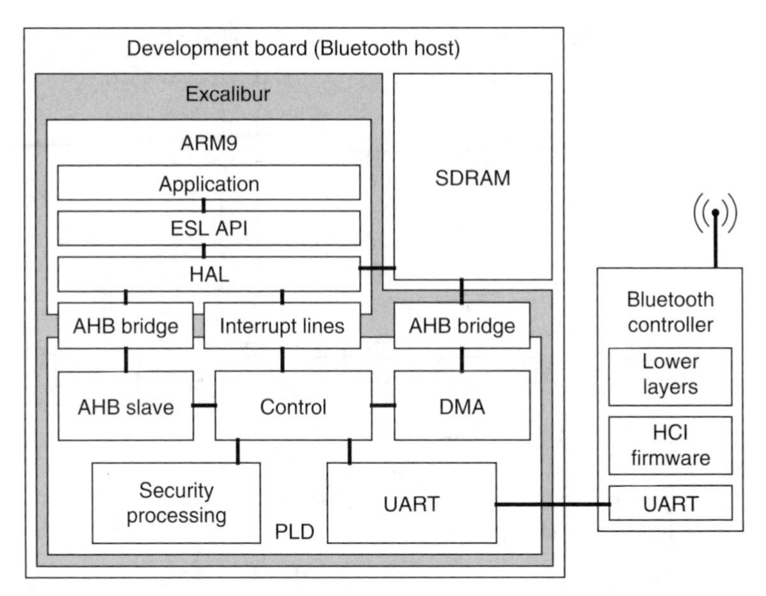

FIGURE 8.7 Architecture of the ESL prototype. Most functionalities of the prototype are implemented as hardware in the PLD. The software on the ARM9 side contains the HAL, the ESL API, the authentication protocols, and the test application.

to PLD communications. The hardware design was captured in VHDL and the software in C.

8.4.1 Security-Processing Hardware Architecture

AES and its ESL modes of operation were chosen to be implemented in hardware for high performance in the prototype. The iterative, 128-bit key AES core published in [48] was used. The core computes the round keys on-the-fly and one encryption round at a clock cycle. It offers high throughput and does not require setup time for switching the key. Due to the feedback loop of the ESL modes of operation, only iterative AES implementations are reasonable choices for a single-core implementation. The on-the-fly key schedule is well suited for the implementation since the processing is constantly altered between encryption and MAC computation with different keys, except in the MAC and the CCM modes. A precomputed schedule would require setup latency and storage for the round keys.

The datapath of the security-processing hardware is presented in Figure 8.8. Input data encoding and MAC value comparisons are performed by ARM9. The internal signals are 128-bit wide, unless specified otherwise. The parameter updated internally is the block counter (Register 5). The load signal sets up the module for reading a new encryption key, MAC key, and nonce. The keys are stored in Register 6 and Register 7, and fed to the AES

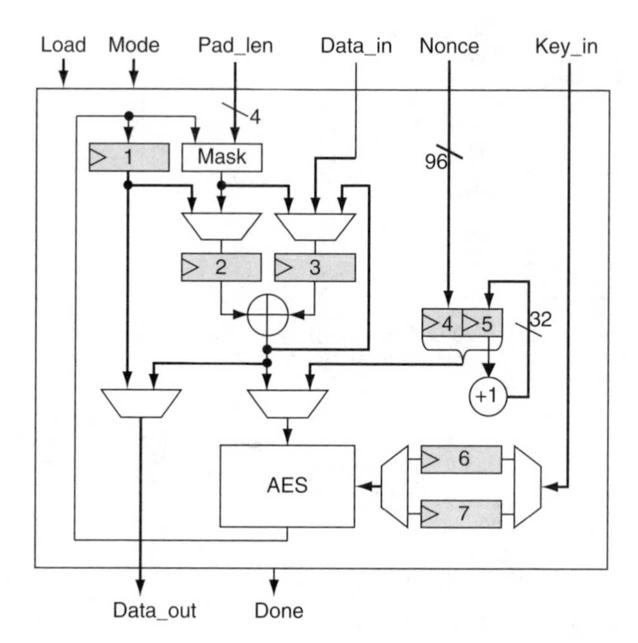

FIGURE 8.8 Datapath architecture of the ESL prototype security-processing hardware.

core in turns by the control logic. The nonce in Register 4 is concatenated with the block counter for the CTR processing. The load port allows also maintaining the old key values and updating only the nonce. The block counter is reset by updating the encryption key or nonce.

A new data block is input through the data_in port and stored in Register 3. The signal mode defines whether the module operates in the MAC, MAC-then-encrypt, encrypt-then-MAC, or CCM mode and sets the module to encrypt or decrypt. The pad_len signal is required in the combined modes for informing the number of padding bytes in the last input block. The data_out port is used to output the encrypted or decrypted data blocks as well as the MAC values.

8.4.1.1 Operation Modes

In the MAC mode, MAC is computed using the MAC key. The chaining value in Register 1, obtained after processing the previous data block, is transferred to Register 2. Initially, the value is zero. After XORing the chaining value with the data block in Register 3, the result is processed by the AES core and the output is written back to Register 1. After the last data block, the contents of Register 1 are output and the register is reset. The MAC verification is carried out in the same way.

In the MAC-then-encrypt mode, first, the hardware performs the MAC computation for a block with the MAC key as described. Then, the operation is switched to the CTR encryption with the encryption key. In CTR, the nonce and the block counter are fed to the AES core. Initially, the block counter is set to zero. The result is written through the mask component to Register 2 and XORed with the data block maintained in Register 3. The XOR result is output and the block counter is incremented. After the last data block, the nonce and the block counter are processed once more and the result is written to Register 3 (through the mask component). The contents of Register 1 are transferred to Register 2 and XORed with Register 3. The result is output as the encrypted MAC.

In the encrypt-then-MAC mode, the processing order of MAC-then-encrypt is inverted. The only difference is that after encrypting a data block the output is also written back to Register 3 for the MAC computation.

In the MAC-then-encrypt decryption, the processing is mainly the same as in the encrypt-then-MAC encryption. After the last data block, the received, encrypted MAC is input through data_in to Register 3. The nonce and the block counter are input to the AES core. XORing the AES output with Register 3 yields the decrypted MAC, which is output for comparison with the earlier output, computed MAC. The encrypt-then-MAC decryption processing is the same as the MAC-then-encrypt encryption processing. The received MAC is not input since it is already in the plaintext form.

The CCM mode operation is similar to the MAC-then-encrypt mode. The encryption and the MAC keys are set to the same value. The difference is that in the CCM mode the block counter starts initially from one and the MAC value is encrypted or decrypted with the block counter value zero.

If the data length is not a multiple of 16 bytes, the last output encrypted or decrypted data block has to be truncated to the original length. This implies the need for the mask component in the encrypt-then-MAC encryption, MAC-then-encrypt decryption, and CCM decryption. Before XORing the last key stream block with the data block, the bytes of the key stream block corresponding to the extra bytes (zeroes) of the data block have to be masked to zero. This way, the XOR result of the input block and the last key stream block, which is used as the input for the MAC computation, has the correct padding (zeroes). The mask logic is implemented with a ROM, containing 16 masking entries of size 16 byte, and an AND gate.

8.4.2 On-Board Communications

As shown in Figure 8.7, the external SDRAM is used for data transfers between ARM9 and PLD in the prototype. The external memory is larger than the fixed-size DPRAM and it can be switched, which makes the ESL implementation scalable for processing larger amounts of data. By sharing SDRAM, the processor does not have to transfer the data from the data memory in SDRAM to DPRAM for the PLD usage. Instead, PLD can access SDRAM directly, which is faster and allows ARM9 to perform other tasks concurrently. The DMA entity was implemented for the purpose in PLD. It accesses the memory via an AHB bridge. An UART entity was implemented in PLD for transferring data between the development board and the Bluetooth controller.

The data transferred through SDRAM consist of HCI commands and data to the Bluetooth controller, HCI events and data from the controller, and the data to or from the security-processing entity. Nonces, ESL keys, and the UART initialization data are also conveyed through the memory. Nonces and keys are only transmitted to the security-processing entity when the values are initialized or changed.

After writing a HCI command or data to SDRAM, ARM9 uses the other AHB bridge for initiating operations in PLD. The control entity receives the processor requests through the AHB slave in Figure 8.7. The slave contains logic for interfacing the AHB as well as control and memory address registers to which ARM9 requests are written from the bus. Depending on the request, the control entity begins an UART transmission of a HCI command or the encryption and decryption of data. ARM9 is interrupted after an operation is finished. The processor reads the reason for the interrupt from the AHB slave.

When a HCI data packet is received from the Bluetooth controller, it is written to SDRAM by DMA, and ARM9 is interrupted. If the packet payload is not encrypted, the processor decodes the packet and gives the ESL payload to the application. Otherwise, it provides the security-processing entity with the keys and the nonce for decrypting the payload. When the payload is decrypted, ARM9 is again interrupted. If a MAC scheme is used, the processor verifies whether the received and the locally computed MAC values match and conveys the data to the application.

Each HCI command has a corresponding event (acknowledgment) with which the Bluetooth controller replies to the command. In addition, the network operations trigger events. For simplicity and removing unnecessary memory accesses, the PLD control entity filters out the events uninteresting to the processor.

8.4.3 Software Interfaces

In Figure 8.7, hardware abstraction layer (HAL) implements the ESL functionalities on the ARM9 side. It constructs HCI commands and data to SDRAM and reads HCI events from SDRAM as well as decodes payload data from the HCI data packets. It also handles the interrupts initiated by PLD. HAL controls the security-processing entity and UART by modifying the memory mapped control and address registers in the AHB slave. HAL allows using the Bluetooth's own security features as well as choosing among the enhanced security features. If an enhanced security mode is used, HAL performs the MAC value comparison on the reception of a HCI data packet.

The ESL API implementation provides procedures for initiation, connection management, and sending and receiving application data using the HAL services. A pseudocode example of the API usage is presented in Figure 8.9.

In the example, first, the Bluetooth device is initialized. It is defined that the standard security features are not used. The InitBd procedure returns the unique Bluetooth address of the controller and the maximum size of the payload for a single transmission. After the initialization, the device is able to operate in the slave mode. Next, the role of the device is switched to the master mode. The PrepareMaster procedure also scans for the devices in the range and returns a list of found device addresses. Before connecting to a device, the parameters for the enhanced security features are set. A successful connection creation returns the handle of the created link. Application data can be sent over the link with a single procedure call. The data size must respect the maximum payload size defined in the initialization. Finally, the connection is closed.

```
void main() {
   ...
   // initialize Bluetooth device:
   // authentication disabled, no link key
   // type defined, no PIN input
   InitBd(FALSE, NULL, NULL, OwnBdAddress,
              payloadMaxSize);

   // set device into master and scan
   // devices nearby
   PrepareMaster(numberOfResponses, deviceArray);

   // set values for Bluetooth's own link key
   // and for the keys of the enhanced
   // encryption and data authentication,
   // choose encrypt-then-MAC
   SetEncMode(NULL, encKey, macKey,
            ENC_THEN_MAC);

   // create connection to the first found
   // device with the security parameters above
   ConnectToBd(deviceArray[0], connectionHandle);

   // send data to the connected device
   TransmitData(connectionHandle, payloadSize,
                payload, nonce);

   // close connection
   Disconnect(connectionHandle);
}
```

FIGURE 8.9 A pseudocode example of using the ESL API implementation. Values for the parameters in italics are returned by the procedure calls.

Even though not used in the example, each procedure returns a Boolean value that informs whether the performed operation was successful or not. For example, if the transmit buffer of the Bluetooth controller is full, the TransmitData procedure fails. Changing the security parameters requires closing the link and calling the InitBd and SetEncMode procedures again. The application must implement a separate call-back procedure for receiving data. The procedure is automatically called by the ESL API implementation on the reception of a data packet. Connection and MAC failures also trigger a call-back procedure, which the application can use for handling the failures. Similar to the standard Bluetooth, the implemented software supports up to seven simultaneously active master–slave connections.

8.4.4 ESL Authentication Protocols

The TLS and AKEP2 authentication protocols were implemented as application software in ARM9 in the ESL prototype. Their performance does not have a significant effect on the overall ESL processing as the protocols are only required during link establishments. The TLS handshake implementation was derived from the software library [49]. Initially, an unprotected Bluetooth connection is established with the procedures presented in Figure 8.9 and used for running either one of the protocols. The authentication messages are treated as regular data. After the protocol run has been finished, the connection is closed and a new, protected connection with the same peer is created using the newly derived ESL keys or PIN code.

8.4.5 Implementation Results and Comparison

The hardware entities were tested in VHDL simulation with ModelSim SE PLUS 5.8d 2004.06 (Mentor Graphics, San Jose, California, USA). The AES hardware was separately verified against the AES software library [50]. The hardware netlist was generated with Precision RTL Synthesis 2003b.41 (Mentor Graphics, San Jose, California, USA), and the netlist was synthesized with Quartus II v4.1 (Altera Corporation, San Jose, California, USA). The software was compiled with ARM Developer Suite 1.2 (ARM Ltd., Cambridge, UK). The complete ESL implementation was tested in practice with a test application between two development boards. The hardware synthesis results on the PLD of EPXA10-F1020-C2 are presented in Table 8.1.

The maximum clock frequency of the prototype hardware is 43.55 MHz. At the maximum frequency, the throughput of the security-processing entity is 214 Mbit/s (26 cycles/block) in the MAC-then-encrypt, encrypt-then-MAC, and CCM modes, and 507 Mbit/s (11 cycles/block) in the MAC mode. Compared with the maximum Bluetooth transmission speeds, negligible latencies are implied by the added processing.

The maximum size of the data payload that can be transmitted or received in a HCI ACL data packet is 2^{16} bytes [1]. However, the buffers of the Ericsson

TABLE 8.1
Resource Consumption of the ESL Prototype Hardware on EPXA10-F1020-C2

Component	LEs	Memory (bits)
Security processing	3,527	43,008
Control	2,175	0
DMA	894	0
AHB slave	440	0
UART	208	0
Total (% of max.)	7,244 (18%)	43,008 (13%)

Bluetooth controller support only 672 byte HCI payload. Regarding this, using 128-bit MACs and assuming that the 96-bit nonce is transmitted in the payload and that the controller transmits at the maximum payload speed of 723.2 kbit/s, it can be computed that the maximum application data throughput of the ESL implementation is 693 kbit/s. The total throughput of the prototype is further limited by the fixed UART implementation of the Ericsson Bluetooth controller. Its highest speed is 460 kbit/s.

Table 8.2 compares the hardware part of the ESL prototype with the hardware implementation of the standard Bluetooth security [51]. The reported throughputs are for the security-processing components of the implementations at the maximum clock frequency (in the combined modes for ESL). Furthermore, Table 8.3 presents comparisons between the cryptographic cores of the standard Bluetooth design (E_0 and SAFER+) and ESL (AES). To evaluate the AES core used in this work, the table includes measures for two other programmable logic designs. A compact and a fast iterative AES design suitable for the feedback modes of ESL are presented in [52] and [53], respectively. All the reference designs were targeted at Xilinx (Xilinx Inc., San Jose, California, USA) field programmable gate arrays (FPGAs) [54], which differ from the Altera PLDs. However, the basic building

TABLE 8.2
Comparison of the ESL Prototype Hardware with an Implementation of the Standard Bluetooth Security

Implementation	FPGA Device	Logic Units	Memory (bits)	Clock (MHz)	Throughput (Mbit/s)
Standard Bluetooth security [51]	XV2600E-FG1156	19,905 slices	38,272	15	15
ESL (this work)	EPXA10-F1020-C2	7,244 LEs	43,008	44	214

TABLE 8.3
Comparison of AES Hardware Implementations with Implementations of the Cryptographic Cores of the Standard Bluetooth Security [Block RAM (BRAM) Is a Dedicated Xilinx Memory Block of Size 18 Kbits]

Implementation	Device	Logic Units	Memory	Clock (MHz)	Throughput (Mbit/s)
E_0 [51]	XV2600E-FG1156	895 slices	0	15	15
SAFER+ [51]	XV2600E-FG1156	4,058 slices	6,272 bits	20	320
AES [52]	XC2V40-6	146 slices	3 BRAMs	123	358[a]
AES [53]	XV1000-BG560-6	2,257 slices	0	127	1,563[a]
AES (this work)	EPXA10-F1020-C2	1,246 LEs	40,960 bits	44	507

[a] For nonfeedback modes of operation.

blocks of Xilinx FPGAs, logic cells (LCs), are close to the Altera LEs. Each programmable Xilinx FPGA component, slice, contains two LCs. Thus, a slice roughly corresponds to two LEs.

Compared with the hardware implementation of the standard Bluetooth security [51], considerably lower LE (LC) consumption and higher encryption throughput were achieved with the ESL prototype. However, part of the ESL control is implemented in the ARM9, which slightly decreases the occupied PLD resources. On the other hand, the ESL implementation includes the additional ARM9 and Bluetooth controller interfaces. The number of memory bits was increased but this can be reduced by modifying the AES core according to the reference AES implementations. Compared with the works of Rouvroy et al. and Standaert et al. [52,53], the core of this work is an average implementation with reasonable resources and throughput. For example, a compact CCM mode implementation using the AES architecture [52] is presented in [55].

Instead of using E_0 for encryption and SAFER+ for authentication and key generation, in ESL all three procedures can use AES when AKEP2 is used for authentication and key agreement. Table 8.3 shows that this reduces the hardware resources and also implies shorter latencies because of the higher performance.

8.5 CONCLUSIONS

In this chapter, an ESL for protecting Bluetooth data links was proposed and implemented. ESL improves the standard security design by replacing the proprietary encryption with an AES-based design and adding cryptographic integrity protection. Furthermore, two authentication protocols are supported for entity authentication and key agreement in different usage scenarios.

All the components used are long-lived, generally considered secure, or they have been proven to be secure. The easy-to-use ESL API offers an application developer simple access to the wireless link and transparent security processing, supporting both ESL and the standard Bluetooth design with safer parameterization. The only security-related task for the application using ESL is nonce management. The prototype implementation showed that ESL implies only a negligible processing latency, which is also lower than that of the standard Bluetooth design. A beneficial aspect of ESL is that its low-level security-processing components are compatible with those of the other significant short-range wireless technologies, which enables efficient resource sharing in devices supporting multiples of these technologies. ESL can also be extended to support the 802.1X authentication framework [24]. This allows interoperability with the IEEE 802.11i WLAN [6] authentication architecture as well as support for a wider range of standard authentication protocols.

REFERENCES

1. Bluetooth Special Interest Group (SIG), Specification of the Bluetooth System, Version 2.0 + EDR, 2004.
2. Official Bluetooth membership web site, Online: http://www.bluetooth.org (visited 2/3/2006).
3. Official Bluetooth wireless info web site, Online: http://bluetooth.com (visited 2/3/2006).
4. Menezes, A.J., van Oorschot, P.C., and Vanstone, S.A., *Handbook of Applied Cryptography*, 5th printing, CRC Press, Boca Raton, FL, 2001.
5. National Institute of Standards and Technology (NIST), Advanced encryption standard, FIPS-197, 2001.
6. Institute of Electrical and Electronics Engineers (IEEE), IEEE standards for information technology—telecommunications and information exchange between systems—local and metropolitan area networks—specific requirements—part 11: wireless LAN medium access control (MAC) and physical layer (PHY) specifications—amendment 6: medium access control (MAC) security enhancements, IEEE 802.11i, 2004.
7. Institute of Electrical and Electronics Engineers (IEEE), IEEE standards for information technology—telecommunications and information exchange between systems—local and metropolitan area networks—specific requirements—part 15.3: wireless medium access control (MAC) and physical layer (PHY) specifications for high-rate wireless personal area networks (WPANs), IEEE 802.15.3, 2003.
8. Institute of Electrical and Electronics Engineers (IEEE), IEEE standards for information technology—telecommunications and information exchange between systems—local and metropolitan area networks—specific requirements—part 15.4: wireless medium access control (MAC) and physical layer (PHY) specifications for low-rate wireless personal area networks (WPANs), IEEE 802.15.4, 2003.
9. ZigBee Alliance, ZigBee Specification, Version 1.0, 2004.

10. Hämäläinen, P. et al., Design and implementation of an enhanced security layer for Bluetooth, in *Proceedings of 8th IEEE International Conference on Telecommunications* (*ConTEL 2005*), Zagreb, Croatia, 2005, p. 575.
11. Institute of Electrical and Electronics Engineers (IEEE), IEEE standards for information technology—telecommunications and information exchange between systems—local and metropolitan area networks—specific requirements—part 15.1: wireless medium access control (MAC) and physical layer (PHY) specifications for wireless personal area networks (WPANs), IEEE 802.15.1, 2005.
12. Muller, T., Bluetooth Security Architecture, Version 1.C.116/1.0, Bluetooth SIG, 1999.
13. Gehrmann, G., Bluetooth Security White Paper, Version 1.01, Bluetooth SIG Security Expert Group, 2002.
14. Morris, S., Recommendations to Early Implementers: Encrypting Broadcast Transmissions in Bluetooth Piconets, Version 1.0, Bluetooth SIG Security Expert Group, 2002.
15. Cylink, SAFER+: Cylink corporation's submission for the advanced encryption standard, presented at the 1st Advanced Encryption Standard Candidate Conference (AES1), Ventura, CA, 1998, Online: http://csrc.nist.gov/CryptoToolkit/aes/round1/conf1/saferpls-slides.pdf (visited 2/9/2006).
16. Borisov, N., Goldberg, I., and Wagner, D., Intercepting mobile communications: the insecurity of 802.11, in *Proceedings of 7th Annual International Conference on Mobile Computing and Networking*, Rome, Italy, 2001, p. 180.
17. Laurie, A. and Laurie, B., Serious flaws in Bluetooth security lead to disclosure of personal data, The Bunker, October 2004, Online: http://www.thebunker.com/security/Bluetooth (visited 2/9/2006).
18. Oates, J., Virus attacks mobiles via Bluetooth, The Register, June 2004, Online: http://www.theregister.co.uk/2004/06/15/symbian_virus (visited 2/9/2006).
19. Whitehouse, O., Bluetooth: Red Fang, Blue Fang, presented at CanSecWest/core04, 2004, Online: http://cansecwest.com/csw04/csw04-Whitehouse.pdf (visited 2/9/2006).
20. Long distance snarf web site, 2004, Online: http://trifinite.org/trifinite_stuff_lds.html (visited 2/9/2006).
21. Jakobsson, M. and Wetzel, S., Security weaknesses in Bluetooth, in *Proceedings of Cryptographer's Track at RSA Conference 2001* (*CT-RSA 2001*), San Francisco, CA, 2001, p. 176.
22. Vainio, J.T., Bluetooth security, 2000, Online: http://www.niksula.cs.hut.fi/~jiitv/bluesec.html (visited 2/9/2006).
23. Shaked, Y. and Wool, A., Cracking the Bluetooth PIN, in *Proceedings of 3rd Conference on Mobile Systems, Applications, and Services* (*MobiSys 2005*), Seattle, WA, 2005, p. 39.
24. Institute of Electrical and Electronics Engineers (IEEE), IEEE standard for local and metropolitan area networks—port-based network access control, IEEE 802.1X-2004, 2004.
25. Gehrmann, G. and Nyberg, K., Enhancements to Bluetooth baseband security, in *Proceedings of Nordic Workshop on Secure IT-Systems* (*NordSec 2001*), Copenhagen, Denmark, 2001, p. 39.
26. Whitehouse, O., War nibbling: Bluetooth insecurity, 2003, Online: http://www.rootsecure.net/content/downloads/pdf/atstake_war_nibbling.pdf (visited 2/9/2006).

27. Levi, A. et al., Relay attacks on Bluetooth authentication and solutions, in *Proceedings of 19th Symposium on Computer and Information Science* (*ISCIS 2004*), Antalaya, Turkey, 2004, p. 278.
28. Kügler, D., Main in the middle attacks on Bluetooth, in *Proceedings of 7th International Financial Cryptography Conference* (*FC '03*), Gosier, Guadeloupe, French West Indies, 2003, p. 149.
29. Hager, C.T. and Midkiff, S.F., An analysis of Bluetooth security vulnerabilities, in *Proceedings of IEEE Wireless Communications and Networking Conference* (*WCNC 2003*), New Orleans, LA, 2003, p. 1825.
30. Hermelin, M. and Nyberg, K., Correlation properties of the Bluetooth combiner, in *Proceedings of International Conference on Information Security and Cryptology* (*ICISC '99*), Seoul, Korea, 2000, p. 17.
31. Fluhrer, S. and Lucks, S., Analysis of the E0 encryption systems, in *Proceedings of 8th Annual International Workshop on Selected Areas in Cryptography* (*SAC 2001*), Toronto, Canada, 2001, p. 38.
32. Lu, Y. and Vaudenay, S., Faster correlation attack on Bluetooth keystream generator E0, in *Proceedings of 24th Annual International Cryptology Conference, Advances in Cryptology* (*CRYPTO 2004*), Santa Barbara, CA, 2004, p. 407.
33. Lu, Y., Meier, W., and Vaudenay, S., The conditional correlation attack: a practical attack on Bluetooth encryption, in *Proceedings of 25th Annual International Cryptology Conference, Advances in Cryptology* (*CRYPTO 2005*), Santa Barbara, CA, 2004, p. 97.
34. Fluhrer, S., Mantin, I., and Shamir, A., Weaknesses in the key scheduling algorithm of RC4, in *Proceedings of 8th Annual International Workshop on Selected Areas in Cryptography*, Toronto, Canada, 2001, p. 1.
35. Bluetooth SIG, Bluetooth SIG presents new specification, and two implementation guides, press release, June 2003, Online: http://www.bluetooth.com/news/sigreleases.asp?A = 2&PID = 870&ARC = 1&ofs = 20 (visited 2/9/2006).
36. Lipmaa, H., Rogaway, P., and Wagner, D., CTR-mode encryption, 2000, Online: http://csrc.nist.gov/CryptoToolkit/modes/workshop1/papers/lipmaa-ctr.pdf (visited 2/9/2006).
37. Bellare, M. et al., A concrete security treatment of symmetric encryption: analysis of the DES modes of operation, 2000, Online: http://www.cs.ucsd.edu/users/mihir/papers/sym-enc.html
38. Dworkin, M., Recommendations for block cipher modes of operation—methods and techniques, NIST special publication 800–38A, 2001.
39. Bellare, M., Killian, J., and Rogaway, P., The security of the cipher block chaining message authentication code, *Journal of Computer and System Sciences*, 61(3), 362, 2000.
40. Krawczyk, H., The order of encryption and authentication for protecting communications (or: how secure is SSL?), in *Proceedings of 21st Annual International Cryptology Conference, Advances in Cryptology* (*CRYPTO 2001*), Santa Barbara, CA, 2001, p. 310.
41. Ferguson, N. and Schneier, B., *Practical Cryptography*, Wiley Publishing, Inc., Indianapolis, IN, 2003.
42. Whiting, D., Housley, R., and Ferguson, N., Counter with CBC-MAC (CCM)—AES mode of operation, 2003, Online: http://csrc.nist.gov/CryptoToolkit/modes/proposedmodes/ccm/ccm.pdf (visited 2/9/2006).

43. Jonsson, J., On the security of CTR+CBC-MAC, in *Proceedings of 9th Annual International Workshop on Selected Areas in Cryptography* (*SAC 2002*), St. John's, Newfoundland, Canada, 2002, p. 76.
44. Dierks, T. and Allen, C., The TLS protocol 1.0, RFC 2246, 1999.
45. Bellare, M. and Rogaway, P., Entity authentication and key distribution, in *Proceedings of Advances in Cryptology* (*CRYPTO '93*), Santa Barbara, CA, 1993, p. 232.
46. Altera Corporation web site, Online: http://www.altera.com (visited 2/10/2006).
47. Ericsson Microelectronics AB, BSK technical documentation, BSK/S10311/1.2, 2001.
48. Hämäläinen, P. et al., Implementation of link security for wireless local area networks, in *Proceedings of IEEE International Conference on Telecommunications* (*ICT 2001*), Bucharest, Romania, 2001, Vol. 1, p. 299.
49. Mozilla.org, Network security services (NSS) software library, Online: http://www.mozilla.org/projects/security/pki/nss (visited 2/10/2006).
50. Gladman, B., AES code, Online: http://fp.gladman.plus.com/AES (visited 2/10/2006).
51. Kitsos, P. et al., Hardware implementation of Bluetooth security, *IEEE Pervasive Computing*, 2(1), 21, 2003.
52. Rouvroy, G. et al., Compact and efficient encryption/decryption module for FPGA implementation of the AES Rijndael very well suited for small embedded applications, in *Proceedings of IEEE International Conference on Information Technology: Coding and Computing* (*ITCC 2004*), Las Vegas, NV, 2004, Vol. 2, p. 583.
53. Standaert, F.-X. et al., A methodology to implement block ciphers in reconfigurable hardware and its application to fast and compact AES Rijndael, in *Proceedings of ACM/SIGDA International Symposium on Field Programmable Gate Arrays* (*FPGA 2003*), Monterey, CA, 2003, p. 216.
54. Xilinx web site, Online: http://www.xilinx.com (visited 2/10/2006).
55. Hämäläinen, P., Hännikäinen, M., and Hämäläinen, T.D., Efficient hardware implementation of security processing for IEEE 802.15.4 wireless networks, in *Proceedings of International Midwest Symposium on Circuits and Systems* (*MWSCAS 2005*), Cincinnati, OH, 2005, p. 484.

9 WLAN Security Processing Architectures

Neil Smyth, Maire McLoone, and John V. McCanny

CONTENTS

9.1 INTRODUCTION

The growth of wireless devices [1,2] is ever-increasing, as are the data transmission bandwidths of the underlying technologies. However, the constraints on battery life and hence processing power are in contradiction to the increased demands and complexity of data processing such as that required by new and more robust security protocols. It is natural for security standards to evolve as more secure methods become available and weaknesses become apparent. This situation necessitates the use of programmable systems to adapt to counteract security weaknesses and provide some degree of future-proofing to prolong the lifetime of products.

IEEE 802.11i [3] is an optional amendment to the IEEE 802.11 standard offering enhanced security at the medium access control (MAC) layer, which is intended to overcome the weaknesses of previous security schemes. These enhancements include: (a) an improved RC4-based scheme for legacy systems [4,5], (b) advanced encryption standard (AES)-based encryption [6,7] in newer wireless local area network (WLAN) devices, and (c) a design that has been integrated with IEEE 802.1x [8,9] to provide a system whereby clients and access points must query an authentication server. The wired equivalent privacy (WEP) security scheme defined in the IEEE 802.11b amendment has been effectively enhanced to form the temporal key integrity protocol (TKIP), designed for use in legacy systems and offered to the consumer by the interim Wi-Fi protected access (WPA) standard as IEEE 802.11i was getting finalized. New 802.11 stations and access points are expected to implement more secure and modern schemes described by IEEE 802.11i, based on AES. This scheme is based on the royalty-free, well understood, and proven counter (CTR) with cipher-block chaining message authentication code (CBC-MAC) protocol (CCMP).

Previous research into cryptographic microprocessor architectures has included extensions to instruction sets to increase the performance of symmetric key [10] and asymmetric key [11] cryptographic algorithms. Moreover, Fiskiran and Lee [12] have developed a data-scalable, general-purpose processor architecture with cryptographic extensions. However, these previous architectures do not accelerate specific cryptographic algorithms (such as AES) as they are generic in nature and do not target specific applications (such as WLAN security). Commercial WLAN security solutions for integration into SoC designs do exist, such as Elliptic Semiconductor's CCM IP core [13] and Helion's 802.11i CCM IP core [14]. These efficient hardware cores integrate into ASIC/FPGA designs, but only perform one WLAN protocol and do not offer the versatility of software. Cavium Networks NITROX processors [15] offer impressive data throughputs and versatility for numerous security applications, but are expensive in terms of area, being single-chip solutions, which are unsuitable for low-power, low-cost, and compact SoC WLAN security solutions.

In this chapter two dedicated WLAN security architectures are proposed. The first is a programmable design that comprises the authors' own primitive RISC processor design [16] and two hardware accelerators, which perform AES and RC4 encryption. The RISC processor is designed not only to execute a standard range of arithmetic and logic instructions, but also dedicated cryptographic instructions that are required to implement WLAN protocols. These include 32-bit cyclic redundancy checks (CRC32) and Michael authentication [3], a packet authentication algorithm developed for IEEE 802.11i.

The WLAN processor has been designed specifically to perform the frame processing requirements of WEP, TKIP, WRAP, and CCMP as specified in draft 3.0 of the IEEE 802.11i standard. It should be noted that WRAP was not adopted in the final IEEE 802.11i standard. The programmability of the processor also provides the ability to manipulate packet types, such as AES-CCM or AES-offset codebook mode (OCB) [17] encapsulation, which can provide functionality for internet protocol security (IPSec) [18].

The second approach evaluates the performance of a fixed-functionality WLAN security design. In contrast to the versatility offered by the microcode-driven programmable processor, this architecture acts as an accelerator with a limited range of functionality, but is intended to be more resource efficient and have a higher throughput. This design is also targeted at IEEE 802.11i applications, but provides a level of generic symmetric cryptography functionality for both the RC4 and AES ciphers. It is the responsibility of the host processor to perform packet manipulation such as header processing or field padding and provide data for the accelerator in such a format that it may be processed.

In Section 9.2 a brief background of the basic operation of an IEEE 802.11 wireless network is provided. Section 9.3 provides a brief description of the RC4 and AES silicon cores used throughout the designs. Section 9.4 provides a summary of the operation and architecture of the programmable WLAN processor. Section 9.5 describes the alternative approach of a fixed-functionality accelerator. The performance of both the WLAN processor and the accelerator device are outlined in Section 9.6. Finally, conclusions are given in Section 9.7.

9.2 BACKGROUND ON IEEE 802.11

WLAN technology is standardized by IEEE 802.11 and in particular, these standards define the MAC and physical (PHY) layers. The original standard has evolved with a number of amendments that adopt new technologies as they become available. The original standard described a communication technology that operates at 1 Mbps, whereas the IEEE 802.11b amendment introduced in 1999 increased the maximum throughput to 11 Mbps. The IEEE 802.11a and IEEE 802.11g amendments have increased the maximum theoretical throughput of WLAN technology to 54 Mbps. A number of manufacturers have produced nonstandard solutions that increase the data throughput further, although these implementations are not generally interoperable.

MAC service data units (MSDUs) containing data payloads are provided to the MAC layer for transmission by the logical link control (LLC) layer. The PHY layer interfaces with the radio and handles packetized data (known as frames) that are modulated for transmission or demodulated when received. This includes control and management frames used to administer the wireless network and data encapsulated into the payload of MAC physical data units

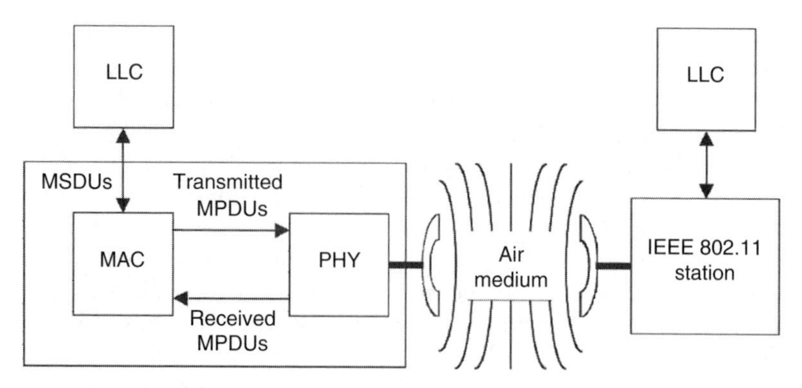

FIGURE 9.1 IEEE 802.11 Device-to-device interface.

(MPDUs). The MAC monitors the activity on the wireless medium to determine if it is inactive and available to transmit data, otherwise the IEEE 802.11 station is configured to receive data. A basic outline of the processing layers in an IEEE 802.11 station is illustrated in Figure 9.1.

WLAN communication disseminates information indiscriminately and therefore does not offer the inherent security of a wired LAN. The optional WEP amendment was introduced to overcome this security weakness. WEP provides a means of confidentiality for the packetized data using the RC4 stream cipher. Authentication is provided through the use of cyclic redundancy checksums. However, WEP has been shown to be a weak security protocol with many flaws [19–21]. Manufacturers have improved the security of WEP by introducing proprietary amendments and enhancements.

The IEEE 802.11i [3] standard for enhanced MAC security was developed to address the need for more robust security as the uptake of wireless communications increases. This optional amendment provides an upgrade path for the RC4-based WEP scheme known as the TKIP, which may be supported by legacy systems already in the field. However, new devices are expected to use the higher security AES block cipher with the CTR and CCMP as described by IETF RFC 3610 [22].

Two AES schemes were proposed for adoption within the IEEE 802.11i standard. The relatively new and licensable wireless robust authentication protocol (WRAP) was proposed as an efficient means to provide confidentiality and authentication using a block cipher. WRAP is based on Rogaway's OCB [17] for block ciphers. The royalty-free CCMP emerged as part of the standard after a number of years of discussion.

All frames transferred to and from the PHY from the MAC layer are composed of header fields, an optional data payload field, and a frame check sequence (FCS) field. The FCS is composed of a CRC32 checksum computed over the data payload. This FCS allows for error detection of those frames received. Cryptographic processing of frames occurs at the MAC layer where the data payload may be encrypted and an integrity check is computed over the header and data payload. The WEP, TKIP, and CCMP security schemes alter only the data payload field and subsequently the FCS field.

IEEE 802.11i requires the use of two basic encryption primitives—AES and RC4. WEP and TKIP use RC4 for confidentiality purposes to transform plaintext to or from ciphertext. Michael (a Feistel-based algorithm) and CRC32 are used to provide authentication in the form of message integrity check (MIC) and integrity check value (ICV) fields in TKIP and WEP frames, respectively. These values are appended to the MPDU of a frame, while initialization vectors and miscellaneous control data are inserted at the start of the MPDU.

WRAP is a licensable scheme based on AES-OCB that was discussed as part of the standardization process, but was not adopted. WRAP is more efficient than CCMP in terms of the AES processing required as authentication does not require an additional block cipher operation and is instead performed in parallel to encryption. CCMP is the AES scheme adopted by IEEE 802.11i and is defined by the CCM algorithm [22], which uses the CTR and CBC-MAC block cipher modes. The predominant reason that CCMP prevailed over WRAP is its status as a royalty-free scheme with a proven record of security.

9.3 CRYPTOGRAPHIC ACCELERATOR CORES

9.3.1 AES

A commercial AES core [23] has been used in both the processor and accelerator approaches to WLAN security processing. This synthesizable verilog core is capable of both encryption and decryption and uses a fixed key length of 128 bits. A 32-bit datapath is used to offer a 44 clock cycle latency when used in electronic code book (ECB) mode. An on-the-fly key scheduler is used to negate the need to store the expanded keyspace. The basic architecture of this core is outlined in Figure 9.2.

9.3.2 RC4

The RC4 core [23] is shown in Figure 9.3. This RC4 core comprises a 256-byte dual-port RAM to store the RC4 state array and control logic to manipulate the contents of this state array. The control logic uses a simple state machine to perform byte swapping operations and permutations on the state array contents. The core must be initialized with a key of up to 256 bits in length, an operation that requires 1152 cycles regardless of the key length or content. The RC4 core produces a stream of pseudorandom bytes that is exclusive-ORed with an input byte stream. The encryption and decryption of a message are identical operations.

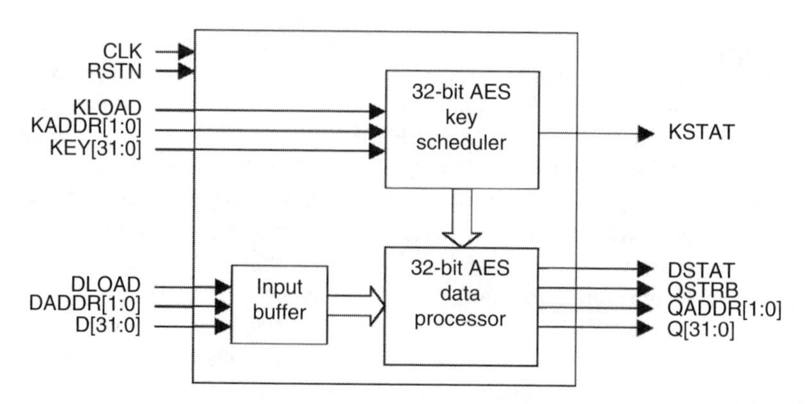

FIGURE 9.2 AES block diagram.

9.4 WLAN SECURITY PROCESSOR ARCHITECTURE

9.4.1 DESIGN OVERVIEW

During the standardization process of IEEE 802.11i, a novel WLAN security processor [16] was proposed. This processor was designed to reduce the risks involved in developing a hardware implementation of a standard that had not yet been finalized by providing a highly flexible architecture, as depicted in Figure 9.4. The design comprises the basic elements of any RISC processor such as an instruction decode unit, an arithmetic and logic unit (ALU), and a barrel shifter. RC4 and AES encryption accelerators were used to provide high-performance encryption that may operate in parallel to the main execution pipeline of the processor. In addition, IEEE 802.11i-specific instructions were provided to enable support for CRC32 checksums and Michael authentication tags.

The use of accelerators allows the intensive encryption algorithms of AES and RC4 to be performed in parallel to other operations, such as data fetch/store to main memory. These accelerators are based on the commercially available AES and RC4 cores [23] described previously. This use of

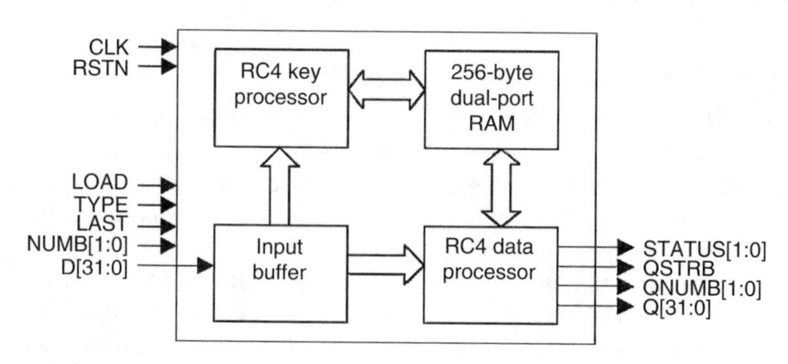

FIGURE 9.3 RC4 block diagram.

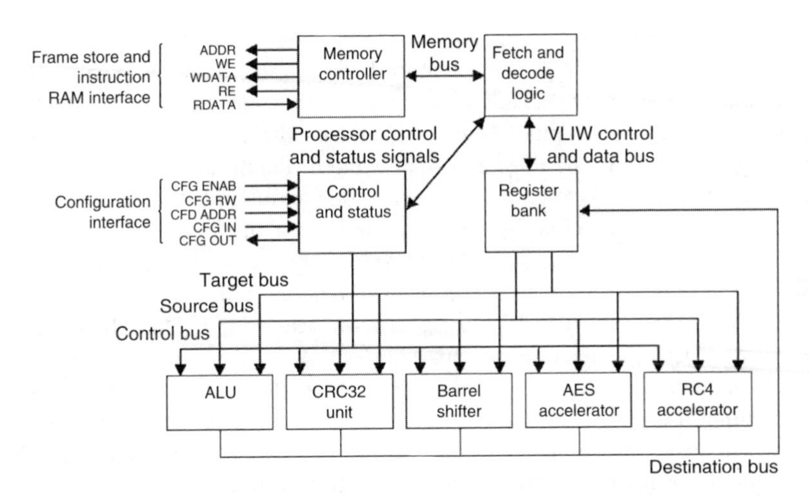

FIGURE 9.4 WLAN security processor block diagram.

high-performance accelerators within the execution pipeline overcomes the reduced throughput inherent in a processor architecture. This is achieved while maintaining a low clock frequency, which aids in reducing power dissipation.

9.4.2 Architecture Description

The processor has been developed as a synthesizable verilog core. Synchronous read RAM is used to efficiently store the microcode that defines the frame encapsulation schemes, all input frames, and all generated output frame data. This RAM is local to the processor and may be defined to use a number of configurations such as separate instruction and data RAM.

Figure 9.4 shows the WLAN processor's two distinct and simple RAM-based interfaces. The first is a 32-bit interface to an external memory containing the instruction and data. The second is a 32-bit interface to the processor's configuration registers, which may be used by the host processor to control the operating parameters. This simple RAM-based interface allows bridging to many commonly used processor buses, such as the ARM advanced peripheral bus (APB) [24].

The register bank is composed of a 32-word register file with a single register window and uses a three-port RAM with two read ports and a write port. The 32-bit instructions are segmented into five components to control how the execution pipeline manipulates contents of the register bank. These components include: (a) an 8-bit instruction code, (b) a 5-bit source register pointer, (c) a 5-bit target register pointer, (d) a 5-bit destination register pointer, and (e) a 9-bit region used to store miscellaneous data.

When processing an instruction, the code word is initially fetched from memory and passed to the decode logic. In the first phase of the three-stage processor pipeline, the 32-bit code word is extended to form a very long instruction word (VLIW) to create control data for the execution pipeline and to provide two register read addresses. In the second clock cycle, the VLIW is passed to the execution pipeline alongside the source and target registers from the register bank. In the third and final cycle of the processing pipeline, the manipulated data is written back to a specified address of the register bank.

In terms of operation the host microprocessor must take control of the WLAN security processor in an 802.11 MAC. The processor is ready to begin security encapsulation once it has been initialized with microcode and the frame to be processed has been written to local data RAM. The WLAN security processor then operates autonomously from the host, requiring only that a number of address pointers are programmed to indicate where the input and output frames are located in local data RAM and that a start command is issued to the processor. When frame encapsulation is complete, an interrupt is generated and the output frame can then be transferred to the PHY.

9.5 WLAN SECURITY ACCELERATOR ARCHITECTURE

9.5.1 Design Overview

A contrasting approach to the reprogrammable WLAN processor is provided by a fixed-functionality solution. This fixed-functionality solution was designed after the IEEE 802.11i standard was finalized and with knowledge of the security schemes that must be supported. Therefore, it was designed to be highly efficient rather than programmable. As such, a second peripheral device to accompany a host microprocessor was realized.

The WEP, TKIP, and CCMP schemes described by IEEE 802.11i have been implemented using specific high-performance hardware acceleration. The encryption and authentication schemes have been accelerated using the same RC4 and AES cores used previously. General-purpose functionality is provided by the ECB, CTR, CBC, and CFB modes of operation for the AES block cipher. This is achieved by reusing the same logic resources used to provide AES-CCM. The RC4 stream cipher may be accessed directly without WEP or TKIP specific functionality. This general-purpose functionality is at the cost of additional silicon resources, but this is felt to be acceptable given the provided benefits in flexibility.

The overall design of the high-performance WLAN security accelerator is shown in Figure 9.5. This design uses the same AES and RC4 encryption components used in the processor design. Rather than controlling these components using a microcode-driven processor pipeline, a number of hardwired finite state machines (FSMs) are used. These FSMs provide fixed functionality for two stream processing pipelines that operate in parallel. This includes an RC4 pipeline that offers key sizes from 8 to 256 bits and an AES pipeline that provides a range of AES schemes using 128-bit keys.

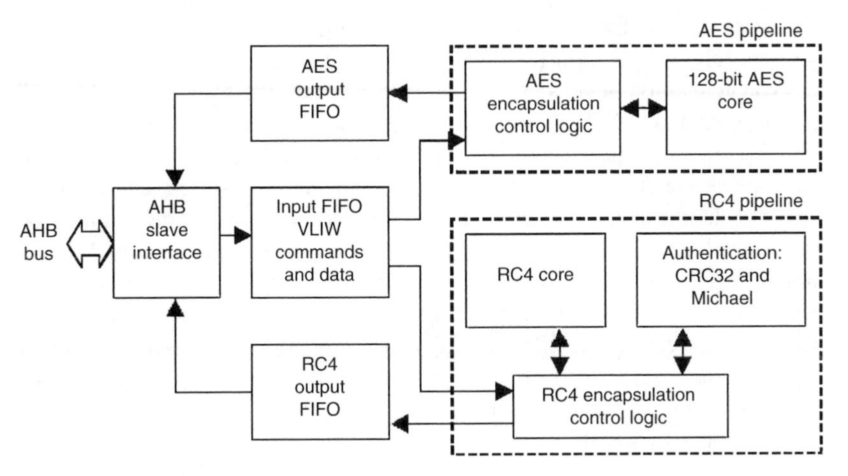

FIGURE 9.5 WLAN security accelerator block diagram.

9.5.2 AHB Slave Interface and Data Queues

The advanced high-performance bus (AHB) system bus [24] has been chosen to provide the necessary memory bandwidth. This open standard provides for burst access between master and slave devices, offering improved memory bandwidth in comparison with APB, which supports only single data transfers. To take advantage of burst access it is necessary to provide input and output buffers.

A single input buffer is used to queue all data and commands to be processed by either pipeline. When that pipeline is free to accept further data and sufficient storage is available at the output buffer, the command and data are pulled from the input FIFO. The accelerator operates at maximum throughput provided the host can maintain the fill level of the input and output buffers so as to prevent stalling. The maximum throughput of the accelerator is detailed later in the description of the two processing pipelines. In addition, a host microprocessor or a dynamic memory allocation (DMA) controller requires less frequent polling of the accelerator's status and fewer interrupts from which to respond. Burst access therefore increases the memory bandwidth available to other peripherals on the bus.

9.5.3 RC4 Processing Pipeline

The RC4 pipeline performs encryption of a byte-oriented packet using the RC4 stream cipher. Encryption and decryption using RC4 are identical processes. In addition, WEP requires a CRC32 checksum to be generated over a packet before RC4 encryption during encapsulation or after RC4 decryption during decapsulation. It is also necessary to encrypt the CRC32 checksum itself.

TKIP is supported by replacing the CRC32 checksum with a 64-bit Michael ICV. The accelerator provides a 2-bit control register to allow the host system to enable either of the two authentication methods or to disable both entirely and perform RC4 encryption only.

The generation of keys is essential for TKIP on a per-frame basis. The key mixing operation of TKIP involves two phases to generate a suitable key for the RC4 stream cipher. These key mixing operations require basic arithmetic and logical operations including XOR, AND, addition, shifting, and rotation. The AES box is also required for byte permutation, an operation that may be performed using a 256-byte lookup table. It should be noted that TKIP would most likely be used for backward compatibility only in a WLAN that also supports CCMP. Therefore, expending silicon resources for this functionality has been determined as too expensive and the decision was taken that it should be performed in firmware by the host microprocessor. This provides some flexibility in the event of changes to the TKIP key selection process brought about by situations such as security weaknesses.

9.5.4 AES Processing Pipeline

The pipeline that supports AES is required to perform CCM as described by IETF RFC 3610 [22]. This scheme requires CTR mode encryption and CBC-MAC authentication. The CCM encryption and decryption processes require only the order of encryption and authentication to be reversed and that only AES encryption functionality be provided.

An important consideration of AES-CCM is the necessity to perform two block encryptions for each 128-bit block—one pass to encrypt and a second to authenticate. If the accelerator simply offered CTR and CBC modes individually, this would require the host system to perform two read and write operations of the entire packet. In order to reduce the required memory bandwidth of the accelerator, AES-CCM is performed in series. This requires the host system to write the message once to the input buffer where the accelerator will then perform two AES passes as necessary on the same 128-bit data block.

The silicon resources used to perform CCM have been employed to provide the ECB, CTR, CBC, and CFB modes of operation. Whereas CCM requires only AES encryption for full functionality, these additional modes require AES decryption and the extra silicon resources this implies. As discussed previously this provides functionality beyond that of IEEE 802.11i, allowing generic acceleration of those applications requiring the AES block cipher and providing some of the general-purpose functionality offered by the WLAN processor approach.

9.6 PERFORMANCE EVALUATION

The resulting APB slave WLAN processor and the AHB slave WLAN accelerator architectures were developed as VERILOG RTL synthesizable cores. Both approaches were synthesized using SYNPLIFY PRO to create netlists for FPGA implementation. The VERILOG RTL includes

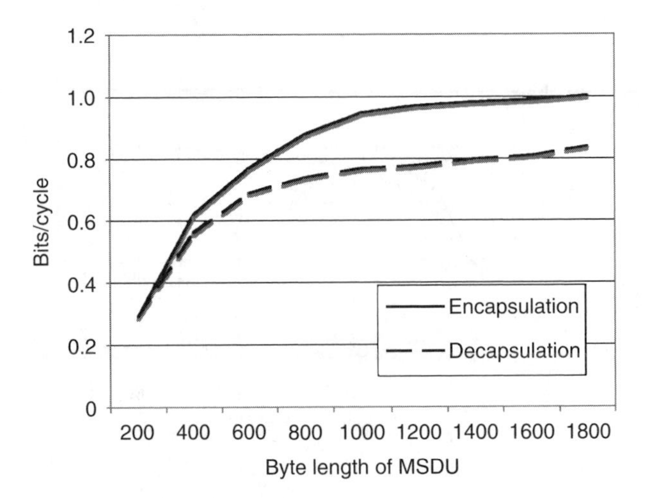

FIGURE 9.6 WEP performance.

compile-time parameters that implement technology-dependant resources, such as RAM. ALTERA QUARTUS II and XILINX foundation series were used to perform place and routing of the netlist onto ALTERA STRATIX and XILINX VIRTEX II devices, respectively. In the case of ASIC implementation, SYNOPSYS DESIGN COMPILER was used to synthesize the cores using TSMC 0.13 μm standard cell libraries under worst-case conditions.

Both devices proposed in this chapter have been described in VERILOG and modelled with a cycle-accurate C^{++} model. They were simulated using MODELTECH MODELSIM and the functionality of the cores was verified against the functional C^{++} model using self-checking testbenches. Test vectors were obtained from various sources, such as the national institute of standards and technology (NIST) [6], the IEEE 802.11 task group I, [3] and the IETF [22], to verify the capability of the cores to perform RC4 and AES encryption, and the various packet encapsulation schemes. The C^{++} model executable of the WLAN security processor allows accurate debugging of microcode and fast-performance evaluations to be made, such as the bits per cycle performance shown in Figure 9.6 through Figure 9.9. This allows microcode to be tested on a hardware model and cycle counts to process MPDUs to be rapidly collated. The following section details the performance of both approaches in terms of data throughput and silicon resources. A comparison to some commercial solutions is also provided.

9.6.1 WLAN Security Processor

RC4 requires an initialization period of 1152 cycles regardless of data payload length. Initialization of coprocessors and the register bank for every new packet or key change can be expected to have a greater effect on small data fields. This is

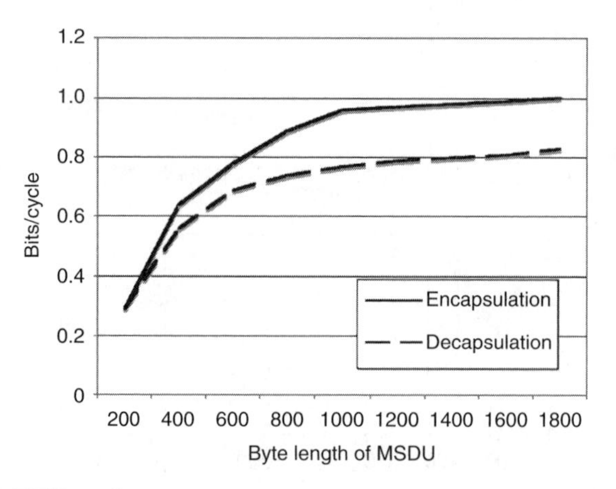

FIGURE 9.7 TKIP performance.

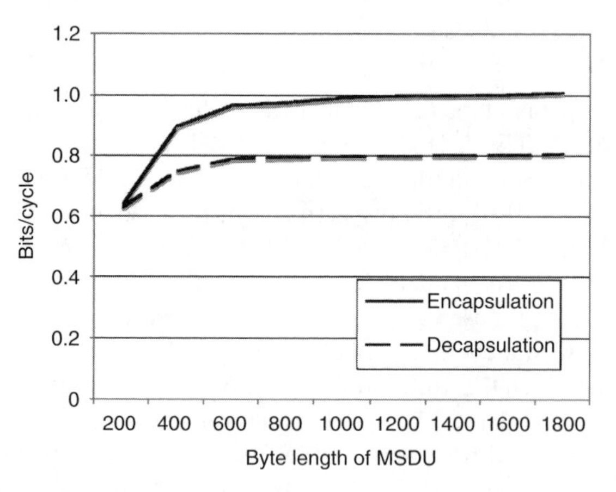

FIGURE 9.8 CCMP performance.

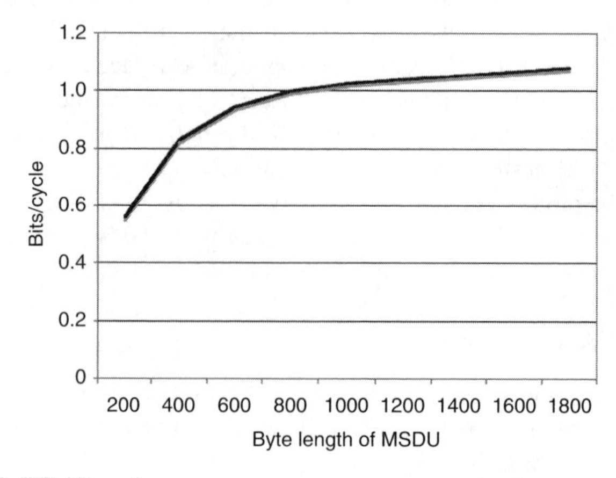

FIGURE 9.9 WRAP performance.

TABLE 9.1
WLAN Security Processor Technology Resource Usage

Technology	Logic Resources	RAM Resources	Timing Constraint
TSMC 130 nm	60.4 k gates	405.8 k gates	250 MHz
XILINX VIRTEX 2–5	3474 slices	15 BRAM	80.3 MHz
ALTERA STRATIX EP1S10F484C5	6873 LE	15 M4K	102.4 MHz

because the number of cycles required for initialization is fixed for each security scheme, hence contributing to larger performance degradation in smaller frames, particularly when RC4 is used. This is illustrated in Figure 9.6 and Figure 9.7.

AES-based CCMP requires less initialization than the RC4-based WEP or TKIP, as illustrated by Figure 9.8. This is largely attributable to the absence of any initialization steps in AES. The degradation in performance of decryption compared with encryption in WEP, TKIP, and CCMP is attributable to the additional processing required for decapsulation and the reordering of processing steps that increases the latency of each step. WRAP encryption and decryption are largely identical and therefore have a similar processing bandwidth as shown by the performance illustrated in Figure 9.9.

WRAP requires half the AES processing that must be performed for CCMP. However, the more complex initialization and data processing must be performed using the general-purpose instructions. The reduced performance of these instructions is offset to some degree by the processor's ability to execute such instructions in parallel to AES encryption. Therefore, although 50% less AES processing is required per data block the more complex encryption and authentication arithmetic requires significant computation. Overall, WRAP offers 10% greater performance throughput than CCMP.

The performance results of the WLAN security processor are illustrated in Table 9.1. The figures quoted are considered worst case (i.e., one 256×8 dual-port RAM for RC4 functionality and one 4096×32 dual-port RAM for packet buffer and instruction memory). The large instruction and frame store RAM may optionally be single port and may be reduced in size depending on the application and specifications.

9.6.2 WLAN Security Accelerator

Performance of the WLAN security accelerator has been measured using WEP, TKIP, and CCMP frame encapsulation. The performance of the AES modes of operation are also presented to illustrate the general-purpose throughput that can be expected from the accelerator. The throughput figures are presented in Table 9.2.

TABLE 9.2
Maximum Data Throughput in Clock Cycles of WLAN Security Accelerator

Operation	Encapsulation	Decapsulation	Notes
WEP	12	12	Encryption and CRC32 authentication of 32-bits
TKIP	12	12	Encryption and Michael authentication of 32-bits
CCMP/AES-CCM	88	92	Encryption and authentication of a 128-bit block
AES-ECB	44	44	Encryption of a 128-bit block
AES-CBC	48	44	Encryption of a 128-bit block
AES-CFB	48	44	Encryption of a 128-bit block
AES-CTR	44	44	Encryption of a 128-bit block

The effect of different frame sizes has more of an effect on those frames to be encapsulated using RC4 in the WEP and TKIP protocols, as shown in Figure 9.10 and Figure 9.11 compared with the performance of CCMP shown in Figure 9.12. This is attributed to the seeding of the RC4 state array that must be performed for every frame. This seeding operation requires 128 clock cycles to initialize the state array and a further 1024 clock cycles to seed the state array with a key.

The CCMP has no initialization cost beyond that of loading the key and initialization vectors. There is typically six blocks of MPDU header fields that are muted before CBC-MAC authentication and are not encrypted. These blocks consume only 48 cycles per block.

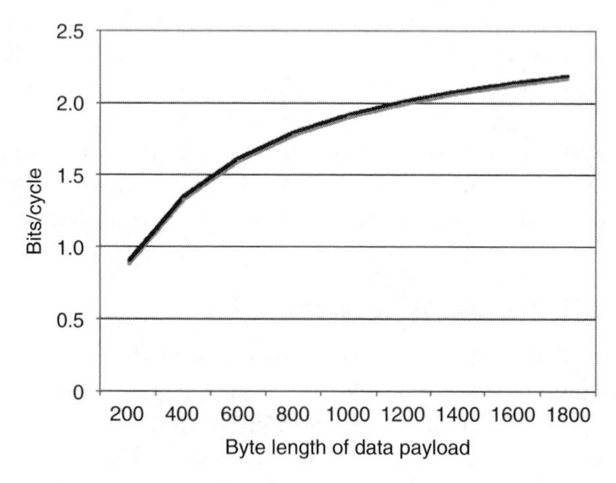

FIGURE 9.10 WEP performance with different frame sizes.

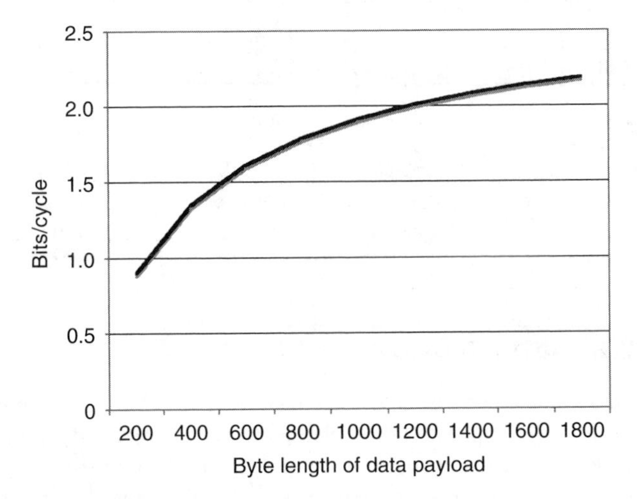

FIGURE 9.11 TKIP performance with different frame sizes.

The resources of the hardwired WLAN accelerator are presented in Table 9.3. All input and output buffers are provided with 64 word FIFOs. It is clear to see that the design of the accelerator approach does not require large and dedicated local RAM resources for instruction space or storing entire MSDU frames before encapsulation. The reduced complexity architecture of the accelerator results in an estimated 33% reduction in logic gates in comparison with the processor approach. Therefore, the accelerator approach results in significantly smaller resource usage.

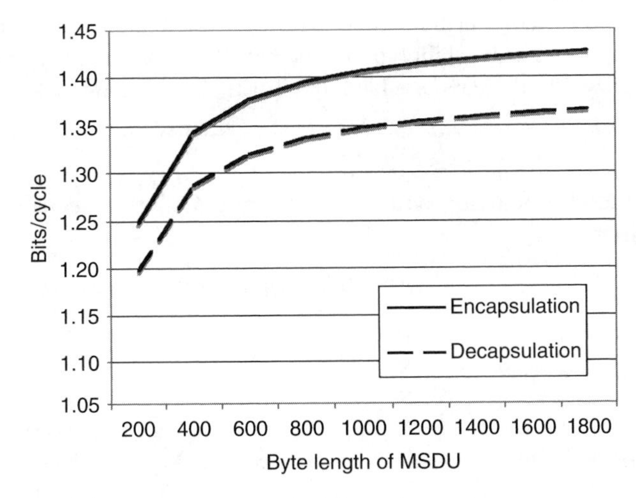

FIGURE 9.12 CCMP performance with different frame sizes.

TABLE 9.3
WLAN Accelerator Resource Metrics

Technology	Logic Resources	RAM Resources	Timing Constraint
TSMC 130 nm	47 k gates	28 k gates	250 MHz
XILINX VIRTEX 2–5	3264 slices	14 BRAM	92.4 MHz
ALTERA STRATIX EP1S10F484C5	6739 LE	8 M4K 23 M512	102.9 MHz

9.6.3 Performance Summary

A comparison of the proposed WLAN processor and WLAN accelerator is shown in Table 9.4. This evaluation also includes a number of commercially available solutions that were previously outlined in Section 9.1.

The specialized and resource-efficient WLAN accelerator offers up to twice the performance of the WLAN processor for both WEP and TKIP while using fewer logic gates. CCMP performance can be sustained in excess of 350 Mbps, 40% faster than that offered by the processor approach. TKIP/WEP key construction, message padding, and header field muting must be performed by the host microprocessor for reasons of future-proofing. Therefore, the host processor cycles consumed to do this must be taken into account in the overall system.

The WLAN Security Processor offers a dedicated solution to wireless security, providing efficient hardware acceleration for the complex operations of encryption and a software-driven execution pipeline to provide versatility. From Table 9.4 it is evident that while performance is more than sufficient for high-speed WLAN the design compares only moderately with currently available solutions in terms of throughput. However, the major advantage of the processor approach is its ability to provide backward compatibility to existing networks as well as those required in future IEEE 802.11i compatible networks.

TABLE 9.4
Performance Comparison with Commercially Available Security Processors

Solution	WEP	TKIP	WRAP	CCMP	Maximum Throughput
WLAN security processor	Y	Y	Y	Y	>275 Mbps
WLAN security accelerator	Y	Y	N	Y	>548 Mbps
Elliptic Semiconductor [13]	Y	Y	N	Y	300–700 Mbps
Helion 802.11 CCM IP core [14]	N	N	N	Y	<2 Gbps
Cavium Networks NITROX processors [15]	N	Y	N	Y	50–10,000 Mbps

A significant advantage of the processor over the accelerator approach is its autonomous operation. The processor approach requires only minimal host processor involvement in writing the data to the relevant address either directly by the host processor or using a DMA controller. The programmable architecture is achieved at the expense of additional logic and RAM resources.

Helion offers an AES-CCM IP core solution with a throughput of up to 2 Gbps and Elliptic Semiconductor provides a CCMP core, which runs at 300–700 Mbps and can optionally support WEP and TKIP functionalities. However, limited design details of these commercial solutions are available. Moreover, since the common operating frequency in commercial MAC/PHY products is 80 MHz and the IEEE 802.11a/g standards only require a throughput of 54 Mbps, these architectures are a very expensive solution for wireless applications and lack the compactness of a dedicated WLAN processor. The proposed processor's power efficiency is lower in comparison with the commercial solutions, due to the supplemental hardware and execution time required to process software.

9.7 CONCLUSION

In this chapter two architectures which can be used to provide WLAN security are described. It has been recognized that there is a security processing gap in wireless devices, caused by the low-power and relatively low processing capabilities of such devices and the demands of complex security protocols on SoC solutions. Cryptographic instructions contained in the instruction set architecture (ISA) of microprocessor technologies can significantly improve the performance of security protocols operating on such microprocessors. Industrially available microprocessor technologies using such techniques include ARM's SecureCore [24] family, MIPS' SmartMIPS, [25], or ARC [26]. Another method to increase throughput is to implement a hardware accelerator block to perform all secure packet processing for a particular application, or provide certain functionality such as AES encryption and map this into a processor's address space as a peripheral device.

The WLAN processor design described in this chapter combines both of these approaches to provide a processor designed specifically to perform efficient cryptographic processing of WLAN frames, with little intervention from the host microprocessor. As the host microprocessor is no longer burdened by performing bulk encryption and packet formatting of 802.11 frames, more processing power can be used to enhance and improve other services on a wireless handset. For example, the user interface may be more feature rich and responsive, there may be less lag experienced when using data services and dedicated hardware can perform cryptographic functions more efficiently than a general-purpose processor thus improving battery life.

The processor approach allows changes to be made to the method of encapsulation while maintaining the efficiency and high throughput of hardware encryption acceleration. Any fluctuations in IEEE 802.11i standards may be overcome by implementing the WLAN security processor into a system where it can be reprogrammed to accommodate any changes to packet structure or security scheme. The security processor offers support for all WLAN protocols and also backward compatibility together with future upgradeability as standards evolve. Moreover, it achieves this extra functionality at a throughput rate required by current 802.11a/g amendments.

In contrast, the more efficient but less versatile fixed-functionality accelerator is targeted at specific WLAN protocols. The accelerator approach offers flexibility by providing high-performance AES and RC4 support. However, this is at the expense of host processor intervention and control to provide any additional computation such as block cipher modes of operation. This flexibility is achieved at the expense of additional silicon gates for control purposes and the provision of AES decryption functionality that is not required for IEEE 802.11i.

ACKNOWLEDGMENT

This work was supported by Amphion Semiconductor Ltd (a subsidiary of Conexant Systems Inc), Belfast, Northern Ireland.

REFERENCES

1. S. Ravi, A. Raghunathan, and M. Sankaradass, Securing Wireless Data: System Architecture Challenges, ISSS '02, Kyoto, Japan, October 2002.
2. S. Gayal and S.A. Vetha Manickam, Wireless LAN Security Today and Tomorrow, Center for Information and Network Security, Pune University, 2002.
3. IEEE 802.11 Wireless LAN Standards, IEEE 802.11 Working Group, Task Group I, URL: http://grouper.ieee.org/groups/802/11, March 2006.
4. B. Schneier, Algorithms and source code in C, *Applied Cryptography: Protocols*, John Wiley & Sons, pp. 397–398, 1996.
5. Alleged RC4 C^{++} source code, URL: http://cryptopp.sourceforge.net/docs/ref, March 2006.
6. National Institute of Standards and Technology, AES (Rijndael) Specification and Information, URL: http://csrc.nist.gov/encryption/aes/rijndael, March 2006.
7. J. Daemen and V. Rijmen, *The Design of Rijndael: AES–The Advanced Encryption Standard*, Springer-Verlag, 2002.
8. J.S. Park and D. Dicoi, WLAN Security: Current and Future, IEEE Internet Computing, URL: http://computer.org/internet/, September/October 2003.
9. J. Williams, Providing for Wireless LAN Security, Part 2, IT Professional, IEEE Internet Computing, URL: http://computer.org/internet/, November/December 2002/2003.
10. J. Burke, J. McDonald, and T. Austin, Architectural support for fast symmetric-key cryptography, Proc. Int. Conf. Architectural Support for Programming Languages and Operating Systems (ASPLOS), pp. 178–189, November 2000.

11. J. Groszschaedl and G.A. Kamendje, Instruction Set Extension for Fast Elliptic Curve Cryptography over Binary Finite Fields GF(2^m), Proc. IEEE Int. Conf. Application-Specific Systems, Architectures and Processors (ASAP), pp. 442–455, June 2003.
12. A.M. Fiskiran and R.B. Lee, PAX: a datapath-scalable minimalist cryptographic processor for mobile environments, *Embedded Cryptographic Hardware: Design and Security*, N. Nedjah and L. de Macedo Mourelle, eds, Nova Science, New York, September 2004.
13. Elliptic Semiconductor, Cryptography IP Cores. URL: http://www.ellipticsemi.com, March 2006.
14. Helion Technology, CCM IP Core. URL: http://www.heliontech.com, March 2006.
15. Cavium Networks, NITROX Processors. URL: http://www.cavium.com, March 2006.
16. N. Smyth, M. McLoone, and J.V. McCanny, WLAN Security Processor, IEEE Transactions on Circuits and Systems I, 53(7): 1506–1520, July 2006.
17. P. Rogaway, M. Bellare, and J. Black, OCB: A block cipher mode of operation for efficient authenticated encryption, *ACM Transactions on Information and System Security (TISSEC)*, 6(3): 365–403, 2003.
18. Internet Engineering Task Force (IETF), Request for Comments (RFC) 2401, Security Architecture for the Internet Protocol, URL: http://www.ietf.org/rfc/rfc2401.txt, March 2006.
19. J.A. LaRosa, WPA: A Key Step Forward in Enterprise-Class Wireless LAN (WLAN) Security, Meetinghouse Data Communications, URL: http://www.meetinghousedata. com/, March 2006.
20. N. Borisov, I. Goldberg, and D.Wagner, Security of the WEP Algorithm, URL: http://www.isaac.cs.berkeley.edu/isaac/wep-faq.html, March 2006.
21. J. Walker, Unsafe at Any Key size: An Analysis of the WEP Encapsulation, IEEE 802.11 Committee, October 2000.
22. Internet Engineering Task Force (IETF), Request for Comments (RFC) 3610, Counter with CBC-MAC, URL: http://www.ietf.org/rfc/rfc3610.txt, March 2006.
23. Amphion Semiconductor Ltd, Cryptography IP Cores. URL: http://www.amphion.com, March 2006.
24. ARM Cores. URL: http://www.arm.com, January 2004.
25. SmartMIPS. URL: http://www.mips.com, January 2004.
26. ARC International, Accelerating Network Processing with Extensions to the User-Customizable ARC Tangent Microprocessor, URL: http://www.arc.com/downloads/downloads-white-papers.html, January 2004.

10 Security Architecture and Implementation of the Universal Mobile Telecommunication System

Paris Kitsos and Nicolas Sklavos

CONTENTS

10.1 INTRODUCTION

In universal mobile telecommunication system (UMTS) networks, security is a vast topic. Access network connection must naturally be secure, but in addition to this, security must be taken into account in many other aspects as well. The various communication models, through fixed, wireless, and satellite networks in both outdoor and indoor environments, lead to situations in which sometimes sensitive information is transferred between different parties and networks.

To achieve efficient and secure roaming among different networks, UMTS supports more complex security mechanisms than the previous mobile systems such as global system for mobile communications (GSM) and digital enhanced cordless telecommunications (DECT). The confidentiality of voice calls is protected in the radio access network (RAN), as is the confidentiality of the transmitted user data. This means that the user has control over the choice of parties with whom he or she wants to communicate. Users also want to know that confidentiality protection is really applied. Therefore, visibility of applied security mechanisms is needed. Privacy of the user's whereabouts is generally appreciated. However, if persistent tracking of users were to occur, they would end up anxious. On the other hand, the privacy of user data is a critical issue when data are transferred through the network. Privacy and confidentiality are largely synonymous in this presentation. Finally, encryption algorithms are stronger. The application of authentication algorithms is stricter and subscriber confidentiality is tighter.

UMTS security architecture is based on three procedures. At the beginning, the user authenticates the network and vice versa. Then, integrity protection of the signaling information is required, and finally the user and signaling data must be confidentiality protected. Note that publicly available cryptographic algorithms are used for encryption and integrity protection. Algorithms for mutual authentication are operator-specific.

Authentication is performed by the authentication and key agreement (AKA) procedure [1]. In this study, the AKA procedure is built on the Rijndael block cipher [2]. In addition to authentication, the AKA procedure also produces the cipher key (CK) and the integrity key (IK). In UMTS, only the encryption mode of the Rijndael block cipher is used [3] as an iterated hash function [4]. The block and key length have been set to 128 bits.

The integrity of signaling information is handled by the message authentication code (MAC) [1] procedure implemented using the one-way hash function $f9$ and the IK [5]. The user and the signaling data confidentiality protection is handled by the data confidentiality [1] procedure implemented using the stream cipher $f8$ and the CK [5].

Both $f8$ and $f9$ algorithms are based on the Kasumi block cipher [6,7]. An implementation of the Kasumi block cipher using feedback logic and negative edge-triggered pipeline [8] is introduced. This implementation makes the critical path shorter, without increasing the latency of the cipher execution. If the clock frequency is determined by the system specifications, the usage of negative edge-triggered pipelining can reduce the clock frequency of its original value for the same data throughput. As a result, power consumption is reduced.

The remainder of this chapter is organized as follows: In Section 10.2, the UMTS security architecture is briefly introduced and in Section 10.3, the proposed system architecture and the hardware implementation are described. In Section 10.4, the hardware implementation results are presented and evaluated. Finally, the chapter ends with some conclusions.

10.2 SECURITY IN UMTS

From the specification point of view, the major scope of the 3rd Generation Partnership Project (3GPP) is to define and maintain the UMTS specifications. In user equipment (UE), all the security tasks are integrated as shown in Figure 10.1. The mobile end-user's terminal-end equipment of the radio interface is officially called user equipment (UE) in the UMTS. From the network point of view, the UE is responsible for those communication functions that are needed at the other end of the radio interface, excluding any end-user applications.

Generally, the cornerstone of the authentication mechanism is a master key K, which is shared between the USIM of the user and the home network database. This is a permanent secret with a length of 128 bits. The key K is never made visible between the two locations. At the time of

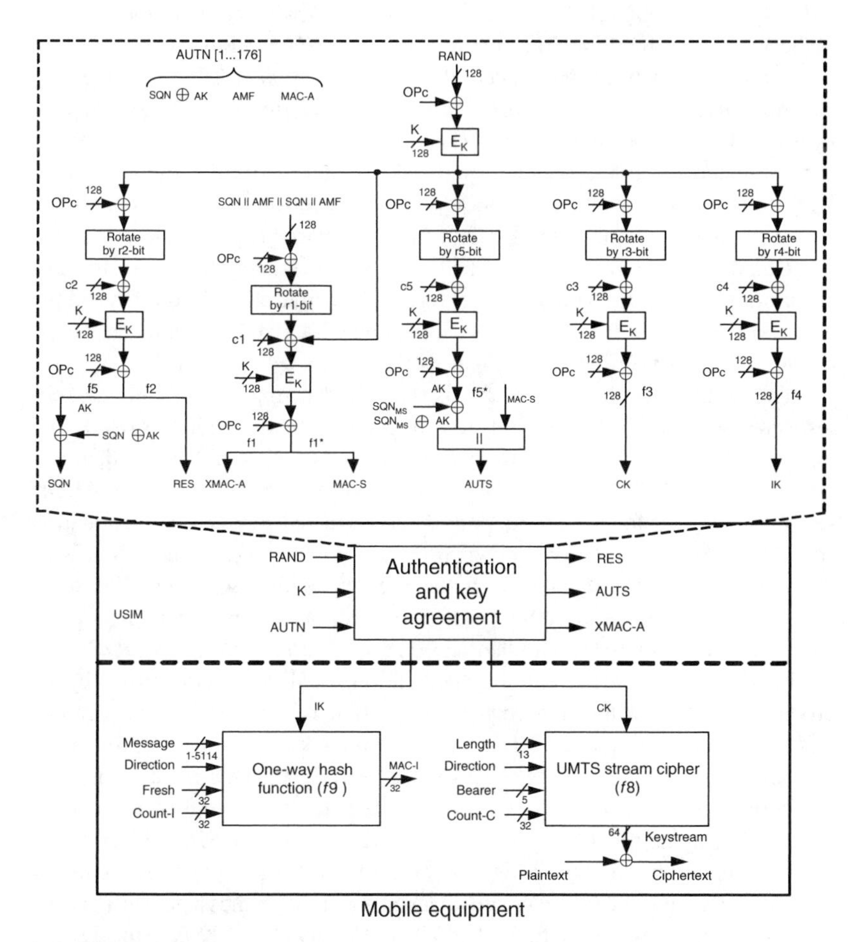

FIGURE 10.1 UMTS security diagram.

authentication, keys for encryption and integrity checking are derived. These are temporary keys with the same length (128-bit). New keys are derived from the permanent key K during every authentication event.

The AKA is integrated in the universal subscriber identity module (USIM). It is the most important security feature of the UMTS security system. The other two basic processes, MAC and data confidentiality, are integrated onto the mobile device.

The authentication procedure can be started after the user is identified in the serving network (SN). Identification occurs when the identity of the user has been transmitted to the visitor location register (VLR) or serving GPRS support node (SGSN). Then, VLR or SGSN sends an authentication data request to the authentication center (AuC) in the home network. Then, the AuC sends to the users the appropriate authentication parameters, authentication token (AUTN) and random challenge (RAND). These parameters with the addition of the secret key (K) [9] are the only information that the AKA module needs to perform the authentication procedure. The AUTN is a 176-bit value that contains three subvalues. The first is the XOR operation product (symbolized by $\oplus$) of the sequence number (SQN) and the anonymity key (AK); the second is the authentication management field (AMF); and the third is the MAC-A.

The main task of the AKA module is the execution of the Rijndael block cipher algorithm (symbolized by E_K). The OPc variable is computed outside the USIM and is stored in USIM. The two necessary parameters for this calculation are the operator variant algorithm configuration field (OP) and the secret key K. The constants c1, c2, c3, c4, and c5 and also the integers r1, r2, r3, r4, and r5 that define the cyclical rotation are specified in [4].

The authentication procedure is executed using the challenge-and-response scheme. Therefore, at the beginning the received RAND and $SQN \oplus AK$ with the secret key K are used to produce SQN and user authentication response (RES). Then the double concatenation (symbolized by $||$) of SQN and AMF is used for the production of the expected message authentication code (XMAC-A) and the resynchronization message code (MAC-S). Then XMAC-A is compared with the received MAC-A and if XMAC-A and MAC-A match, the authentication procedure continues. If these parameters are different, a "user authentication reject" is sent back to VLR/SGSN.

In all cases, for network authentication, USIM should verify if the produced SQN is within the correct range. If it is within the correct range, USIM continues to generate RES. RES is sent back to the VLR/SGSN and it is compared with the expected response XRES. If they match, the user authentication is passed, otherwise it is failed. If SQN is out of the correct range, the authentication procedure is failed, and USIM generates a resynchronization token (AUTS). AUTS is the concatenation of $SQN_{MS} \oplus AK$ and MAC-S. A synchronization failure message with AUTS as parameter is sent back to VLR/SGSN. Finally, the CK and the IK are generated by RAND and the secret key K.

The signaling information that is sent between the user and the network is integrity protected by the integrity algorithm $f9$. Practically, the integrity algorithm is a block cipher that works as one-way hash function. $f9$ appends the MAC-I to messages to ensure that they are generated at the claimed identity. Based on the input parameters such as message, direction, fresh, count-I, and IK, the hash function ($f9$) computes MAC-I. As mentioned earlier, IK is the 128-bit IK produced in USIM. Message is the input bit-stream and direction is a 1-bit input, which indicates the direction of transmission. Fresh is a 32-bit random number and count-I is a 32-bit time variant. MAC-I is appended to the message and is sent over the radio access link. The receiver computes the corresponding XMAC-I and verifies the data integrity of the message by comparing it with the received MAC-I.

User data and some signaling information are considered sensitive and must be confidentiality protected. Based on the input parameters such as length, direction, bearer, count-C, and CK, the stream cipher ($f8$) computes the output keystream block, which is used to encrypt or decrypt the input plaintext or ciphertext and produce the output ciphertext or plaintext. As mentioned earlier, CK is the 128-bit confidentiality key produced in USIM. Length is the number of the plaintext bits and direction is a 1-bit input that indicates the direction of transmission (uplink or downlink). The bearer is a 5-bit input and count-C is a 32-bit time variant. This type of encryption has the advantage that the mask data can be generated even before the actual plaintext is known. In this case, final encryption is a fast operation.

10.3 SECURITY ARCHITECTURE AND HARDWARE IMPLEMENTATION

For UMTS security the architecture illustrated in Figure 10.2 is proposed. This system is a part of UE. It consists of two fragments. The first fragment is integrated in USIM and implements the AKA. The second fragment is integrated in the mobile device and implements the stream cipher ($f8$ algorithm) and the one-way hash function ($f9$ algorithm).

The proposed system architecture is supported by a control unit, which coordinates all system operations and processes. In the mobile device fragment, a common 64-bit data bus is used for the internal data transfers. The algorithms' appropriate keys are stored in the RAM. The required parameters of the $f8$ and $f9$ algorithms are stored in the ROM. Through the I/Os interface, the proposed system transfers data from and to the external environment.

In the USIM fragment, a data bus of 8 bits is used for all the internal data transfers. The appropriate parameters are stored in RAM, whereas the fixed constants ci and integers ri are stored in ROM. The I/Os unit is used for USIM communication with the rest of the system. The controller synchronizes the USIM operation. The three basic units of the system architecture are described in detail subsequently.

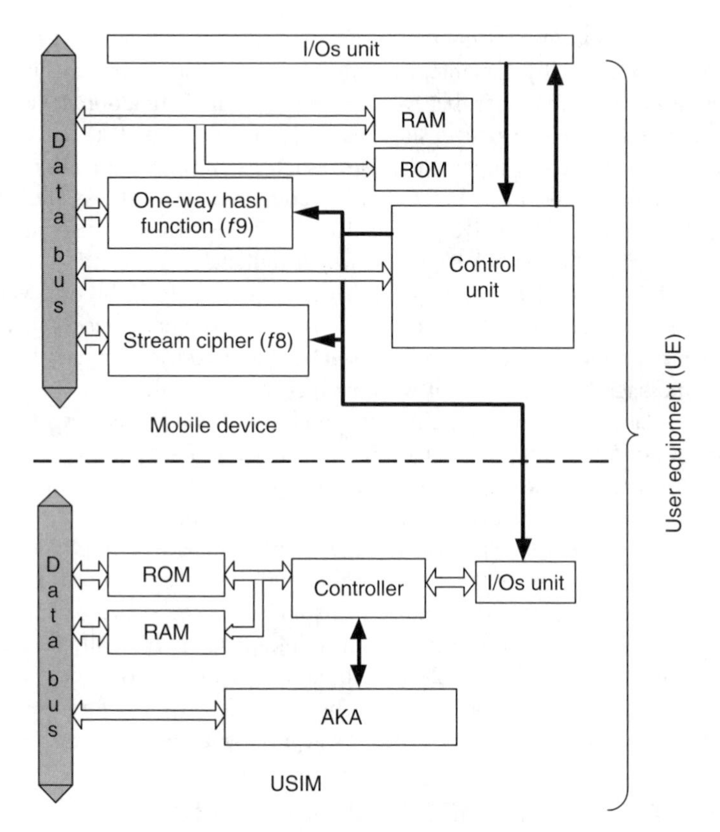

FIGURE 10.2 UMTS security system hardware architecture.

10.3.1 Authentication and Key Agreement Unit

A typical implementation of the AKA procedure needs six Rijndael modules (Figure 10.1). In the proposed system design, as shown in Figure 10.3 only three Rijndael modules are used. In this way, a significant hardware resources reduction is achieved, which is important in applications with strict area limitations. In addition, from the performance measurements following, it is obvious that the proposed architecture performs better than the UMTS specifications demands.

In many devices, the requirement of portability places several restrictions on the consumed power dissipation. To meet these restrictions, special attention has been given to the design of the proposed architecture (Figure 10.2). It is well known that a reduction in hardware resources results in a total active capacitance reduction. Therefore, the overall system power consumption is reduced. In addition, the use of on-chip storage resources reduces the active capacitance significantly. This is achieved using internal memory blocks.

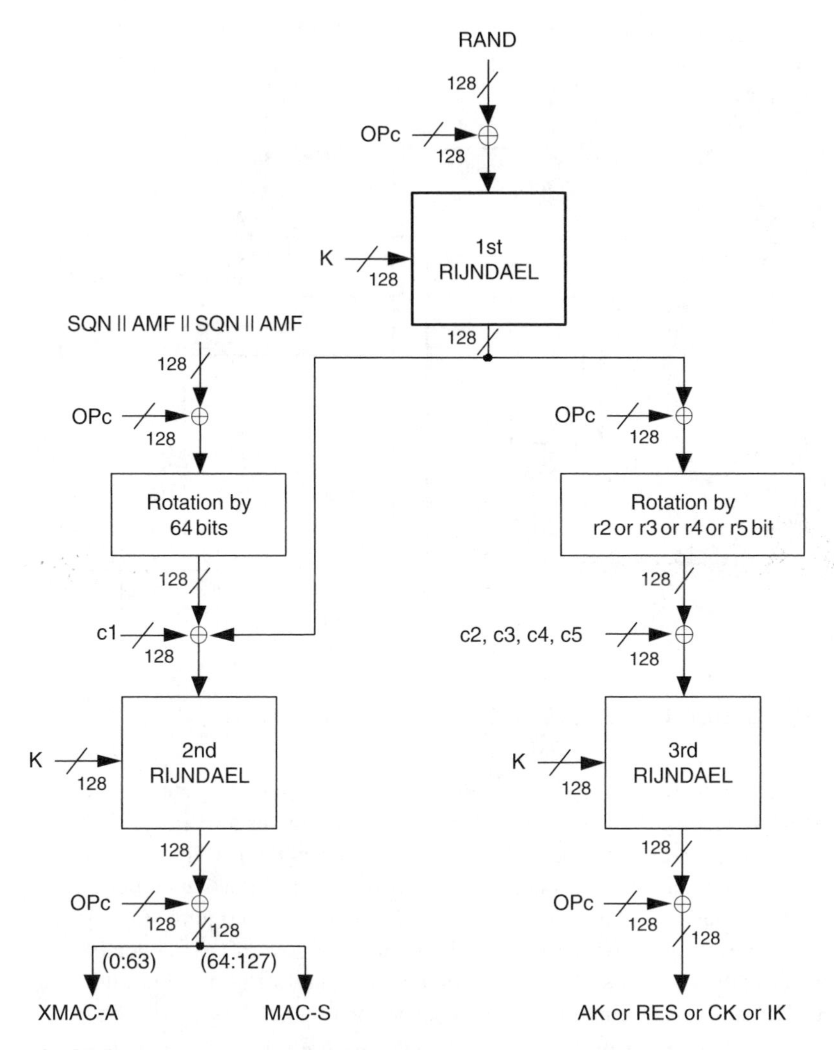

FIGURE 10.3 Authentication and key agreement implementation.

The constants ci and the integers ri are stored and accessed from the 20-byte ROM blocks. The OPc value is stored and accessed from the RAM. The required authentication parameters are produced on the output of the 2nd Rijndael and the 3rd Rijndael modules.

10.3.1.1 Rijndael Block Cipher

The proposed hardware implementation of the Rijndael block cipher is shown in Figure 10.4. This is similar to the implementation in [10], but uses less hardware resources.

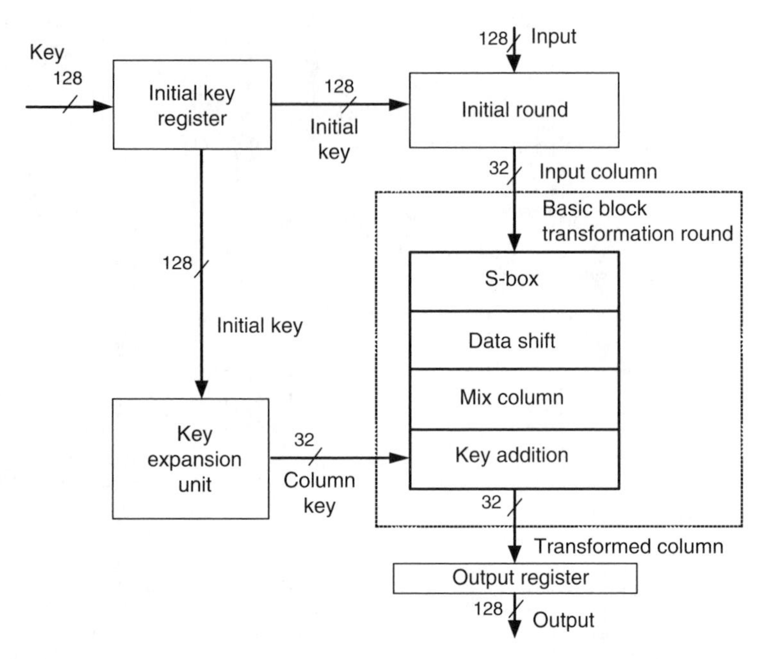

FIGURE 10.4 Rijndael block cipher hardware implementation.

The transformations of the algorithm architecture operate on the intermediate result called state. The state can be pictured as a rectangular array of bytes. This array has four rows. The number of columns (Nb) is equal to the block length divided by 32. The key is also considered as a rectangular array with the same number of rows as state. The number of columns (Nk) is equal to the key length divided by 32. The number of rounds (Nr) depends on the values Nb and Nk. For block and key length equal to 128 bits, both the values of Nb and Nk are equal to 4 and Nr is defined as 10.

The proposed Rijndael architecture consists of the key expansion unit, the basic block transformation round, the initial round, and the appropriate registers. Forty-one clock cycles are needed for the completion of a 128-bit plaintext transformation.

The basic block transformation round is composed of four building blocks: S-boxes, data shift, mix column, and key addition. To achieve high-speed performance, the S-boxes are implemented using ROM blocks.

In general, FPGAs devices have internal ROM (RAM) blocks available. In the proposed implementation, four [256×8]-bit ROM blocks were used. The S-box delay time, implemented by ROM blocks, is 12.8 ns.

The S-boxes require the implementation of two different mathematical functions: (i) the multiplicative inverse of each byte of the state in the finite field GF(2^8) and (ii) an affine mapping transformation over GF(2). The most known very large scale integration (VLSI) architecture for the multiplicative inverse in GF(2^m) uses arrays of basic inversion block cells [11–13]. This design has time and area requirements, with complexity varying from O(m^2) to O(m^4) [11–13]. The execution of the multiplicative inverse in GF(2^m) needs a number of cycles per inversion in the range between m and $3m+2$ [12,13]. These values are unacceptable for high-speed implementation of a cryptographic algorithm. The multiplicative inverse function produces a byte, which is the input of the affine mapping transformation function. This function is defined by the following equation:

$$\begin{aligned} Out(i) = {} & \text{In}[i]xor\text{In}[(i+4)\bmod 8]xor\text{In}[(i+5)\bmod 8]xor \\ & \text{In}[(i+6)\bmod 8]xor\text{In}[(i+7)\bmod 8]xor\, C(i), \end{aligned} \tag{10.1}$$

where In[i] is the ith bit of the input byte and $C(i)$ is the ith bit of the byte constant C (={01100011}), as the algorithm specifications define. The round keys are calculated on the fly by the key expansion unit. Therefore, the keys production procedure has no additional time delay cost on the Rijndael critical path.

10.3.2 Message Authentication Code and Confidentiality Protection Units

The one-way hash function $f9$ [5] with the use of the IK computes MAC-I of a message. The length of the message may be between 1 and 20,000 bits. The $f9$ algorithm is based on a Kasumi block cipher, in a variant version of the cipher block chaining-message authentication code mode (CBC-MAC) standard as defined in ISO 9797 [14].

First, the initial values for the hash function ($f9$) unit according to the input parameters (Figure 10.5) are derived. Second, it computes the padding string (PSi), which is one input of the basic feedback loop. The maximum message length is 20,000 bits (length $\leq$20,000) and the maximum number of the PSi is 313 ($\approx$20,000/64). This value denotes the number of the iteration loops.

This implementation uses two registers to hold variables A and B. After each loop operation, the registers A and B are updated and the new values are the new input of the Kasumi module. Finally, the Kasumi cipher is executed for one more time using a modified IK and the content of the register B. After this operation, in the leftmost 32 bits of register B, the MAC-I value is stored.

The confidentiality algorithm $f8$ [5] is a stream cipher that encrypts or decrypts a block of data between 1 and 20,000 bits in length using a confidentiality key. The stream cipher $f8$ is based on a Kasumi cipher in the form of output feedback mode (OFB) [15]. It generates the keystream in multiples of 64 bits. The proposed cipher ($f8$) implementation is depicted in Figure 10.6.

First, the initial Kasumi values are padded to make a 64-bit input and the input block number (BLKCNT) is computed. During the initialization process

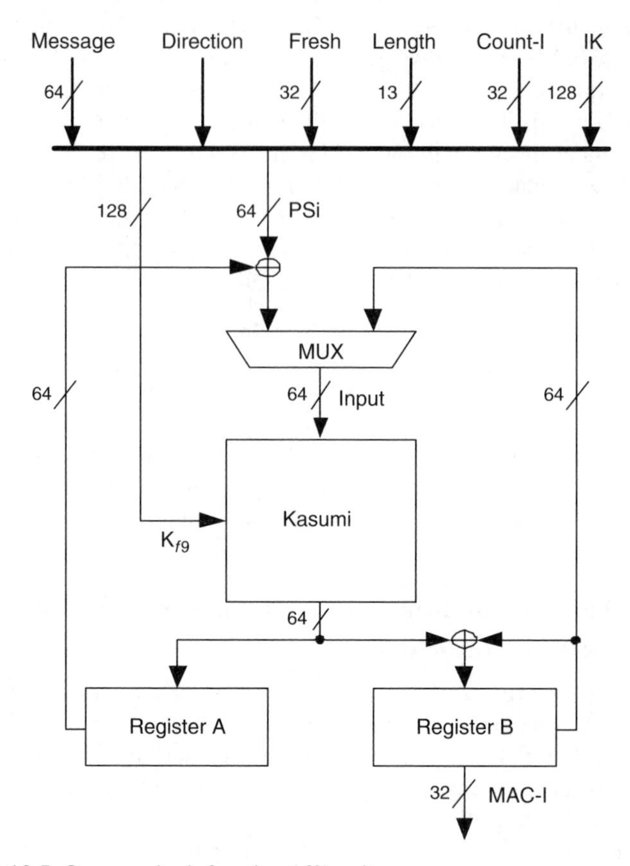

FIGURE 10.5 One-way hash function ($f9$) unit.

(first-loop execution), the MUX subunit selects the initial input and the Kasumi module produces the initial key stream (KS) using the modified CK.

This initial KS is stored in a register and is used for the next iterations. After the initial iteration, in all the next iterations, MUX selects the input produced with the XORing of the previous KS and the CK is used by the Kasumi module. The block count (BLKCNT) counter is set initially to 0 and is increased by 1 after each iteration. The maximum value of the counter is (length/64) rounded up to the nearest integer, which is the number of iterations. The input length defines the plaintext or ciphertext length (number of bits).

10.3.2.1 Kasumi Block Cipher

The Kasumi block cipher [5,6] is used both by the one-way hash function ($f9$) and by the stream cipher ($f8$). Kasumi comprises the Feistel networks [16]

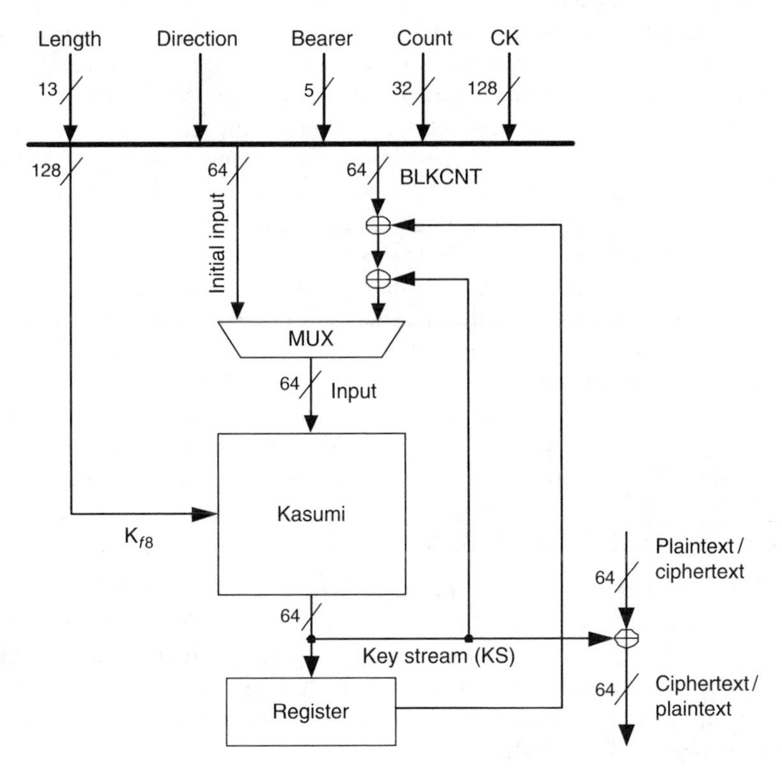

FIGURE 10.6 Stream cipher ($f8$) unit.

operation with eight rounds and operates on a 64-bit input. It produces a 64-bit output according to a 128-bit ciphering key. Let us denote each round as f_i. f_i has two different forms depending on whether it is an even round or an odd round. On the even rounds, the computation procedures are executed in reverse order comparing with the odd rounds.

Let us denote the odd rounds as odd round cell (ORC) and the even rounds as even round cell (ERC). For the implementation of the complete algorithm, the data are applied in a repeated manner to the two basic cells, one ORC and one ERC. This implementation structure is depicted in Figure 10.7.

The round keys are computed by the key expansion unit. The total key schedule is constituted by hardwired right shifters that produce many subkeys, while the remaining are generated by bitwise XOR operation with predefined constants. About forty 16-bit subkeys are generated in total. Many of them are used more than once to generate the round keys with the appropriate concatenations. The round keys are precomputed and stored in a 64×16-bit RAM memory. Therefore, the system does not need to generate the keys for the decryption mode.

To increase the cipher performance, the inner-round pipeline technique is used. A negative edge-triggered register is used for the pipeline inside the rounds. In a conventional pipeline with positive edge-triggered register design between the rounds, the data are transferred between two successive registers in one clock period. The use of this technique (negative edge-triggered pipeline) results in a significant reduction of the round critical path delays. The negative edge pipeline register is inserted in the FO function (Figure 10.7b), which is roughly in the middle of the round datapath. This technique has been recently proposed for low-power and high-speed designs [8]. The execution time of each round is one system clock cycle. To synchronize the processing datapaths, similar registers are inserted in the left and right branches of each round (Figure 10.7b). The result of this insertion is the reduction to roughly half of the clock period.

As in the Rijndael cipher, for the Kasumi cipher implementation the S-boxes were designed using ROM blocks. Therefore, the proposed Kasumi block cipher implementation performance is high, as the UMTS standard demands.

10.4 SYSTEM EVALUATION

The VLSI synthesizes results of the proposed UMTS security hardware implementation they follow. The whole system (Figure 10.2) was captured using VHDL. This code was synthesized, placed, and routed using Xilinx FPGA devices [17]. The VHDL code was simulated and verified using the "implementors" test data [18–20]. The two system fragments are implemented in two separate FPGA devices.

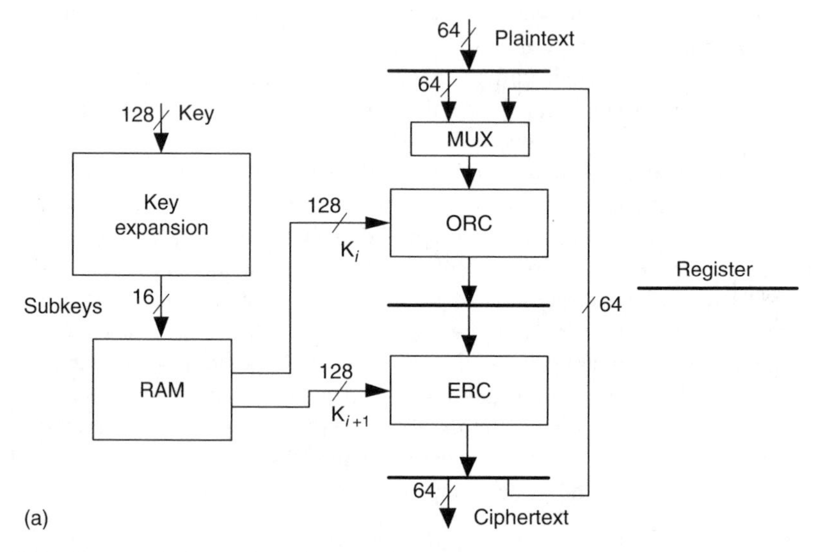

FIGURE 10.7 (a) Kasumi block cipher hardware implementation.

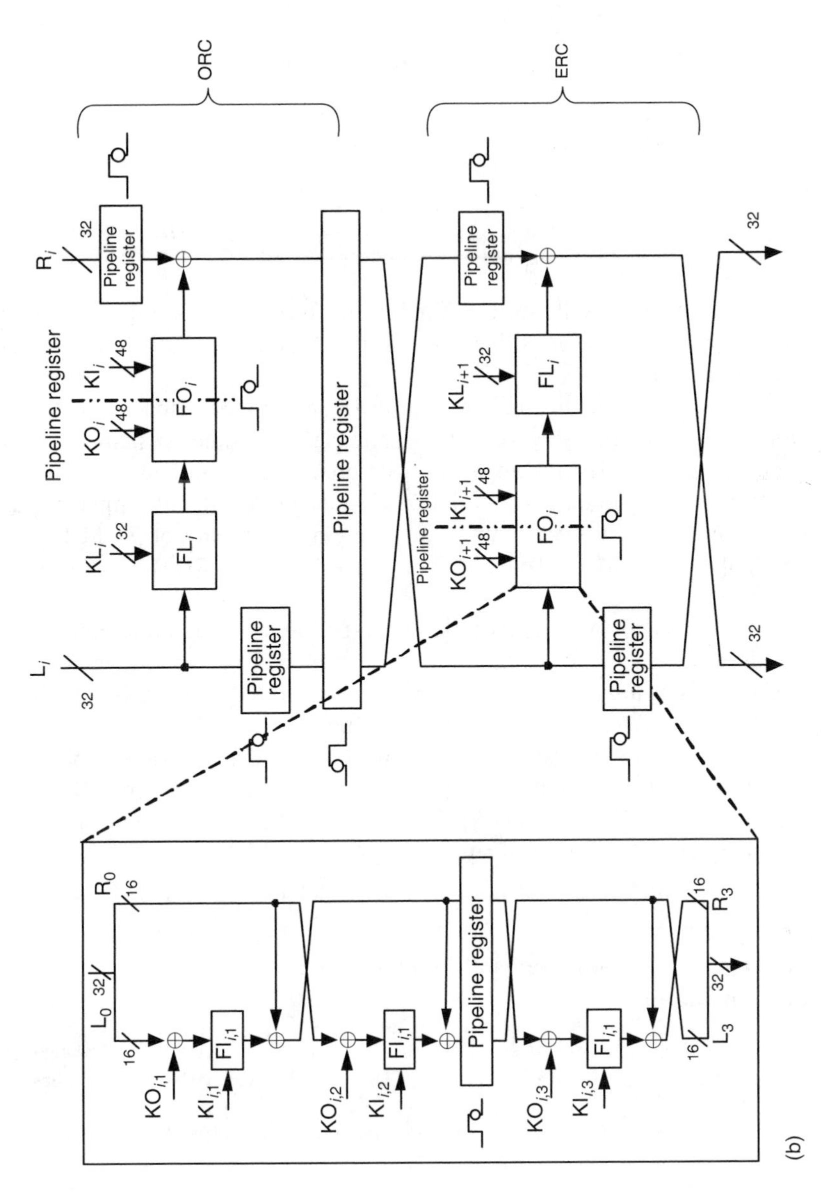

FIGURE 10.7 (Continued) (b) Rounds of Kasumi hardware implementation.

TABLE 10.1
Rijndael Module and AKA Unit Implementation Synthesis Results

	Rijndael Module	AKA Unit
Function generators	2387	7390
Configurable logic blocks	1194	3820
D flip-flops	715	2212
Frequency: F (MHz)	78	70
Throughput (Mbps)	243	218

The synthesis results for the Rijndael module and AKA unit implementations are illustrated in Table 10.1. The FPGA device XILINX V400E-FG676 was used.

The performance measurements of the Rijndael block cipher implementation are shown in Table 10.2. Measurements from other designs are also included in this table for comparison with previously published works.

For the completion of a 128-bit block transformation, both implementations need 41 clock cycles. With the operating frequency of 78 MHz, the throughput is 244 Mbps. To build four S-boxes, 4×2048 ROM bits storage is needed.

The full 3GPP AKA algorithm set can be implemented on an integrated circuit card, equipped with an 8-bit microprocessor, ROM, and RAM modules. From the available hardware resources, 6-KB ROM and ~200-byte RAM are needed for the Rijndael function. Running with a 3.25 MHz clock, the implementation must produce AK, XMAC-A, RES, CK, and IK in <500 ms. The Rijndael should produce the 128-bit output values in <50 ms [4]. From the results in Table 10.1 and Table 10.2, it is obvious that both proposed architectures meet these requirements.

TABLE 10.2
Performance Measurements of Rijndael Block Cipher Implementation

Architecture	Device	CLBs	Frequency (MHz)	Throughput (Mbps)
[21]	XCV1000 BG560	5,302/10,992	14.1/31.8	300/1940
[22]	XILINX (not specified)	5,673	—	353
[23]	Xilinx Virtex	2,902	25.9	331
[24]	ASIC Approach	3.96 mm^2	100	910
[25]	Altera APEX1K4001	845 LE	—	750 (best)
Proposed	XCV200 EFG456	1603	78.3	244

The proposed Rijndael implementation uses 128-bit data and key blocks. This implementation is slightly slower (~10%) in terms of throughput than other earlier works [21–25]. A high-speed reprogrammable Rijndael design is shown in [26]. However, this is not a drawback. Rijndael implementation performs efficiently according to the UMTS specifications. With a 78 MHz clock frequency, the output is produced in 0.52 μs. With the clock frequency of 3.25 MHz, the Rijndael implementation produces the output in 12.6 μs compared with the 50 ms that the UMTS demands.

The major advantage of the proposed implementation is the minimized covered area resources. This is important in applications with strict area limitations. Only 1-KB ROM is used in contrast to 6 KB that 3GPP specifies. There is also no need for RAM blocks.

In the proposed AKA procedure implementation, only 1044-byte ROM and 128-bit RAM are needed to store the OPc values. Therefore, the required memory locations are less than the locations that 3GPP specifies (8-KB ROM and 300-byte RAM). In the proposed design, the parameters AK, XMAC-A, RES, CK, and IK are calculated in 76 μs, whereas UMTS specifies 500 ms at a clock frequency of 3.25 MHz. This reduction is important because the AKA unit can process more messages at the same time.

The synthesis results for the Kasumi module are presented in Table 10.3 The FPGA device XCV300E-8BG432 was used. This device has 16-KB internal ROM, divided into 32 blocks. In the proposed design, 688-byte ROM is used. With a 105 MHz clock frequency, a throughput up to 840 Mbps is achieved.

In [27,28], a two-round architecture with a conventional approach was proposed. Combinational logic and lookup table (LUT) are used to implement the Kasumi S-boxes. In [29], two implementation versions are proposed. The first is the low-power version (Type1) and the second is the high-performance version (Type2) with the usage of four-stage pipeline. Moreover, in [30] a hardware implementation that reduces the hardware resources is presented. Two synthesis results are given. The first (Synth1) was made with the speed grade −6 and the second (Synth2) was made with a speed grade −8. Finally, in [31] a small hardware architecture is presented. The proposed Kasumi

TABLE 10.3
Kasumi Module, *f*8 Unit and *f*9 Unit Synthesis Results

	Kasumi Module	*f*8 Unit	*f*9 Unit	Total
Function generators	1819	2064	2278	4342
Configurable logic blocks	910	1042	1158	2200
D flip-flops	1570	1788	1973	3761
Frequency: F (MHz)	105	104	104	104

TABLE 10.4
Kasumi Time Performance Comparisons

Architecture	Device	Frequency (MHz)	Throughput (Mbps)
Kasumi_comb in [27,28]	XCV300E-6BG432	20.88	167.04
Kasumi_LUT in [27,28]	XCV200E-6FG456	35.35	70.70
Kasumi_type1 in [29]	—	20	110
Kasumi_type2 in [29]	—	60	410
Kasumi_synth1 in [30]	XCV300E-8BG432	33.14	265.12
Kasumi_synth2 in [30]	XCV300E-6BG432	28.38	227.04
Kasumi in [31]	XCV300E-8-BG432	42	393
Proposed	XCV300E-8BG432	105	840

architectures outperform all the earlier (two rounds) implementations in terms of time performance as shown in Table 10.4.

The synthesis results for the one-way hash function ($f9$) unit and the stream cipher ($f8$) unit are also illustrated in Table 10.3. Both units were integrated in the same FPGA device.

Table 10.5 shows the comparisons between the proposed $f8$ and $f9$ units and the previous designs. It is obvious that the proposed implementations perform better in terms of performance.

TABLE 10.5
***f*8 and *f*9 Time Performance Comparisons**

Architecture	Device	Frequency (MHz)	Throughput (Mbps)
$f8$_Comb in [27]	XCV300E-6BG432	20.52	162.1
$f8$_LUT in [27]	XCV200E-6FG432	33.14	261.8
$f8$_Comb in [28]	XCV300E-6BG432	16.93	135
$f8$_LUT in [28]	XCV600E-6FG432	46.56	372
$f8$_type1 in [29]	—	19.5	154
$f8$_type2 in [29]	—	52	411
$f8$_synth1 in [30]	XCV300E-8BG432	30.12	240.96
$f8$_synth2 in [30]	XCV300E-8BG432	25.80	206.40
Proposed	XCV300E-8BG432	104	822
$f9$_Comb in [27]	XCV300E-6BG432	20.68	165.44
$f9$_LUT in [27]	XCV200E-6FG432	20.19	161.52
$f9$_Comb in [28]	XCV300E-6BG432	16.70	134
$f9$_LUT in [28]	XCV600E-6FG432	35.34	340
$f9$_synth1 in [29]	XCV300E-8BG432	30.12	240.96
$f9$_synth2 in [29]	XCV300E-8BG432	25.80	206.40
Proposed	XCV300E-8BG432	104	822

The proposed *f*8 implementation achieves 75.4% (combinational-based S-boxes) and 760% (ROM-based S-boxes) higher throughput than the implementation in [27]. The MAC computation time in the proposed *f*9 implementation gives ~88% less average time. In addition, the *f*8 and *f*9 implementation covered area is half of the covered area in [27]. In addition, in [28] an area-efficient architecture for the *f*8 was proposed. In this architecture, the Type 1 *f*8 implementation operates with 19.5 MHz, whereas Type 2 operates with 52 MHz clock frequency. Table 10.5 shows that the proposed implementation provides much higher throughput than the Type 1 and Type 2 implementations.

10.5 CONCLUSION

A hardware implementation of the UMTS security system was presented in this chapter. The introduced system supports the AKA, user data, signaling information confidentiality protection, and the signaling information integrity. With the proposed designs, a major hardware resources reduction is achieved. The proposed AKA implementation executes the procedure within 76 μs compared with the 500 ms that UMTS specifies. The main architectural units of the system are based on the Rijndael and Kasumi block ciphers. An efficient Rijndael architecture is proposed to reduce the required hardware resources. The proposed Kasumi architecture reduces the hardware resources and power consumption. It uses feedback logic and positive–negative edge-triggered pipeline to make the critical path shorter without increasing the latency of cipher execution.

REFERENCES

1. 3GPP TS 33.102 V4.2.0, Technical specification group services and system aspects, 3G security. Security Architecture, September 2001.
2. Daemen, J. and Rijmen, V., AES Proposal: Rijndael, available at http://csrc.nist.gov/encryption/aes/round2/AESAlgs/Rijndael/Rijndael.pdf
3. 3GPP TS 35.206 V4.0.0, Technical specification group services and system aspects, 3G security, specification of the MILENAGE algorithm set: An example algorithm set for the 3GPP authentication and key generation functions f1, f1*, f2, f3, f4, f5 and f5*, Document 2: Algorithm Specification, April 2001.
4. 3GPP TR 35.909 V4.0.0, Technical specification group services and system aspects, 3G security, specification of the MILENAGE algorithm set: An example algorithm set for the 3GPP authentication and key generation functions f1, f1*, f2, f3, f4, f5 and f5*, Document 5: Summary and Results of Design and Evaluation, April 2001.
5. *f*8 and *f*9 Specification. Specification of the 3GPP confidentiality and integrity algorithms, Document 1, ETSI/SAGE, September 2000.
6. KASUMI specification. Specification of the 3GPP confidentiality and integrity algorithms, Document 2, ETSI/SAGE, December 1999.

7. 3GPP KASUMI evaluation report. Security algorithms group of Experts (SAGE), report on the evaluation of 3GPP standard confidentiality and integrity algorithms, SAGE version 2.0.
8. Strollo, A.G.M., Napoli, E., and Cimino, C., Analysis of power dissipation in double edge-triggered flip-flops, *IEEE Transaction on Very Large Scale Integration (VLSI) Systems,* 8(5): 624–629, 2000.
9. Dohmen, J.R. and Olaussen, L., UMTS Authentication and Key Agreement, Graduate thesis, Agder University College—2001, online available at http://siving.hia.no/ikt01/ikt6400/jrdohm99
10. Sklavos, N. and Koufopavlou, O., Architectures and VLSI implementations of the AES-proposal Rijndael, *IEEE Transaction on Computers,* 51(12): 1454–1455, 2002.
11. Brunner, H. et al., On computing multiplicative inverses in GF(2^m), *IEEE Transactions on Computers,* 42(8): 1010–1015, 1993.
12. Wang, C.C. et al., VLSI architectures for computing multiplications and inverses in GF(2^m), *IEEE Transactions on Computers,* C-34(8): 709–717, 1985.
13. Araki, K., Fujita, I., and Morisue, K., Fast inverters over finite field based on Euclid's algorithm, *Transaction IEICE*, E-72(11): 1230–1234, 1989.
14. ISO/IEC 9797–1:1999 (E). Information technology—security techniques—message authentication codes (MACs)—Part 1.
15. Recommendation for block cipher modes of operation. Methods and techniques, National Institute of Standards and Technology (NIST). Available at http://csrc.nist.gov/encryption/modes/Recommendation/Modes01.pdf
16. Menezes, A.J., Oorschot, P.C., and Vanstone, S.A., *Handbook of Applied Cryptography,* 1996, CRC Press, October 1996.
17. Xilinx, San Jose, California, USA, Virtex, www.xilinx.com
18. 3GPP TS 35.207 V4.0.0, Technical specification group services and system aspects, 3G security, specification of the MILENAGE algorithm set: An example algorithm set for the 3GPP authentication and key generation functions f1, f1*, f2, f3, f4, f5 and f5*, Document 3: 'Implementors' Test Data, April 2001.
19. 3GPP TS 35.208 V4.0.0, Technical specification group services and system aspects, 3G security, specification of the MILENAGE algorithm set: An example algorithm set for the 3GPP authentication and key generation functions f1, f1*, f2, f3, f4, f5 and f5*, Document 4: Design Conformance Test Data, April 2001.
20. Design conformance test data. Specification of the 3GPP confidentiality and integrity algorithms, Document 4, ETSI/SAGE, December 1999.
21. Elbirt, A.J. et al., An FPGA based performance evaluation of the AES block cipher candidate algorithm finalists, in *Proc. of the 3rd Advanced Encryption Standard (AES) Candidate Conference*, New York, USA, April 13–14, 2000.
22. Dandalis, A., Prasanna, V.K., and Rolim, J.D.P., A comparative study of performance of AES final candidates using FPGAs, in *Proc. of the 3rd Advanced Encryption Standard (AES) Candidate Conference*, New York, USA, April 13–14, 2000.
23. Gaj, K. and Chodowiec, P., Comparison of the hardware performance of the AES candidates using reconfigurable hardware, in *Proc. of the 3rd Advanced Encryption Standard (AES) Candidate Conference*, New York, USA, April 13–14, 2000.
24. Kuo, H. and Verbauwhede, I., Architectural optimization for a 1.82 Gbits/sec VLSI implementation of the AES Rijndael algorithm, in *Proc. of the CHES 2001*, France, May 14–16, 2001.

25. Fischer, V. and Drutarovsky, M., Two methods of Rijndael implementation in reconfigurable hardware, in *Proc. of the CHES 2001*, France, May 14–16, 2001.
26. Hodjat, A. and Verbauwhede, I., Area-throughput trade-offs for fully pipelined 30 to 70 Gbits/s AES processors, *IEEE Transaction on Computers*, 55(4): 569–572, 2006.
27. Verbauwhede, I., Schaumont, P., and Kuo, H., Design and performance testing of a 2.29-GB/s Rijndael processor, *IEEE Journal of Solid-State Circuits*, 38(3): 569–572, 2003.
28. Marinis, K. et al., On the hardware implementation of the 3GPP confidentiality and integrity algorithms, in *Proc. of the 4th International Conference for the Information Security, ISC 2001*, Malaga, Spain, pp. 248–265, October 1–3, 2001.
29. Marinis, K. et al., An area optimized hardware implementation of the 3GPP confidentiality and integrity algorithms, in *Proc. of the 8th Conference on Optimization of Electrical and Electronic Equipment, OPTIM 2002*, Brasov, Romania, May 16–17, 2002.
30. Kim, H. et al., Hardware implementation of 3GPP KASUMI crypto algorithm, in *Proc. of the 2002 International Technical Conference on Circuits/Systems, Computers and Communications* (*ITC-CSCC*), Phuket, Thailand, 1: pp. 317–320, July 16–19, 2002.
31. Satoh, A. and Morioka, S., Small and high-speed hardware architectures for the 3GPP standard cipher KASUMI, in *Proc. of the 5th International Conference Information Security, ISC 2002*, Sao Paulo, Brazil, September 30–October 2, 2002.

11 Wireless Application Protocol Security Processor: Privacy, Authentication, and Data Integrity

Nicolas Sklavos

CONTENTS

11.1 INTRODUCTION

Wireless communications have become an attractive and interesting sector for the provision of electronic services. Mobile networks are available almost anytime, anywhere and the user's acceptance of wireless devices is high. One of the major scopes of the wireless protocols and especially of wireless application protocol (WAP) [1] is to bring the Internet applications to mobile devices [2–5]. Security is a key issue in the world of electronic communication, especially for services with a sensitive purpose such as electronic commerce and online banking. New ciphers have been developed [6] to support the networks' defense against the increasing range of attacks. Optimizations of the security layer specifications have also been published in the last year [7]. However, the time overhead due to data encryption should not impose an intolerant penalty on the communication process. Almost all the ciphers of today have an important drawback: the slowness of their operation, due to the mathematic and logic transformations. Because software implementations are too slow, even running in fast processors, the use of specific hardware modules seems to be the only reasonable solution for the requested high performance. Different designs have been proposed for the hardware implementation of ciphers [8–22]. These works present the implementation of one cipher in a hardware device at a time. Modern applied cryptography in the wireless communications networks demands powerful encryption engines with special purposes of privacy, authentication, and integrity. In order to support these security needs efficiently, hardware security engines have to be implemented in a single hardware module because of the restricted availability of resources for mobile devices. It is necessary that a set of ciphers is integrated in the same chip for the security layers of wireless protocols. The implementations of ciphers in different hardware devices, one for each algorithm, have proved to be insufficient and forbidden solutions in the wireless world.

An efficient architecture for WAP security layer implementation is proposed in this chapter. The introduced system supports six different ciphers for both architecture and security purposes: IDEA, DES, RSA, Diffie–Hellman, MD-5, and SHA-1, in the same hardware module. The proposed architecture has been implemented in an field programmable gate array (FPGA) device. The synthesis results prove that the performance of integrated ciphers is high and better than implementations with separate ciphers [8–22], in most of the cases. The IDEA architecture uses a modified transformation round, which minimizes the allocated area resources by ~30% compared with the other works [13–15]. The proposed DES implementation performs better, with a range from 200% to 400%, than in the other related studies [16–19]. In particular, the proposed DES architecture has been designed with a slight modification to operate alternatively as an authorized user verification unit. The introduced reconfigurable authentication unit performs efficiently for both RSA and Diffie–Hellman and decreases the covered area by ~70% in total compared with the case of two separate hardware implementations, one for each cipher. The proposed reconfigurable integrity unit minimizes the allocated area resources and performs efficiently for two different operation modes: SHA-1 and MD5. The performance of both SHA-1 and MD5 is better at ~50%–300% compared with the conventional implementations [20–22]. The proposed system can be applied as a powerful solution for the wireless transport layer security (WTLS) implementation in wireless networks. It can be used for both server providers and mobile devices. Furthermore, the system can be used as a powerful security core in wireless communications supporting privacy, authentication, and integrity. The high-speed performance and the minimized resources make this architecture suitable and practical, in spite of the limitations that wireless networks impose.

This work is organized as follows: in Section 11.2 the WTLS is introduced. In Section 11.3 the proposed architecture is presented. Section 11.4 is dedicated to verification and testing. The synthesis results of the FPGA implementation and comparisons with other published works are given in Section 11.5. Finally, in Section 11.6 conclusions and observations are presented.

11.2 WIRELESS TRANSPORT LAYER SECURITY

11.2.1 WAP STANDARD

Mobile networks of our days do not provide the desired flexibility when added services are about to be introduced. Often it has proved to be a complicated and lengthy task to launch such services. WAP offers an efficient solution to this problem by adopting Internet capabilities as a powerful service platform in the wireless communication world. This is based on the fact that the Internet has proven to be an easy and efficient way of delivering services to millions of wired users.

WAP is a completely new concept. It was specified by an industry consortium, the WAP Forum. This forum was founded in December 1997 by Ericsson, Motorola, Nokia, and Unwired Planet. The primary goal of WAP is to bring Internet content and advanced data services to handheld devices and other wireless terminals. At the same time, the main attempt of the WAP Forum is to create a global wireless protocol specification that could be applied across all wireless existing or new technologies. A great number of applications and services are intended to be enabled in a wide range of wireless bearer networks and unwired devices. Generally, it is important to note that the already existing standards will be embraced and extended. WAP specifications have been developed for the above-described goals to be accomplished according to the design principles of the WAP.

Today, the Internet is used mainly from personal computers and all the common requested services such as links, searches, and downloads are easily accessed by every user, even in the case of an amateur user. The user-friendly interface, provided by the software tools, supports all the described requests. Providing Internet and web services on the wireless communication protocol or network presents many challenges to mobile service providers and application developers.

The types of both wireless devices and communication network set limitations on the range and the kind of the services provided. Wireless networks have fundamental restrictions in the available spectrum, power consumption, and mobility. They also have less bandwidth and more latency than the wired networks. Connection stability and predictable availability are also less. Furthermore, possible improvements in the bandwidth rate in the wireless communication world have power consumption penalty in the limited battery life of handheld devices.

Low-power consumption is not the only limitation that mobile phones tend to have. In contrast with the personal computers, mobile phones have less powerful CPUs and less available memory of any kind. The small size of screens in addition to the different way of data input, voice input, and a smaller keypad requests a different user interface than the one used in computers. The applied wireless protocol has to overcome the network and device limitations and at the same time support an acceptable level of services to satisfy the requests and needed applications.

The kind of services to the users is similar to the one that is used on the Internet, with a slight difference in the way the provided services are accessed. In the wireless world, the user needs an interface utility of the services, at any given point of time, without having to use search engines, special links, or full download capabilities. The supported applications and services must be in a comparatively high level because they have to attract a wide range of users. The provided services vary from banking accounts and products on sale to gambling and ticketing operations. Many other information-type applications, such as news and weather forecasts, are also in wide use. New and different kinds of applications are introduced by WAP such as voice mail, faxes, and e-mails in mobile phones. Today, the WAP forum is conducting work in several areas that will facilitate mobile value added services (VAS), such as persistent storage, smart cards, and user agent profiles.

11.2.2 WAP Architecture Overview

The WAP standard defines an application environment and an application protocol. The application environment consists of the markup language, WML, and a programming language, the WMLScript. Since WAP-supported applications and services can be downloaded on demand and discarded when no longer needed, the application environment also allows dynamic extension of the terminal's user interface.

The actual application protocol architecture provides a scalable and extensible application development environment for mobile communication devices. This is achieved through a layered architecture of the protocol stack. WAP is designed as a layered-type protocol to be extensible, flexible, and scalable. Based on the open system interconnection model (OSI), the WAP stack is basically divided into five essential layers. These are (i) wireless application environment (WAE), (ii) Wireless session protocol (WSP), (iii) wireless transaction protocol (WTP), (iv) WTLS, and (v) wireless datagram protocol (WDP), which are described in detail by the specifications of the WAP forum [1]. Each layer of WAP provides a set of functions and services to other applications through well-defined interfaces. This means that each of the five building layers of WAP defines a well-specified interface to the one mentioned here and makes the lower layers invisible to it. The protocol stack isolates the applications from the bearer so that one application can be executed or run regardless of the actual transport service used.

In addition to the application environment and the application protocol, the WAP standard also defines a technology known as wireless telephony application (WTA) specification. It is a telecom-oriented technology that allows WAP to be integrated with the advanced services in the telecom network, such as intelligent networks. Combined with the browser-based user interface of WAP, the WTA allows new intelligent network services to be introduced to users without modifying the terminals in any way.

Figure 11.1 shows the WAP stack and how it relates to the protocols on the Internet.

The differences between the WAP and the commonly used Internet protocol stack have been proved to be the most important parts for enabling wireless access to mobile devices. The WAP stack does not map directly onto other stacks but a comparison between them could take place. As illustrated in Figure 11.1, the kind of functionality that is provided by HTML and Java in the Internet is incorporated in the WAE layer of WAP and the wireless

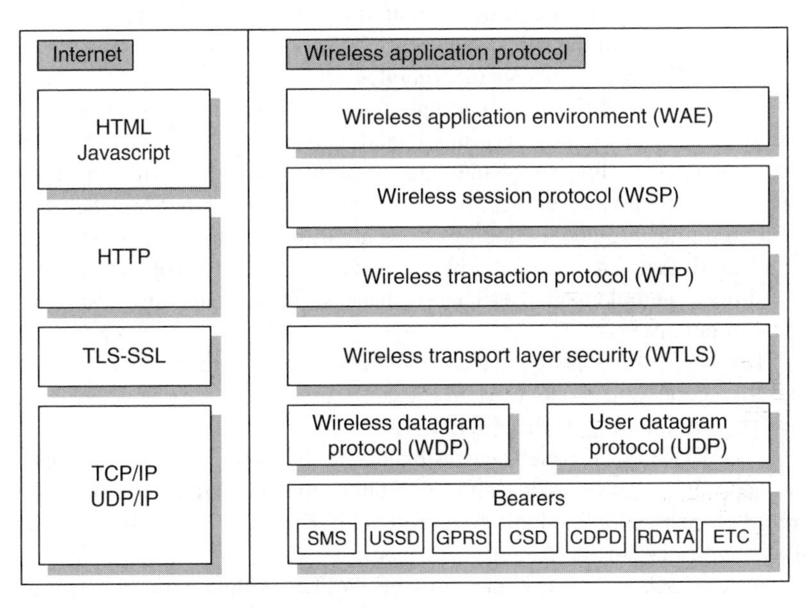

FIGURE 11.1 WAP stack architecture.

session layer (WSP). At the same time, WSP and WTP layers include the functionality provided by HTTP. The transport layer in the wired communications is usually either TCP or user datagram protocol (UDP), whereas in the wireless world it is UDP over IP where it is possible. WDP is provided for networks that cannot support IP at the network layer.

Transport layer security (TLS) is basically used in Internet applications to ensure and guarantee the desired security level. WTLS is its equivalent in the unwired communications world. This layer includes a specification set that implements options for authentication and encryption and is optimized for use in the mobile environment.

11.2.3 Wireless Transport Layer Security

WTLS, as its name indicates, is dedicated to the security layer of the WAP. Security is needed for the WAP to safely support the provided services, particularly the sensitive ones such as online banking and e-commerce.

WTLS is based on the TLS 1.0, which is the security standard founded in the Internet, but is optimized for narrowband communication channels. Furthermore, a number of modifications and changes were needed according to the nature of the wireless networks. One basic difference, compared with the Internet, is that in wireless transmission, support for both datagram and connection-oriented transport layer protocols is required. On the other hand, there are many specifications in the used encryption algorithms because of the limited processing power and memory. Furthermore, the bandwidth limit is a fact that must be taken care of in addition to all the other restrictions of handheld devices in the supported security.

The WTLS security layer incorporates new features such as wide range in the selection of the used encryption algorithms, multi-operation mode encryption, dynamic key refreshing, and handshake. The WAP has also been optimized for low-bandwidth bearer networks with relatively long latency. Flexible encryption algorithms with high performance have been chosen to provide a wide available algorithm set to the users. It has to be cleared that, in a mobile phone device, only a small set of these algorithms is implemented efficiently because of the hardware restrictions and limitations.

The WTLS layer is optional and can be used with both the connectionless and the connection mode WAP stack configuration. In case it is used, it is always placed on top of the WDP layer.

The WTLS layer, which accepts data from the upper WAP layers, applies the appropriate compression and encryption and then transmits them. The record protocol is divided into four protocol clients. The WTLS layer stack is illustrated in Figure 11.2.

In this section, every part of the WTLS architecture is described briefly. During the handshake, all the related parameters are agreed on.

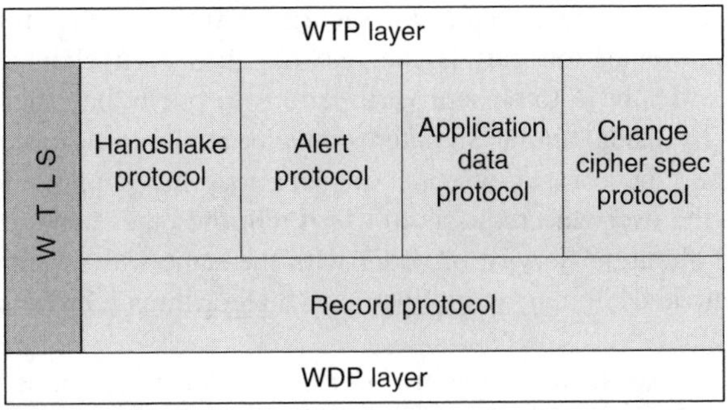

WTP layer

W T L S

Handshake protocol | Alert protocol | Application data protocol | Change cipher spec protocol

Record protocol

WDP layer

FIGURE 11.2 WTLS internal architecture.

Such parameters specify the used protocol versions, the characteristics of the encryption algorithms, and the established connection properties. Different handshake types can take place alternatively, according to the connection needs. The alert protocol provides all the appropriate alert messages that are sent each time in the established secure state. Three different types of alert messages are specified: warning, critical, and fatal. Depending on the type of the alert message, connections may result in termination, initiation of the exchange, or may even continue. Further procedures could be accomplished according to the error level. The change cipher spec is sent peer to peer either by the client or by the server. The appropriate messages of this protocol are sent after the handshake to set the values in the current and the pending states of the receiver. Finally, the application data protocol is involved with all the appropriate processes and then applied to the pure transmitted data amount.

Security in every communication protocol or network is composed of separate security entities of special purposes. In WTLS, three different encryption operations have been defined: privacy, authentications, and data integrity. These operations are fully supported by the WTLS layer, with the availability of selection between different encryption algorithms in each kind of security process.

With the rule of privacy, an applied transfer method that ensures a private end-to-end transfer is defined. Privacy is also called bulk encryption in the literature. The sender and the recipient have to use a commonly known cryptographic algorithm to encrypt and decrypt the transmitted data. They also have to know the used key, on which the operation of the encryption algorithm is based. The key is determined as a block of data known only to both parties of the communications and nobody else. It is obvious that the used algorithm should be resistant to all known cryptanalysis attacks. The encryption algorithms that are proposed to be used in WTLS are IDEA, DES, and RC5.

Authentication is described as the security unit that ensures if the identity of every communication part is the one that has been claimed. The two contacting parties have to present verifications to prove their identities. This is achieved by using digital signatures or electronic certificates. After the authentication, the service provider is sure that the supported service is available to the user who requests to use it. On the other hand, the user can be confident about the service provider with the same way of authentication. The RSA, Diffie–Hellman, and elliptic curve algorithms serve authentication in WAP.

Integrity is used to verify that the transmitted data have not been modified in all the travel through the network. In different words, integrity secures the reliability of the information. This could be guaranteed by calculating checksums from the original information to be sent. One plain checksum is not enough. More sender-related information mixed into these calculations is needed. This information is signed with the user's digital signature. Hash functions are the most common methods used for integrity. In WTLS, the SHA and the MD5 hash functions are introduced by the specifications.

11.3 PROPOSED WTLS ARCHITECTURE

11.3.1 Proposed Crypto-Processor Architecture

The proposed crypto-processor architecture, for the WTLS hardware implementation is illustrated in Figure 11.3. The introduced system has been designed like a typical processor with datapath, memory, I/O interface, and control unit [23,24].

Six different ciphers are supported by the proposed crypto-processor. DES and IDEA algorithms are selected for the bulk encryption unit. The reconfigurable integrity unit performs efficiently in two different operation modes, for SHA-1 and MD5 hash functions. The operations of both RSA and Diffie–Hellman are performed by the reconfigurable authentication unit. An extra security scheme is also supported by the proposed crypto-processor. A reconfigurable logic block, in combination with the modified DES unit, implements the authorized user verification unit. A common data bus of 64 bits and a 32-bit address bus are used for the internal data transfer purposes. Two different storage units have also been integrated. The appropriate algorithm keys are stored and loaded in the RAM blocks, whereas all the transformed data are kept as long as it is necessary in the transformed data registers. A common bus interface unit, which supports 32-bit input data and 32-bit address buses, has also been implemented, for the crypto-processor to communicate efficiently with the external environment. This environment may be a general purpose processor or a special CPU.

It has to be mentioned that WAP is intended to be applied mainly in mobile devices. Because of their hardware integration limitations, only a set

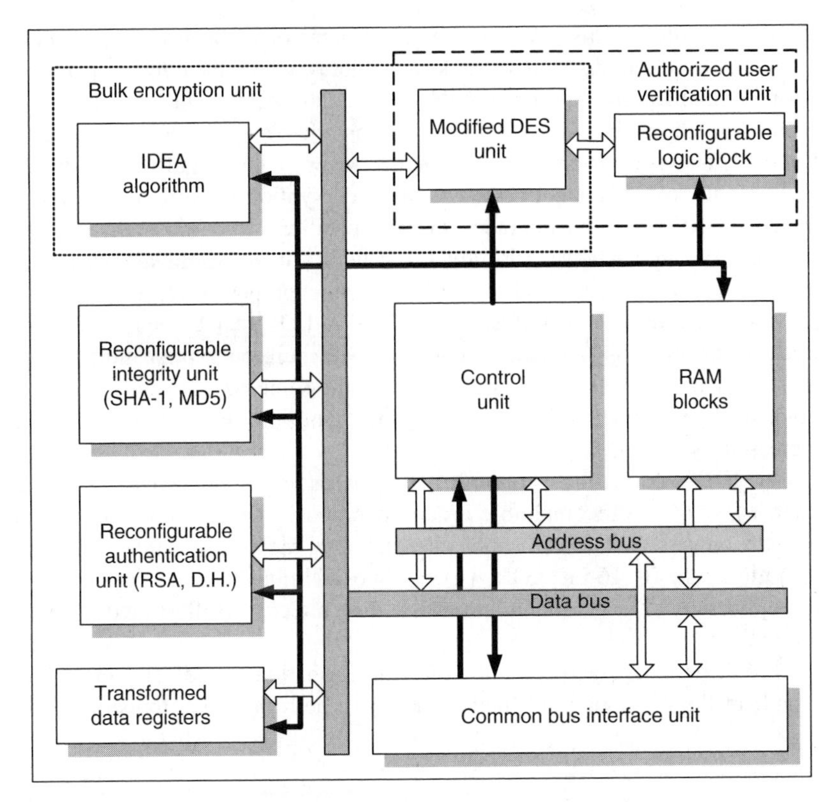

FIGURE 11.3 Proposed architecture for a crypto-processor.

of ciphers and not all specified by WTLS can be implemented in a flexible embedded system [25,26]. The ciphers set in the proposed crypto-processor were selected by considering security and hardware implementations parameters. As presented in the rest of the chapter, according to our study, the integration units of the selected ciphers ensure the highest provided security and the best hardware performance at the same time.

11.3.2 Bulk Encryption Unit

In the proposed crypto-processor, the bulk encryption unit provides the capability of selection between two ciphers: IDEA and DES. According to the opinion of the security experts, IDEA is one of the most secure block algorithms available to the public [27]. On the other hand, DES has been established as the data encryption standard and has proved to be a flexible design for VLSI implementations. These algorithms have been preferred, for several reasons, instead of RC5, which is the third cipher, specified by WTLS. RC5 is a trademark of RSA data security [28] and it is expected to be patented in the near future. This cipher has a set of parameters, such as key rounds, operation word lengths, and secret key variables, which have to be specified during initialization. All the possible set values seem to be available and usable in theory, but many of them in practice may be forbidden [29]. Detailed analysis is needed to prove the security level and the hardware performance of each set. Unlike the other encryption algorithms, the parameterized RC5 permits upgrades in the operation, with the main goal to increase the supported security level, but with a major disadvantage in the performance and vice versa. In order that the proposed crypto-processor provides alternative capabilities, both IDEA and DES have been integrated. Because of the large amount of data, which a bulk encryption unit has to transform, these two ciphers can operate at the same time, or only one at a time. In the case of parallel operation, the crypto-processor performance is increased by a great factor.

In IDEA [30], the required data confusion is achieved by using three different and incompatible group operations. These group components operate on pairs of 16-bit subblocks and mix them. The three algebraic operations are (i) 16-bit XOR, (ii) modulo addition 2^{16}, and (iii) modulo multiplication 2^{16} [31]. The proposed IDEA architecture is illustrated in detail in Figure 11.4.

IDEA defines four or eight basic transformation rounds and one half-round. In the previous published works [15], both the basic transformation

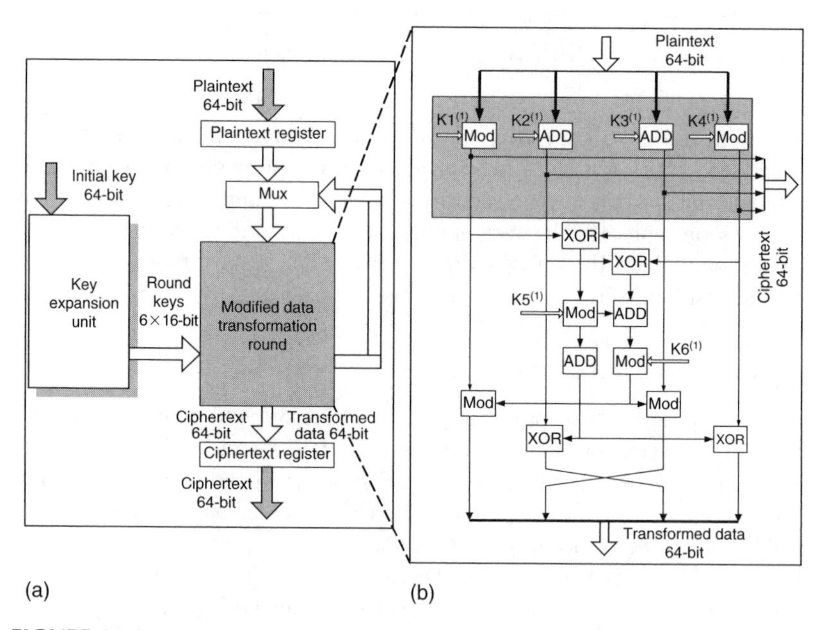

FIGURE 11.4 (a) IDEA architecture; (b) modified data transformation round.

round and the half-round have been integrated separately. Such an implementation approach has a major cost in the covered area resources. The proposed IDEA architecture is based on a feedback logic operation mode. As Figure 11.2 illustrates, the specified IDEA transformation round architecture is partially modified (modified data round). It operates efficiently for both the basic and the half-round, as the specifications demand [30]. By this proposed architecture, the allocated resources of IDEA implementation are reduced to about two modulo multipliers area, which are the fundamental covered area components. It has been measured that by the proposed architecture, 40% area reduction is achieved compared with the design approach applied in [15]. The analytical synthesis results and detailed comparisons are given in Section 11.5.

Furthermore, in the proposed IDEA architecture the key expansion unit has also been integrated and the appropriate round keys are generated on the fly. By the key expansion unit integration, the required dynamic key refreshing of WTLS is achieved, with no extra time delay. In addition, the introduced design has no time delay in the initialization process for the key setup like in [15]. According to our study, the separate integrated key expansion unit has a cost of ~7%–10% in the total IDEA implementation area resources. In the implementation in [15] the key expansion unit has not been integrated. In our work, such an approach is forbidden because of the WTLS specified key refreshing. In the proposed architecture, the key refreshing process is supported during data transformation. A design methodology like [15] causes a great time delay cost for every new key setup. In this way, the system performance in [15] is dramatically decreased. Analytical time delay cost measurements for both operations (key refresh and initialization) are presented in Section 11.5.

DES cipher operation is supported by a 64-bit key [27,32,33]. The computation of the key schedule is clearly described by the standards specifications [34]. The algorithm defines 16 rounds. Each data round uses a different key comprising 48 bits from the initial input key (64 bits). The total key schedule process was analyzed and according to our study it is proved that a certain combinational shift register can produce every round key. The key expansion unit design in the proposed DES architecture is built on 16 different shift registers and not with a full rolling design technique used in other published works [16–19]. With this applied technique (shift registers), DES performance is increased by ~170% compared with any architecture with a key expansion unit, built on the defined key scheduling logical components [16–19]. The only drawback of the proposed architecture is the 8% increased covered area resources. According to DES specifications, the decryption keys are the same as those used in encryption mode, if they are processed in the reverse order. For this reason, RAM blocks are used for the encryption keys to be stored after their generation. RAM blocks equal to 16×48 bits are allocated in total. In this way, the system does not generate extra keys for the decryption mode and no extra time for the generation of decryption keys is needed.

The common DES architecture has been slightly modified and the proposed architecture, in cooperation with a reconfigurable logic block, operates as a bulk cipher and as an authorized user verification unit. This unit is analyzed in detail in the following section.

11.3.3 Authorized User Verification Unit

In order to have only authorized users accessing to multiuser handheld devices and provided services, personal identification numbers (passwords) are used. In order not to allocate many extra resources for the implementation of an authorized user verification unit, a UNIX method for password verification [35] could be adopted in wireless devices. This method has the advantage that it is based on DES cipher, specified by WTLS. This means that it can be applied efficiently to the proposed crypto-processor with minimized extra covered area resources. The proposed architecture for the authorized user verification unit is illustrated in Figure 11.5.

The random number generator could be implemented by using the well-known random or pseudorandom generation techniques [27]. In the proposed architecture (Figure 11.3), the generation of the salt is based on the system clock. In this way, no extra allocated area resources for the integration of the random number generator are needed, compared with the resources that the implementation of the published techniques [27] demand. After 25 iterations of data transformation, both the encrypted password and the random salt are stored in the storage unit.

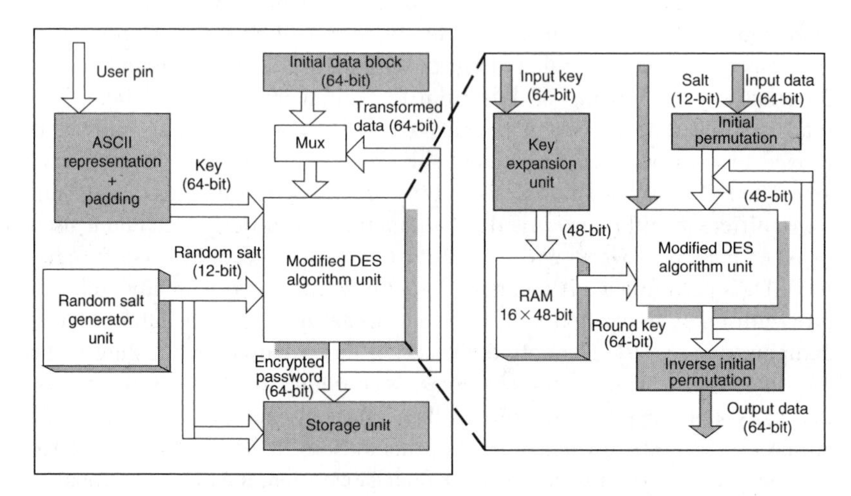

FIGURE 11.5 Proposed architecture for authorized user verification unit.

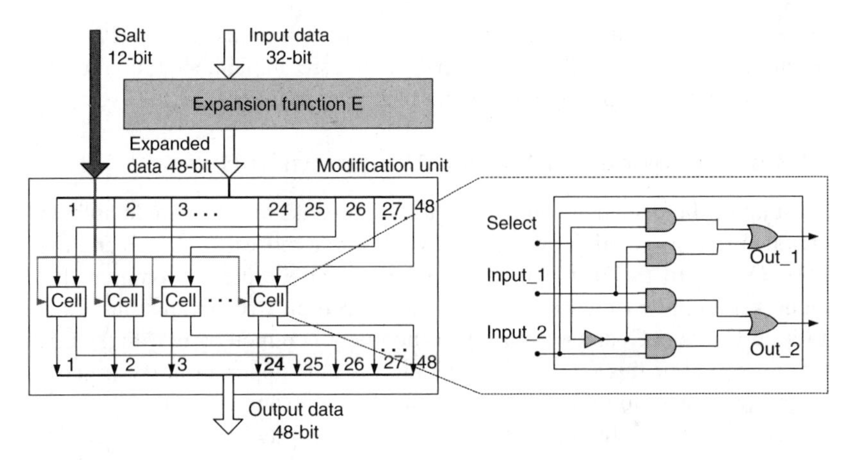

FIGURE 11.6 Proposed architecture for a modified function E.

Particularly, the salt is used to support the modified behavior of DES expansion function E, providing 4096 different operation modes. In this way, the security of the authorized user authentication unit is increased dramatically by a factor of 2^{12} ($=4096$). In systems with eight-character PINs, an attacker has to use a database of $2^{64} \times 2^{12}$ possible passwords for function E variations. In the architectural level, the desired variable operation of function E is achieved by the proposed modified function E. The architecture of modified function E is illustrated in Figure 11.6.

The modified function E is based on a dynamic combinational circuit, which is called modification unit. This unit consists of 24 similar cells. By using the 12-bit salt, 2^{12} ($=4096$) different modification cases on the expanded data are achieved. Each bit of the salt is used in two basic cells. For example, salt bit (k) is used for basic shells (k) and ($k+12$). In the proposed architecture, every bit of modification unit output is predetermined by a pair of input data bits (expanded data). The basic cell determines the appropriate modification by multiplexing every pair of input bits, with the salt bit as the select input (Figure 11.4). If the select is equal to logic one, the two associated bits are swapped. Otherwise (select is zero), no modification takes place in the certain pair of bits. The security strength of the encrypted password is augmented with the use of the randomly generated salt, and so any possible dictionary attacks become less effective.

The proposed DES architecture (Figure 11.3) is used alternatively as the original DES cipher core with the appropriate commands of the crypto-processor control unit. With the help of the applied design (Figure 11.4) that was described earlier, the proposed crypto-processor is supported with one more extra security scheme (authorized user verification) and the provided security of the system becomes higher. The major advantage of the proposed authorized user verification unit is that it uses only 2% extra resources compared with the original DES core implementation for bulk encryption, with no performance penalty at the same time.

11.3.4 Reconfigurable Message Authentication Unit

To support the demanded authentication in WAP, a reconfigurable authentication unit is proposed. It operates in two different modes: RSA [36] and D.H. [37]. Owing to the fact that their major operations are in common, both ciphers are implemented in the same unit, based on a reconfigurable design. Elliptic-Curve (EC) algorithm [38] has no major common parts with the other two ciphers. For this reason, EC operation is not supported in the proposed crypto-processor, to minimize the allocated area resources.

The reconfigurable authentication unit is presented in Figure 11.7. This proposed architecture is reconfigurable in the sense that it performs efficiently for both RSA and D.H. on to the user selection and it is not predefined by the crypto-processor (Figure 11.3).

This proposed reconfigurable unit is based on the array multiplier. The most widely known algorithm for both encryption and decryption processes of RSA to be performed is the square and multiply algorithm [39]. A number of works have been published reporting systolic array architectures

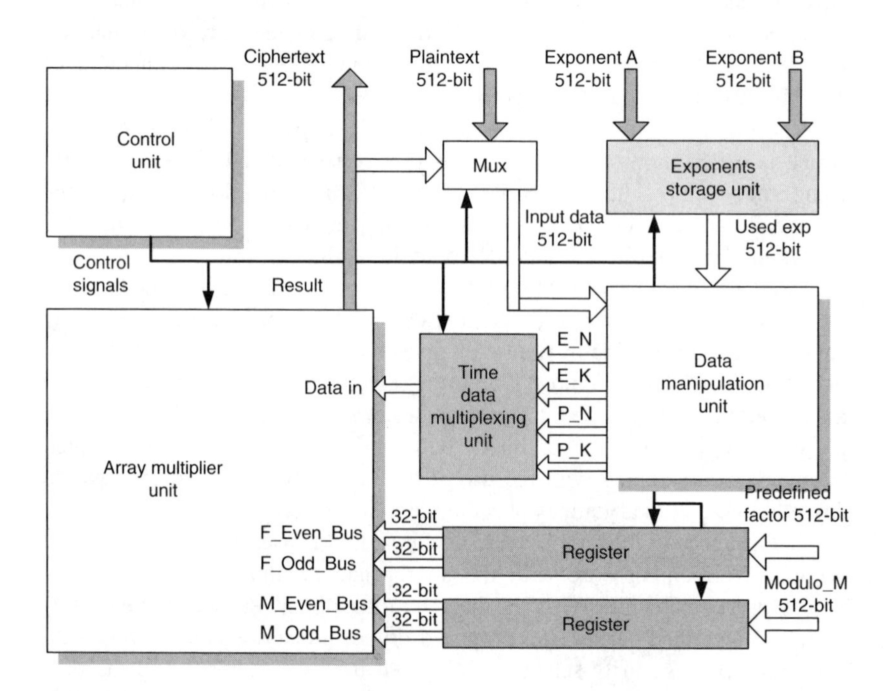

FIGURE 11.7 Proposed architecture for a reconfigurable authentication unit.

for modular multiplication. One of the most well-known modular multiplication algorithms is the Montgomery algorithm [40]. Different architectures have been published [9–12,41] with alternative implementation criteria (performance, covered area, and run time), for modular multiplication applied to RSA hardware integration. Based on the work of [40] and in combination with the previous works [42,43] on systolic multiplication, Blum and Paar [9] introduced a systolic array modular multiplication suitable for hardware implementations. The proposed reconfigurable authentication unit, which is illustrated in Figure 11.6, is based on the square and multiply algorithm [39,40] and uses the modular multiplication systolic array architecture proposed in [9].

The supported plaintext and keyword length is equal to 512 bits, defined by the WTLS specifications. RSA uses one exponent (A) of 512 bits, whereas D.H. architecture is based on two 512-bit exponents (A, B). Both the algorithms are based on modular multiplications on an input modular base (M). The applied multiplier architecture of [9] demands an extra precomputational factor input. (More details on the multiplier operation could be found in [9].)

D.H. operation is based on the same array multiplier unit used in RSA. The only basic difference is that D.H. uses two exponents compared with the one used in RSA. This results in the doubling of the number of clock cycles for a produced cipher of D.H. compared with the RSA requested time for a complete encryption or decryption process. Although the operating frequency is common for both operation modes, the proposed reconfigurable authentication unit decreases the covered area compared with the case of two separate implementations, one for each cipher, by ~70%. Furthermore, the proposed WTLS crypto-processor implementation (Figure 11.3) is able, without sacrificing the system performance or using extra resources, to provide two alternative operation modes for authentication (RSA and D.H.). The analytical synthesis results of the proposed implementation of the reconfigurable authentication unit are given in Section 11.5.

11.3.5 Reconfigurable Integrity Unit

The proposed architecture for the implementation of the reconfigurable integrity unit is presented in Figure 11.8.

This proposed architecture is reconfigurable in the sense that it operates efficiently for both SHA-1 [44] and MD5 [45]. The used operation mode (SHA-1 or MD5) is based on the users needs each time. First, in the padder, the input data are padded to be a multiple of a 512-bit block as both MD5 and SHA-1 specifications define. The padding process is exactly the same for both SHA-1 and MD5 hash functions, according to the two algorithm specifications. The 512-bit padded data blocks are divided into sixteen 32-bit words in the data manipulation unit. Then, these words are processed in order, according to each algorithm's specified data manipulation procedures.

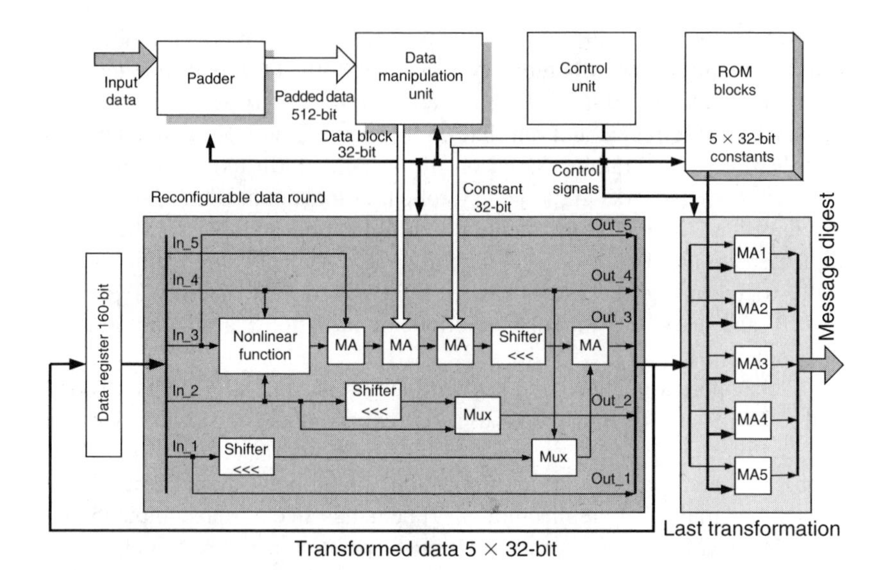

FIGURE 11.8 Proposed architecture for a reconfigurable integrity unit.

SHA-1 demands 8 × 32-bit ROM, whereas MD5 requires 68 × 32-bit ROM blocks, for the specified constants of these hash functions to be stored. The reconfigurable data round is the most critical component of the proposed reconfigurable integrity unit architecture. It has been designed as a mix of both MD5 and SHA-1 specified transformation rounds. The MA (modulo adder) component denotes modulo adder 2^{32}. The nonlinear function is a combination of mathematical functions and digital logic. It performs in two different ways for SHA-1 and MD5 operation modes. The multiplexers and also the shifters (left circular shift) operate according to the control unit commands for the two different operation modes (SHA-1, MD5) to be performed. MD5 defines 64 data transformation rounds, whereas SHA-1 specifies 80 rounds. Finally, the last transformation modifies the data. This unit consists of five modulo adders 2^{32}, where modulo additions between the five data inputs and the five 32-bit constants are performed in parallel. The 160-bit SHA-1 message digest is obtained by concatenating the 32-bit outputs of all the modulo adders. In the case of MD5 operation, the 128-bit message digest is equal to the concatenation of the first four modulo adders' 32-bit outputs (MA1 to MA4).

Our proposed reconfigurable integrity unit implementation needs ~7%–10% extra covered area resources compared with a separate SHA-1 or MD5 implementation. The critical path of this proposed unit (Figure 11.8) is defined by the out_3 data arrival time of the reconfigurable data round. The achieved frequency for both SHA-1 and MD5 operation modes is equal to 70 MHz for the proposed implementation. It is important to note that the achieved frequency is reduced by ~2% compared with one of the SHA-1 and MD5 implementations in the case of two separate hardware devices.

11.4 VERIFICATION AND TESTING

The proposed crypto-processor architecture (Figure 11.1) has been captured using VHDL. All the internal components of the design were synthesized, placed, and routed using a XILINX FPGA device [46]. The system was then simulated again for the verification of the correct functionality. To verify the right operation of the developed system, the test board of XILINX (XSA board) was used. This board is shown in Figure 11.9.

Initially, the developed architecture is downloaded to the FPGA device of this board by the parallel computer port. Then, the required VHDL code that permits emulation of the developed architecture is created and downloaded to the FPGA device too. The values of the input or output signals were monitored with the help of a logic analyzer, which is connected to the whole board structure. (The PS/2 connector and the CRT port are not used in this case.) The test scenarios that are applied to the board, to verify the correct functionality of the system, are provided by the cipher standards. In addition, during the test procedure a great number of test vectors were used to verify the right operation of the received FPGA device samples. These test vectors were mostly selected in a random manner, but some special values of the

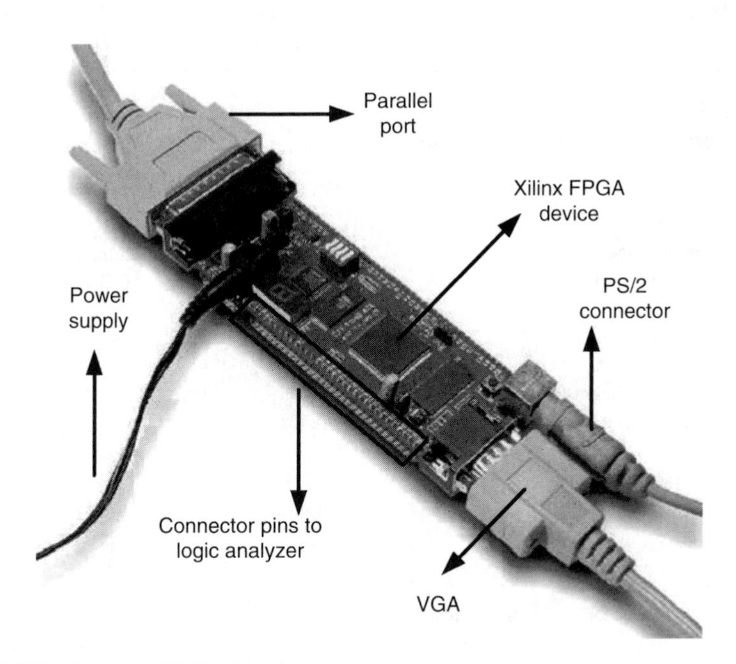

FIGURE 11.9 Used FPGA board.

input data have been included (e.g., "FFF...FFF," "000...000") to ensure maximum test coverage.

The proposed crypto-processor architecture passed all these test vectors correctly. In addition, the VLSI synthesis results of the crypto-processor hardware implementation are given in Section 11.5.

11.5 SYNTHESIS RESULTS AND EVALUATION

The synthesis results of the proposed crypto-processor are shown in Table 11.1.

In the FPGA-proposed implementation RAM, blocks are used for both keys and specified constants storage. Many FPGAs provide embedded RAM, whereas external RAM blocks can also be used, in cases where internal RAM is not available. In such implementations, the switching time of the RAM is a factor that has to be considered in the total performance timing measurements. The proposed crypto-processor requirements for RAM blocks are 128 bits for IDEA, 2048 bits for the reconfigurable authentication unit (RSA and D.H.), 16×64 bits for DES, and finally 72×32 bits for the reconfigurable integrity unit (8×32 bits for SHA-1 and $4 \times 16 \times 32$ bits for MD5). In Table 11.2 and Table 11.3, performance comparisons of the proposed crypto-processor with other related works are presented.

The proposed ciphers' implementations are compared with the best implementations published in the technical literature. The introduced work

TABLE 11.1
Implementation Synthesis Analysis

Hardware Device System Component	FPGA Device (XILINX Virtex v2000ebg560)			
	Covered Area			
	CLBs	FGs	DFFs	*F* (MHz)
IDEA algorithm	1,852	3,104	380	50
DES unit	341	682	170	85
Reconfigurable integrity unit (SHA-1, MD5)	1,653	2,905	1,049	70
Reconfigurable authentication unit (RSA, D.H.)	5,421	8,303	4,056	41
Bus interface	242	413	453	70
Reconfigurable logic block	117	546	387	70
Data registers	1,952	0	680	70
Control unit	1,576	3,200	1,200	70
Crypto-processor	14,154	20,153	8,375	—

Note: D flip-flops, DFFs; Configurable logic blocks, CLBs; Function generators, FGs; Frequency, *F* (MHz).

TABLE 11.2
Encryption Algorithms' Performance Comparison I

Architectures	Performance of Implementations		
	Area (CLBs)	F (MHz)	Throughput (Mbps)
IDEA [14]	40.561 gates	8 kHz	—
IDEA [15] 1.2 μm	108 mm^2	25	177
IDEA [13]	24,442,878	82,150	1,166,600
IDEA proposed	1,852	50	711
RSA [9] (512-bit)	25,553,413	45.6	1,446
RSA [10] 0.8 μm (512-bit)	77,988 gates	50	24 kbps
RSA [11] 0.5 μm (1024-bit)	105,000 gates	40	20 kbps
RSA [8] 2 μm (512-bit)	23,000 optimum cost	77	300 kbps
RSA [12] 1 μm (512-bit)	75,000 gates	25	100 kbps
Reconfigurable authentication unit (RSA, D.H.)	5,421	41	1.1 RSA 0.5 D.H.
DES [19]	11.1×11.1 mm^2	105	10 Gps
DES [18]	50 k transistor	250	1 Gbps
DES [17]	—	—	11.6
DES [16]	262,433,741	251,811	99,148,184
DES proposed	341	85	245

of [13] achieves throughput values almost equal to the proposed IDEA design. The basic drawback of this system is the doubled covered area, compared with the proposed FPGA implementation. The presented work in [14] operates with very low frequency compared with the proposed one, whereas no other information of the system throughput and the needed clock cycles for the encryption/decryption process is given in [14]. These omissions in the

TABLE 11.3
Encryption Algorithms' Performance Comparison II

Architectures	Performance of Implementations		
	Area (CLBs)	F (MHz)	Throughput (Mbps)
SHA-1 [20]	1,004	43	119
MD5 [21]	—	300	256
MD5 [20]	1,004	43	146
MD5 [22]	8,804,763	2171.4	165,354
Reconfigurable integrity unit (SHA-1, MD5)	1,653	70	442 SHA-1 551 MD5

reported synthesis results of [14] do not ensure a detailed comparison of this work with the proposed and the other conventional IDEA architectures [13,15]. In [15], the round keys are generated internally using the basic transformation round architecture. It has to be stated that according to the architecture of the transformation round [15], four round keys can be generated during one clock cycle at maximum. The generation of the 104 specified subkeys demands $104/4 = 26$ extra clock cycles during initialization, and $104 \times 16 = 1664$ bits of allocated RAM blocks are needed for key storage. This time delay has to be considered in the total system performance [15], in addition to the achieved 177 Mbps throughput. If the used initial key is refreshed N times, as WAP specifies, the extra needed time is equal to $N \times 26$ clock cycles. This time delay decreases the performance of such an implementation dramatically [15]. In the proposed IDEA implementation, the key expansion unit has also been integrated separately and supports the on-the-fly key generation. Only 128-bit RAM blocks are used for the initial key storage. In addition, the integrated key expansion unit of the proposed IDEA architecture supports the dynamic specified key refreshing of WTLS with no delay penalty.

The proposed reconfigurable authentication unit has an operating frequency equal to 57 MHz, for both operation modes (RSA and D.H.). The RSA algorithm performance is in general data-dependent. The performance values are illustrated in Table 11.2 and have been measured for 512-bit key and plaintext blocks specified by WTLS. Different test vectors were used to measure the average value of the performance. This was done due to the fact that RSA performance is dependent on the number of logic ≪1≫ values, that the input key vector may have. The proposed reconfigurable authentication unit has almost the same performance compared with the introduced work of [9]. Although, in the proposed implementation the same array multiplier with [9] is used, our reconfigurable authentication unit allocates 18% more area resources than the conventional. This extra area is allocated for the control logic and needed registers, for the proposed unit to perform efficiently for the two different modes (RSA and D.H.). The other compared architectures [8,10–12] have worst performance than the proposed, although this result is slightly unfair. The implementation technology libraries of these works are somehow dated (2, 1, 0.6 μm), making the comparison aspects difficult. It is possible that these works [8,11,12] would achieve better performance if they are upgraded to currently available CMOS. On the other side, the D.H. performance is estimated to be half in terms of throughput compared with the RSA. This is due to the fact that D.H. operation needs double the number of performed multiplications, based on the two used exponents, compared with the RSA. As not many hardware implementations of D.H. have been published until now [47], the comparative study of D.H. performance is a difficult process. In the only well-known work of D.H. implementation [47], it is claimed that by using GF multipliers, the performance of this cipher could be increased by ~33%. However, no detailed synthesis results about the operating frequency and covered area of the implementation are given in this work [47].

For DES hardware implementation, different works have been proposed, which focus on different implementation aspects [8,16–19]. The goals of these implementations vary from research in the key expansion process strength to programmable and parallelism designs. The proposed DES implementation is 400% faster than the best implementation presented in [16]. Nevertheless, this work [16] discusses a universal key-search machine based on fast DES architectures. The achieved results related to the key search on the DES key expansion process in [16] are very good but the used DES implementations in this work are inferior in performance. A case study of exploiting parallelism in hardware implementation of DES is introduced in the work of [17], but with low performance. The proposed DES implementation has better performance at ~430%–880% compared with the work of [17]. The major goal of the designs [18,19] is high performance. The throughput of the implementation [19] remains the best reported in the technical literature until now. The work [18] employs a novel methodology for the design of GaAs architectures and has an operating frequency three times better than the proposed implementation.

As Table 11.3 presents, the SHA-1 operation mode of the proposed reconfigurable integrity unit is better at ~260% in terms of throughput than the one in [20]. The proposed reconfigurable unit area resources are ~1.6 times more than in the same work [20]. The proposed reconfigurable integrity unit, in the case of MD5 operation mode, has better throughput than the conventional implementations [20–22], by ~260%, 115%, and 55% respectively. It has to be mentioned that the work of [21] provides a performance estimation of a theoretical MD5 hardware implementation and does not report implementation results in detail. The estimations of [21] are still important for the readers and researchers. The first implementation of [22] allocates less covered area than the proposed. This is a physical result of this design rather than a disadvantage of the proposed reconfigurable integrity unit. In this work [22], the specified MD5 processes of both padding and data manipulation have not been integrated. The proposed reconfigurable integrity unit supports both these fundamental units of MD5 specifications. Their integration has resulted in a low increase in the covered area resources. The second implementation of [22] uses full-step architecture. Although this design approach achieves doubled performance compared with the first implementation of the same work [22], the covered area resources are increased by a factor equal to 6. The proposed reconfigurable integrity unit has better area-delay product than the two implementations of [22]. The architecture introduced in [20] has been designed like a typical digital processor with data and address buses. This work requires 206 and 255 clock cycles to perform 64 rounds of MD5 and 80 rounds of SHA-1, respectively, with 59 MHz clock frequency.

Our proposed unit, based on the full rolling (feedback) technique, requires for SHA-1 and MD5 operation modes 81 and 65 clock cycles respectively, with a clock frequency of upto 70 MHz. The shared used arithmetic units in [20], supported by data and address buses, are a design technique with not very good performance, compared with the applied technique in the proposed reconfigurable integrity unit. This is because of the fact that the architecture of [20] requires many clock cycles. However, in [20] components can be added and removed from the system easily (scalability) and the system performance can be increased using more arithmetic units (exploiting parallelism). In the case of WAP integrity unit, the proposed full rolling loop architecture is a design with better performance for both MD5 and SHA-1 operation modes.

The major advantage of both the reconfigurable authentication unit and the reconfigurable integrity unit is that each one ensures the operation of two ciphers, RSA–D.H. and MD5–SHA-1, respectively, but they allocate at ~40%–60% minimized area resources compared with two separate implementations of each pair of ciphers. This is a major issue in mobile communications where many limitations exist in the area resources and in the available memory. In addition, these reconfigurable units have high operation frequency. Both authentication and integrity units achieve throughput compatible and in many cases better than the other separate conventional implementations. In the case of bulk encryption specified by the WAP, ciphers have no commonality in their architecture. This reason makes every design approach inefficient for a reconfigurable bulk encryption unit. That is why ciphers for bulk encryption have been designed as separate cores.

The main scope of the design of this proposed crypto-processor architecture is to achieve the best balance as possible between the implementation parameters such as bandwidth, allocated area, energy, and so on. In the earlier sections, the design criteria of each separate unit of the proposed architecture are analyzed in detail. It has to be mentioned that the achieved performance is superior to today's WTLS specifications and it is estimated that it could efficiently satisfy future upgrades. Concerning area and energy issues, optimizations and better synthesis results could be achieved by excluding security features or possibly one of the integrated units such as authentication unit, in cases where the applications demand. Generally, such an approach would result in reducing the supported security level of the proposed crypto-processor, and it is not recommended from the security point of view.

The proposed architecture can also be used as a powerful security core, in wireless communication networks of any kind, supporting bulk encryption, authentication, and data integrity. This means that wireless networks with no specific security requirements could adopt the powerful WTLS as an alternative flexible crypto-processor.

11.6 CONCLUSION

An efficient architecture for WTLS implementation was proposed in this chapter. All the WTLS-specified encryption units are supported by the introduced system, which guarantees a high level of security strength at the same time. The proposed architecture performs efficiently for a great set of ciphers: IDEA, RSA, D.H., DES, MD5, and SHA-1, integrated in the same hardware module. In addition, an authorized user verification has also been implemented. The proposed architecture operation is mainly based on two reconfigurable designed units. With this applied technique, the allocated area resources have been minimized to a great extent compared with other conventional implementations. The introduced system has been integrated in an FPGA hardware device and has been tested in real-time conditions, using an FPGA board. The synthesis results prove that the system has compatible (for RSA and D.H. operation modes) and better performance (for IDEA, DES, MD5, and SHA-1) compared with previously published works. The architecture proposed for IDEA is based on a modified transformation round, which minimizes the allocated area resources by ~40%. DES implementation performs better, with a range from 200% to 400%, compared with the conventional works. The reconfigurable integrity unit has better performance at ~50%–300% compared with the other conventional architectures for both MD5 and SHA-1 operation modes.

The proposed architecture is a flexible solution for WAP security layer implementation. The introduced system can be applied efficiently in both servers and mobile devices of WTLS wireless networks. The implementation of the proposed architecture achieves high-speed performance and minimized area resources, supporting six ciphers operation.

REFERENCES

1. WAP Forum: Wireless Application Protocol Architecture Specification and Wireless Transport Layer Security, www.wapforum.org, 2002.
2. G.J. Hwang, J.C.R. Tseng, and Y.S. Huang, I-WAP: An Intelligent WAP Site Management System, *IEEE Transactions on Mobile Computing*, Vol. 1, No. 2, April–June 2002.
3. N. Leavitt, Will WAP Deliver the Wireless Internet? In *Proceedings of IEEE Computer*, pp. 16–20, May 2000.
4. T. Lewis, Why WAP May Never Get Off the Ground, in *Proceedings of IEEE Computer*, pp. 110–112, August 2000.
5. M. Metter and R. Colomb, WAP Enabling Existing HTML Applications, in *Proceedings of the First Australian User Interface Conference*, pp. 49–57, Australia, 31 January–2 February 2000.
6. Advanced Encryption Standard, http://csrc.nist.gov/CryptoToolkit/aes/, 2003.
7. R. Zhang and K. Chen, Improvements on the WTLS Protocol to Avoid Denial of Service Attacks, Computers and Security, Elsevier, Vol. 24, No. 1, pp. 76–82, 2005.

8. S.S. Raghuram and C. Chakrabarti, A Programmable Processor for Cryptography, in *Proceedings of IEEE International Symposium on Circuits and Systems (ISCAS '00)*, May 28–31, Switzerland, 2000.
9. T. Blum and C. Paar, High–Radix Montgomery Modular Exponentiation on Reconfigurable Hardware, *IEEE Transactions on Computers*, Vol. 50, No. 7, 2001.
10. P.S. Chen, S.A. Hwang, and C.W. Wu, A Systolic RSA Public Key Cryptosystem, in *Proceedings of International Symposium of Circuit and System (ISCAS '96)* 1996, Vol. 4, pp. 408–411.
11. S. Ishii, K. Ohyama, and K. Yamanaka, A Single-Chip RSA Processor Implemented in a 0.5 μm Rule Gate Array, in *Proceedings of 7th Annual IEEE International ASIC Conference Exhibit*, 1994, pp. 433–436, Rochester, NY, September 19–23, 1994.
12. Holger Orup, A 100 kbits/s Single Chip Modular Exponentiation Processor, in HOT Chips VI, Symposium Record, pp. 53–59, 1994.
13. O.Y.H. Cheung, K.H. Tsoi, P.H.W. Leong, and M.P. Leong, Tradeoffs in Parallel and Serial Implementations of the International Data Encryption Algorithm, in *Proceedings of CHES 2001*, LNCS 2162, pp. 333–337, Springer-Verlag, 2001.
14. Walter Sachs and Stefan Wolter, Specification and Implementation of a Crypto-Coprocessor for ISDN, in *Proceedings of IEEE International Symposium on Circuits and Systems (ISCAS '00)*, Vol. 1, pp. 275–278, Switzerland, May 28–31, 2000.
15. R. Zimmermann, A. Curiger, H. Bonnenberg, H. Kaeslin, N. Felber, and W. Fichtner, A 177 Mb/s VLSI Implementation of the International Data Encryption Algorithm, *IEEE Journal of Solid State Circuits*, Vol. 29, No. 3, March 1994.
16. J. Kaps and C. Paar, Fast DES Implementations for FPGAs and Its Application to a Universal Key-Search Machine, in *5th Annual Workshop on Selected Areas in Cryptography (SAC '98)*, August 17–18, Ontario, Canada.
17. A.G. Broscius and J.M. Smith, Exploiting Parallelism in Hardware Implementation of the DES, in *Advances in Cryptology: CRYPTO-91 Proceedings*, Springer-Verlag, pp. 367–376, 1992.
18. H. Eberle, A High-Speed DES Implementation for Network Applications, in *Proceedings of 12th Annual International Cryptology Conference, CRYPTO '92*, Santa Barbara, August 16–20, 1992.
19. D.C. Wilcox, L.G. Pierson, P.J. Robertson, E.L. Witzke, and C. Gass, A DES ASIC Suitable for Network Encryption at 10 GPS and Beyond, in *Proceedings of CHESS '99*, LNCS 1717, pp. 37–48, 1999.
20. S. Dominikus, A Hardware Implementation of MD4-Family Hash Algorithms, in *Proceedings of IEEE International Conference on Electronics Circuits and Systems (ICECS '02)*, Vol. 3, pp. 1143–1146, Croatia, September 15–18, 2002.
21. J.D. Touch, Performance Analysis of MD5, in *Proceedings of ACM SIGCOMM '95*, Cambridge, Massachusetts, 1995.
22. J. Deepakumara, H.M. Heys, and R. Venkatesan, FPGA Implementation of MD5 Hash Algorithm, in *Proceedings of IEEE Canadian Conference on Electrical and Computer Engineering (CCECE 2001)*, Toronto, Ontario, May 2001.
23. J.M. Rabaey, *Digital Integrated Circuits*, Prentice Hall, Englewood Cliffs, NJ, 1996.
24. R. Rajsuman, *System-on-a-Chip, Design and Test,* Artech House, Boston, 2002.
25. S. Jormalainen and J. Laine, Security in WTLS, http://www.hut.fi/~jtlaine2/wtls/, 2002.

26. I. Goldberg and D. Wagner, Architectural Considerations for Cryptanalytic Hardware, Chapter 10 of *Cracking DES: Secrets of Encryption Research, Wiretap Politics and Chip Design*, O'Reilly, July 1998.
27. Bruce Schneier, *Applied Cryptography—Protocols, Algorithms and Source Code in C*, 2nd ed., John Wiley and Sons, New York, 1996.
28. Ronald L. Rivest, The RC5 Encryption Algorithm, Proceedings of the 1994 Leuven, Workshop on Fast Software Encryption (Springer 1995), pp. 86–96.
29. B.S. Kaliski Jr. and Y.L. Yin, On the security of the RC5 Encryption Algorithm, RSA Laboratories Technical Report TR-602, September 1998.
30. X. Lai and J.L. Massey, A Proposal for a New Block Encryption Standard, in *Proceedings of Eurocrypt '90*, pp. 389–404, Aarhus, Denmark, May 21–24, 1990.
31. D.R. Stinson, *Cryptography: Theory and Practice*, CRC Press LLC, 1995.
32. A. Menezes, P. van Oorchot, and S. Vanstone, *Handbook of Applied Cryptography*, CRC Press, Inc, October 1997.
33. A. Curiger, H. Bonnenberg, and H. Kaeslin, Regular VLSI Architectures for Multiplication Modulo (2^n+1), *IEEE Solid-State Circuits*, Vol. 26, No. 7, pp. 990–994, July 1991.
34. Data Encryption Standard, Federal Information Processing Standard (FIPS) 46, National Bureau of Standards, 1977.
35. F.T. Gramp and R.H. Morris, UNIX Operation System Security, *AT&T Bell Laboratories Technical Journal*, Vol. 63, No. 8 (part 2), October 1984.
36. R. Rivest, A. Shamir, and L. Adleman, A Method for Obtaining Digital Signatures and Public Key Cryptosystems, *Comm. ACM*, Vol. 21, pp. 120–126, February 1976.
37. W. Diffie and M.E. Hellman: New Directions in Cryptography, *IEEE Transactions on Information Theory*, Vol. IT-22, pp. 644–654, 1976.
38. IEEE P1363, Standard Specifications for Public-Key Cryptography (Draft Version 8), October 1998.
39. D.E. Knuth, *The Art of Computer Programming. Vol. 2: Seminumerical Algorithms*. Addison-Wesley, Reading, Massachusetts, 2nd ed., 1981.
40. P. Montgomery, Modular Multiplication with Trial Division, *Mathematics of Computation*, Vol. 44, pp. 519–521, April 1985.
41. C.D. Walter, Systolic Modular Multiplication, *IEEE Transactions on Computers*, Vol. 42, No. 3, pp. 376, March 1993.
42. T. Blum and C. Paar, Montgomery Modular Exponentiation on Reconfigurable Hardware, in *Proceedings of the 14th Symposium on Computer Arithmetic*, pp. 70–77, 1999.
43. T. Blum, Modular Exponentiation on Reconfigurable Hardware, Master thesis, Electrical and Computer Engineering Department, Worcester Polytechnic Institute, May 1999.
44. SHA-1 Standard National Institute of Standards and Technology (NIST), Secure Hash Standard, FIPS PUB 180–1, www.itl.nist.gov/fipspubs/fip180-1.htm, 2001.
45. R.L. Rivest, The MD5 Message Digest Algorithm, RFC 1321, MIT LCS and RSA Data Security, Inc., April 1992.
46. Xilinx Inc., San Jose, California, Virtex, 2.5 V Field Programmable Gate Arrays, 2002.

47. C.N. Zhang, M. Deng, and R. Mason, Two Improved Algorithms and Hardware Implementations for Key Distributing Using Extended Programmable Cellular Automate, in *14th Annual Computer Security Applications Conference*, Phoenix, Arizona, 1998.
48. SSL Protocol Specifications, www.netscape.com/eng/ssl3, 2002.

12 Binary Algorithms for Multiplicative Inversion

Erkay Savas

CONTENTS

The basic arithmetic operations (i.e., addition, multiplication, and inversion) in prime and binary extension fields, GF(p) and GF(2^n), have several applications in cryptography, such as RSA algorithm, Diffie–Hellman key exchange algorithm [1], the U.S. federal Digital Signature Standard [2], and also elliptic and hyperelliptic curve cryptography [3,4]. Efficient calculation of multiplicative inverses of elements in both fields is of utmost importance since inversion is the most time-consuming operation in hyperelliptic curve cryptography when affine coordinates are selected [5–10].

The majority of the currently employed inversion algorithms used to compute inverses in both GF(p) and GF(2^n) have their roots in the Euclidean algorithm reported by Euclid in his *Elements* [11]. The Euclidean algorithm provides a simple and efficient means for computing the greatest common divisor (GCD) denoted $gcd(u,v)$ of two positive integers u and v, without finding their factorizations. In many cryptographic applications, the extended version of the Euclidean algorithm plays an important role. In addition to the GCD, the extended Euclidean algorithm (EEA) returns two unique integers s and r. Using these integers the GCD may be expressed as a linear combination of u and v

$$us + vr = gcd(u,v).$$

If u and v are relatively prime, it immediately follows that

$$vr = 1(\bmod u).$$

Hence, the EEA provides an efficient method to compute modular inverse v^{-1} $(\bmod u) = r$. The EEA can easily be modified to compute multiplicative inverses in binary extension fields, GF(2^n), as shown in [12]. The major difficulty in EEA is that it requires many integer division operations, which are considered to be very expensive in cryptography. Therefore, a binary extended Euclidean algorithm (b-EEA), that is attributed to Penk [13], was proposed. The binary algorithm is remarkably suitable for implementation in digital systems (both hardware and software), since it does not require long integer divisions, but only relies on basic addition and subtraction operations. Many inversion algorithms proposed in literature, both for GF(p) and GF(2^n), have similarities to the b-EEA. However, they differ in the number of basic operations they require to compute inverses.

The organization of the chapter is as follows: in Section 12.2, we present four different algorithms proposed for GF(p) and discuss and compare their performances from the perspectives of both software and hardware implementations. We perform the same treatment for inversion algorithms for GF(2^n) in Section 12.3. We demonstrate that more possibilities exist for inversion in GF(2^n) than GF(p) such as systolic array implementation in this section. Finally, we summarize our findings in Section 12.4.

12.1 INVERSION ALGORITHMS FOR PRIME FIELDS GF(p)

12.1.1 BINARY EEA

Algorithm 1 [13] directly computes $a^{-1} \bmod p$ by maintaining the invariants

$$sa + xp = v, \quad ra + yp = u,$$

where x and y are not computed. In the iteration before the last one, $u = v = 1$, and thus the last iteration results in $u = 1$ and $v = 0$. Consequently, we have $ra + yp = 1$, indicating r is the inverse of $a \bmod p$. However, since r is allowed to be negative and r can become larger than p in the last iteration, Step 13 and Step 14 are necessary.

Algorithm 1. Binary Extended Euclidean Algorithm

Input: $a \in [1, p-1]$ and p is prime number

Output: $r \in [1, p-1]$ where $r = a^{-1} (\bmod p)$

1: $u := p$, $v := a$, $r := 0$, and $s := 1$
2: **while** $v > 0$ **do**
3: **while** u is even **do**
4: $u := u/2$
5: **if** r is even **then** $r := r/2$ **else** $r := (r+p)/2$
6: **end while**
7: **while** v is even **do**
8: $v := u/2$
9: **if** s is even **then** $s := s/2$ **else** $s := (s+p)/2$
10: **end while**
11: **if** $u > v$ is even **then** $u := u - v$, $r := r - s$ **else** $v := v - u$, $s := s - r$
12: **end while**
13: **if** $r \geq p$ **then** $r := r - p (\bmod p)$
14: **if** $r < 0$ **then** $r := r + p (\bmod p)$
15: **return** r

A division algorithm to compute $b/a \bmod p$ can be directly obtained by substituting $s := b$ for $s := 1$ in Step 1 of Algorithm 1. Since the operations applied to r and s mostly right shifts and subtractions and occasionally additions with the modulus, it is easier to control the magnitude of r and s in Algorithm 1 than in the Montgomery division algorithm, as can be seen in the next section.

12.1.2 Montgomery Inversion Algorithm

The Montgomery inversion algorithm, as defined in [5], computes

$$b = a^{-1} 2^n (\bmod p), \qquad (12.1)$$

given $a < p$, where p is a prime number and $n = [\log_2 p]$. The algorithm consists of two phases: the output of Phase I is the integer r such that $r = a^{-1} 2^k (\bmod p)$, where $n \leq k \leq 2n$, and Phase II is a correction step and can be modified, as shown in [8], to calculate a slightly different inverse that is a more precise definition of the Montgomery inverse

$$b = MonInv(a2^n) = a^{-1} 2^n (\bmod p). \qquad (12.2)$$

This new definition is more suitable since it takes an integer in so-called residue domain and yields its multiplicative inverse, again in the residue domain.

Algorithm 2. Montgomery Inversion Algorithm—Phase I

Input: $a \in [1, p-1]$ and p is prime number

Output: $r \in [1, p-1]$ and k, where $r = a^{-1}2^k \bmod p$ and $n \leq k \leq 2n$

1: $u := p$, $v := a$, $r := 0$, $s := 1$, and $k := 1$ (loop index)
2: **while** $v > 0$ **do**
3: **if** u is even **then** $u := u/2$, $s := 2s$
4: **else if** v is even **then** $v := v/2$, $r := 2r$
5: **else if** $u > v$ **then** $u := (u - v)/2$, $r := r + s$, $s := 2s$
6: **else** $v := (v - u)/2$, $s := r + s$, $r := 2r$
7: $k := k + 1$
8: **end while**
9: **if** $r \geq p$ **then** $r := r - p(\bmod p)$ **end if**
10: **return** $r := p - r$ and k

Algorithm 2 is in fact the first phase of the Montgomery inversion algorithm and the Montgomery inverse of $a \bmod p$ (i.e., $a^{-1}2^{2n} \bmod p$) can be obtained applying $2n - k$ repeated multiplications of r by 2 after the first phase. In every iteration the following operation must also be performed

if $r \geq p$ **then** $r := r - p$.

There are three important theorems about the algorithm, which have already been proven in [5].

Theorem 1 *If $p > a > 0$, then the intermediate values r, s, u, and v in the Montgomery inversion algorithm are always in the range $[1, 2p - 1]$.*

Theorem 2 *If a and p are relatively prime, p is odd, and $p > a > 0$, then the number of iterations in the first phase of Montgomery inversion algorithm is at least n and at most $2n$, where n is the number of bits in p.*

Theorem 3 *If p and a are relatively prime, p is odd, and $p > a > 0$, then Phase I of Montgomery inversion algorithm returns $a^{-1}2^k$ (mod p).*

A Montgomery division algorithm to compute $(b2^n)/(a2^n) \bmod p$ that can be obtained from the Montgomery inversion algorithm requires substantial modifications in the steps of Algorithm 2. First modification must be made to Step 1 by substituting $s := b2^n$ for $s := 1$. Consequently, the variables r and s become large numbers in the early iterations and furthermore the operations in Step 3 through Step 6 can even result in larger values of r and s. Therefore, after the operations in Step 3 through Step 6, r and s must be reduced ($\bmod p$) if they become larger than p. For example, Step 3 must be modified as

if u is even **then** $u := u/2, s := 2s \bmod p$.

As a result, the advantage of the Montgomery division algorithm over the classical method of invert-and-multiply (i.e., $b/a \bmod p \equiv b(a^{-1} \bmod p)$ is not obvious, if it is not worse. And finally, the second phase of the Montgomery inversion must also be modified. Given residue numbers $a2^n \bmod p$ and $b2^n \bmod p$, the first phase of the Montgomery division yields $b/a \cdot 2^k \bmod p$, where $2n \geq k \geq n$. The residue form of the division, $b/a \cdot 2^n \bmod p$, can be obtained by applying $k-n$ repeated division of the form

if r is even **then** $r := r/2$ **else** $r := (r + p)/2$.

12.1.3 Left-Shift Binary Inversion Algorithm

In Algorithm 1 and Algorithm 2 variables u or v are shifted to the right in every iteration. The algorithm proposed in [14] given as Algorithm 3 computes inversion by employing left-shift operations.

Algorithm 3. Left-Shift Inversion Algorithm

Input: $a \in [1, p-1]$ and p is prime number

Output: $r \in [1, p-1]$ where $r = a^{-1} \bmod p$

1: $u := p, v := a, r := 0$, and $s := 1, c_u := 0, c_v := 0$
2: **while** $u \neq \pm 2^{c_u}$ **and** $v \neq \pm 2^{c_v}$ **do**
3: **if** $(u_n = 0$ **and** $u_{n-1} = 0)$ **or** $(u_n = 1$ **and** $u_{n-1} = 1)$ **and** $(u_{n-2}$ **or**, ..., u_1 **or** $u_0 = 1)$ **then**
4: **if** $c_u \geq c_v$ **then** $u := 2u, r := 2r, c_u := c_u + 1$ **else** $u := 2u$, $s := s/2$, $c_u := c_u + 1$
5: **else if** $(v_n = 0$ **and** $v_{n-1} = 0)$ **or** $(v_n = 1$ **and** $v_{n-1} = 1)$ **and** $(v_{n-2}$ **or**, ..., v_1 **or** $v_0 = 1)$ **then**
6: **if** $c_v \geq c_u$ **then** $v := 2v, s := 2s, c_v := c_v + 1$ **else** $v := 2v$, $r := r/2, c_v := c_v + 1$
7: **else**
8: **if** $v_n = u_n$ **then**
9: **if** $c_u \leq c_v$ **then** $u := u - v, r := r - s$ **else** $v := v - u, s := s - r$
10: **else**
11: **if** $c_u \leq c_v$ **then** $u := u + v, r := r + s$ **else** $v := v + u, s := s + r$
12: **end if**
13: **end if**
14: **end while**
15: **if** $(v = \pm 2^{c_v})$ **then** $r := s, u_n := v_n$
16: **if** $(u_n = 1$ and $r < 0)$ **then** $r := -r$ **else** $r := p - r$
17: **if** $(r < 0)$ **then** $r := r + p$
18: **return** r

Algorithm 3 is designed to be implemented in hardware. It eliminates the need for the integer comparison operation such as $u > v$ that is common in other binary inversion algorithms. Instead, it includes relatively less expensive operations such as ORing of bits of u and v that can be implemented using a tree of OR gates. However, it requires comparisons such as $u \neq \pm 2^{c_u}$ that is probably more expensive to implement than the simple comparison $v \neq 0$ used in other algorithms. The main advantage of Algorithm 3, however, is the fact that it requires fewer number of addition operations on average comparing with other right-shift algorithms at the expense of more shift operations. Since shift operations are likely to be much less expensive than additions, Algorithm 3 can be more efficient when implemented in hardware even though the average number of iterations is higher than those of Algorithm 1 and Algorithm 2. In addition, the control circuit of Algorithm 3 is likely to be more complicated than others.

12.1.4 Binary EEA with Brent–Kung Technique

All algorithms presented here except Algorithm 3 require the operation of comparing two integers, that is, $u > v$. To remove the comparison operation, one can adopt the idea of Brent and Kung [15], which introduces a new variable, δ, to represent the difference between the bit lengths of u and v. Algorithm 4 proposed in [16] uses this idea.

Algorithm 4. Binary EEA with Brent–Kung Technique

Input: $a \in [1, p-1]$ and p is prime number

Output: $r \in [1, p-1]$ where $r = a^{-1} \bmod p$

1: $u := p, v := a, r := 0, s := 1$, and $\delta := 0$
2: **while** $v \neq 0$ **do**
3: **if** v is even **then**
4: $v := v/2$, $\delta := \delta - 1$
5: **else**
6: **if** $\delta < 0$ **then** $u \leftrightarrow v, r \leftrightarrow s, \delta := -\delta$
7: $k := 1$
8: **if** $((u+v) \bmod 4 \neq 0)$ **then** $k := -1$ **else** $\delta := \delta - 1$
9: $v := (v + k \cdot u)/2$, $s := (s + k \cdot r)$
10: **end if**
11: $s := (s + s_0 \cdot p)/2$
12: **end while**
13: **if** $u \neq 1$ **then** $r := p - r$
14: **return** r

The symbol $\leftrightarrow$ indicates the swap of values between two variables. The algorithm is a slightly different version of b-EEA and uses the following properties to compute multiplicative inversion. If v is even and u is odd, then $gcd(u,v) = gcd(u,v/2)$. If both are odd, then 4 divides either $v+u$ or $v-u$. If 4 divides $v+u$, then $gcd(u,v) = gcd(u,(u+v)/4)$. Otherwise, $gcd(u,v) = gcd(u,(v-u)/4)$. In both cases, $|u+v|/4, |v-u|/4 \leq \max(|u/2|, |v/2|)$. Thus, if $v > u$, $(u+v)/4$ or $(v-u)/4$ decrements the bit length of v by 1. If $v < u$, the bit length of v may not be decremented. This may result in negative values of v. On the other hand, when $\delta < 0$, the swap of variables is applied, which is required for the convergence of the algorithm.

Algorithm 4 requires, on average, higher number of iterations to complete as explained in the next section. On the other hand, it is easy to convert it to a division algorithm by substituting $s := b$ for $s := 1$ in Step 1 to compute $b/a \bmod p$. However, the real advantage of the algorithm is the fact that it needs no comparison of integers. Eliminating the integer comparison operation may facilitate using carry-free arithmetic, where the comparison is expensive, but shift and addition operations can be executed efficiently.

12.1.5 Comparison of Binary Inversion Algorithms for GF(p)

Many inversion algorithms proposed in the literature are originally designed to be implemented in software on general-purpose processors. Recently, there has also been considerable interest to design new algorithms favoring hardware implementations, since software implementations are far from achieving the time requirements of elliptic curve cryptography. The aim of this section is to provide a fair comparison of different inversion algorithms from the perspective of both hardware and software implementations and to give designers leverage in choosing the appropriate algorithm for the intended application.

However, it is difficult to derive criteria to assess different algorithms since many details factor in on their efficiency. The best thing one can do in this circumstance is to count average number of iterations and average number of operations such as addition and shifting. Each iteration incorporates different combinations of iterations such as additions, shifts, additions followed by shift operations, and so on. Which combination is executed in a particular iteration is determined using certain conditional check operations such as parity check (e.g., Step 3 and Step 7 of Algorithm 1), comparison (e.g., Step 11 of Algorithm 1), and so on. The expected number of these operations and certain combinations of these operations determine the complexity of the algorithm. Since software and hardware implementations adopt different ways to execute these operations, we separately inspect software and hardware cases.

The software implementation of an inversion algorithm running on the datapath of a simple general-purpose processor executes the operations in a sequential manner since main functional units are not duplicated. For example, Step 5 of Algorithm 2 has two additions and shift operations that can be done in parallel. However, a typical software implementation on a simple reduced instruction set computer (RISC) processor fails to take advantage of these types of concurrencies. Under these assumptions, we ran four algorithms presented above 10,000 times with different precisions, counted the number of shift and addition operations, and demonstrated the results in Figure 12.1.

As can be observed from the figure, while Algorithm 1 requires the fewest number of shift operations, Algorithm 3 requires the fewest number of additions, and Algorithm 1 and Algorithm 2 are comparable in terms of total number of addition and shift operations. Algorithm 4 is apparently not suitable for software implementations. Algorithm 3 is also not suitable for software implementation because shift and addition operations are usually of equal cost in software. From the software implementation aspect, the total number of iterations is usually more important. Furthermore, the control flow of Algorithm 3 poses certain difficulties in software implementation. Therefore, there are two algorithms suitable for software: Algorithm 1 and Algorithm 2. Algorithm 1 has slightly lower total operation count than Algorithm 2. However, Algorithm 2 usually performs better than Algorithm 1 because of the unaccounted factors such as memory efficiency and more comparison operations. For example,

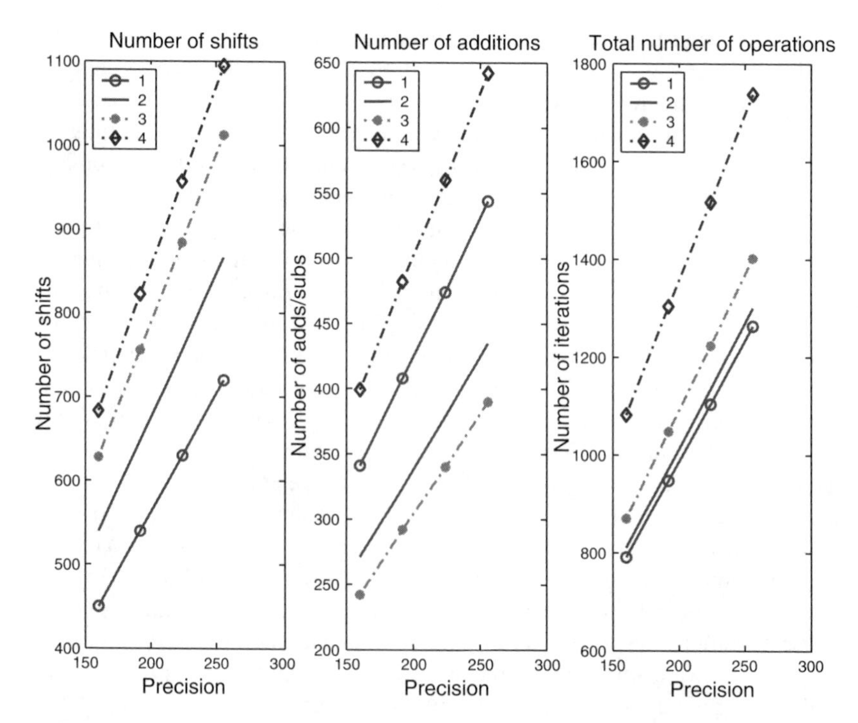

FIGURE 12.1 Comparison of four algorithms in terms of number of operations with respect to software implementation.

Algorithm 1 requires parity check of r or s frequently. In addition, the fifth multiprecision variable, the modulus p, needs to be loaded from memory, which is not necessary in the first phase of Algorithm 2.* Consequently, one can safely conclude that Algorithm 2 is the best choice for software implementation.

To compare different inversion algorithms from the perspective of hardware implementations, we take a different approach by taking into account the operations that can be performed in parallel. We assume that it is possible to employ more than one functional unit such as adders and shifters. We can, thus, classify the operations to count as follows: (i) standalone shift operations that cannot be executed along with an addition and (ii) addition operations that are basically addition or subtraction operations. For example, shifts in Step 3 or Step 4 of Algorithm 2 are standalone shift operations, while $(u-v)/2$ in Step 5 of the same algorithm is considered as an addition operation. Although the latter has also a shift following the subtraction, it is considered as an addition operation since this shift can be incorporated into an adder while designing the hardware. In addition, assuming that we can employ as many adders or shifters as we need, we consider operations that can be executed simultaneously by different units working in parallel, as only one operation. In case two additions and two shifts are executed in the same iteration in parallel, we count them as a single addition and shift operation, respectively. For example, Step 3 of Algorithm 2 is counted as a single shift operation.

Under these assumptions, we compared four algorithms. Excluding parity check and integer comparison, we counted the number of standalone shift operations and additions by running these four algorithms 10,000 times (with 100 different integers whose inverses to be calculated for 100 different primes). In Figure 12.2, the statistics obtained from this experiment are displayed.

From the figure, the number of addition operations in Algorithm 3 is much fewer than those in the other three algorithms. In total number of operations, Algorithm 2 absolutely compares favorably with the others. However, Algorithm 3 may perform better where the shift operations are much less expensive than additions. On the other hand, since there are more complicated conditional checks in Algorithm 3 than Algorithm 2, a better comparison based on actual hardware implementations of both algorithms is needed to determine which one is more efficient by considering the other factors such as area requirements and critical path delays of the actual designs also. As also shown in the figure, Algorithm 4 performs poorly comparing with the other three algorithms. The algorithm does not use a comparison operation used in Algorithm 1 and Algorithm 2 to accelerate the convergence process. Consequently, it converges slowly resulting in higher number of iterations on average. Since it does not use a comparison, it can be profitably used in implementations where carry-free arithmetic is employed, examples of which are given in [17,18].

---

* In fact, all inversion algorithms need to perform many load and store instructions from memory. These memory-access instructions have significant effect on performance.

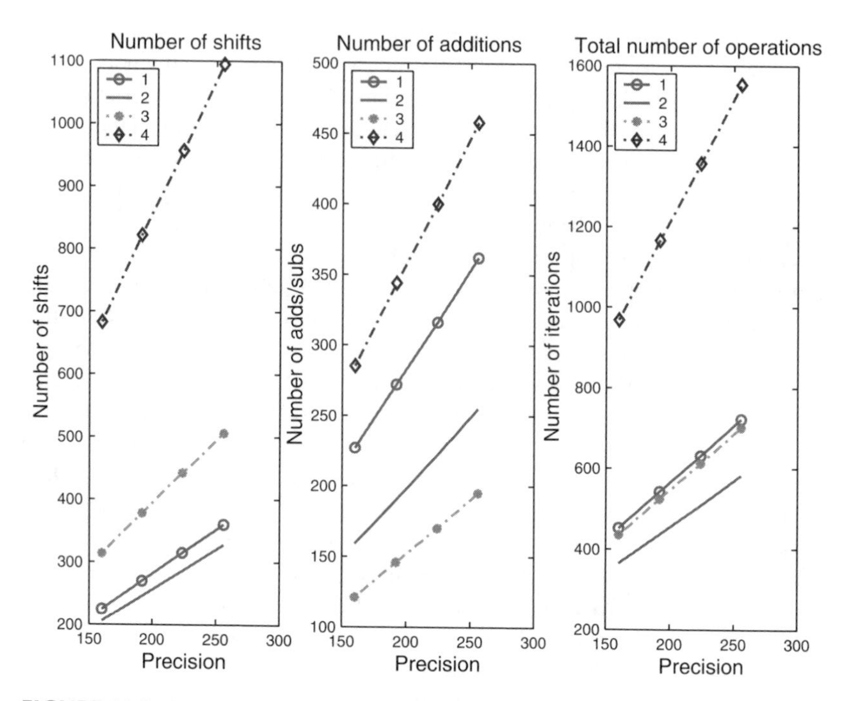

FIGURE 12.2 Comparison of four algorithms in terms of number of operations with respect to hardware implementation.

12.1.6 Other Algorithms for Inversion

Beside the four algorithms described and analyzed earlier, other alternative algorithms for inversion have also been proposed in the literature [7,8,10,19–22] for GF(p) inversion. Some of these algorithms can be considered as slightly different versions of the algorithms described here. For instance, the inversion algorithms in [20] and [21] are the same as Algorithm 1. On the other hand, some algorithms, even though are variations of Algorithm 2, incorporate clever tricks to accelerate the computations. We outline two algorithms in this category in the following:

- *Inversion with Multibit Shifting*. The algorithms in [19,22] proposed for hardware implementations take advantage of so-called multibit shifting technique which allows to shift the variables u and v more than 1 bit in one clock cycle whenever it is possible. Although the technique is originally proposed for Algorithm 2, it can also be used in Algorithm 1 and Algorithm 4. In Step 3 and Step 4 of Algorithm 2, instead of a simple parity check on the variables u and v, the three least significant bits of these variables are checked. This check allows, for example, shifting these variables 3 bits to the right at once if the three least significant bits of the variable are all zero. Checking 3 bits instead of a parity check costs insignificant amount of time and area overhead in hardware. The number of iterations is therefore reduced. The overall effect of the technique is reported in [19,22] as about 15% to 20% decrease in the number of clock cycles in actual hardware implementations of the algorithm. The multibit shifting idea can be extended to Step 5 and Step 6 of Algorithm 2 by slightly modifying the subtraction operations $u - v$ and $v - u$. For example, $u - v$ can be changed as $u - 2v$ or $u - 3v$ when possible. This accelerates the convergence of u to 1, hence decreasing the number of iterations on average. However, this technique also complicates the operations in an iteration. Therefore, its effect needs to be further analyzed.

- *Kobayashi's Word-Based Algorithm Suitable for Software*. The algorithm [7] based on Algorithm 2 proposes a different variant suitable for software implementations. The algorithm treats the variables as multiword integers where each word is w bits, that is usually the word size of the underlying general-purpose processor. The algorithm proposes a major modification in the first phase of the algorithm by nesting an inner "for loop" within the while loop. The inner for loop is executed w times and only the least and most significant words of the four variables, u, v, r, and s, are involved in the computations. These computations are the same as those of Algorithm 2. Therefore, the inner loop consists of only single precision operations. When the inner loop exits, a couple of operations are performed on multiword variables. This technique combined with a postprocessing technique proposed for the second phase of Algorithm 2 reportedly provides almost 5.5 times speedup.

12.2 INVERSION ALGORITHMS FOR BINARY EXTENSION FIELDS GF(2^n)

Although prime fields GF(p) and binary extension fields GF(2^n) are quite dissimilar mathematical structures, many inversion algorithms based on EEA proposed for computing inverses in GF(p) also work for GF(2^n) with only minor modifications. In this section, we describe and analyze inversion algorithms for GF(2^n). In addition, we also explain new possibilities for binary extension fields such as systolic array computation of inversions, which is not possible for prime fields. First, we start by introducing a notation used in all algorithms in this section. Let

$$p(x) = x^n + p_{n-1}x^{n-1} + p_{n-2}x^{n-2} + \cdots + p_1x + p_0$$

be an irreducible polynomial over GF(2) that is used to construct the binary extension field GF(2^n). An element of GF(2^n) can be represented as a polynomial

$$a(x) = a_{n-1}x^{n-1} + a_{n-2}x^{n-2} + \cdots + a_1x + a_0$$

whose coefficients a_is are from {0,1}. Then arithmetic on the elements in GF(2^n) is regular polynomial arithmetic modulo irreducible polynomial $p(x)$, where operations on coefficients are performed modulo 2.

12.2.1 Binary EEA for Computing Inverses in GF(2^n)

The b-EEA for GF(2^n) is quite similar to Algorithm 1, except that binary polynomials are used instead of integers, both of which are represented in the same manner in digital systems. Therefore, the b-EEA for GF(2^n) can be easily obtained from Algorithm 1 by using the following modifications:

- Replace all addition and subtraction operations involving u, v, r, s, and p with addition operation in GF(2^n).
- Replace all divisions by 2 with division by x and multiplication by 2 with multiplication by x. However, both division and multiplication by 2 are implemented as right and left shifts in digital systems, respectively. Therefore, there is no need to do any modification in the implementation since division and multiplication by x are implemented in the same manner.
- Replace the parity checks in Algorithm 1 with checks on the constant term of polynomials. Again, there is no need to do any modification in the implementation since both checks are done identically.

Algorithm 5. Binary Extended Euclidean Algorithm for GF(2^n)

Input: $a(x)$ and $p(x)$ irreducible polynomial, where $deg(a(x)) < deg(p(x))$

Output: $r(x)$ where $r(x) = a(x)^{-1} \bmod p(x)$ and $deg(r(x)) < deg(p(x))$

1: $u(x) := p(x)$, $v(x) := a(x)$, $r(x) := 0$, and $s(x) := 1$
2: **while** $v \neq 0$ **do**
3: **while** $u_0 = 0$ **do**
4: $u(x) := u(x)/x$
5: **if** $r_0 = 0$ **then** $r(x) := r(x)/x$ **else** $r(x) := (r(x) + p(x))/x$
6: **end while**
7: **while** $v_0 = 0$ **do**
8: $v(x) := u(x)/x$
9: **if** $s_0 = 0$ **then** $s(x) := s(x)/x$ **else** $s(x) := (s(x) + p(x))/x$
10: **end while**
11: **if** $deg(u(x)) > deg(v(x))$ **then** $u(x) := (u(x) + v(x))$, $r(x) := r(x) + s(x)$
12: **else** $v(x) := (v(x) + u(x))$, $s(x) = s(x) + r(x)$
13: **end if**
14: **end while**
15: **return** $r(x) \bmod p(x)$

- Replace integer comparison operation with degree comparison of two polynomials. However, comparing binary representation of two polynomials as integers also works.

The b-EEA for GF(2^n) maintains similar invariants to those of Algorithm 1. Similarly, a division algorithm to compute $b(x)/a(x) \bmod p(x)$ can be directly obtained by substituting $s(x) := b(x)$ for $s(x) := 1$ in Step 1 of Algorithm 5.

12.2.2 Montgomery Inversion Algorithm for GF(2^n)

Similar to b-EEA, the Montgomery inversion algorithm for GF(2^n) can be obtained by performing the same modifications proposed for b-EEA. The first phase of the algorithm computes $r(x) = a(x)^{-1}x^k \bmod p(x)$, where $n+1 \leq k \leq deg(a(x)) + n + 1$ and n is the degree of the irreducible polynomial $p(x)$. Moreover, if $deg(p(x)) > deg(a(x)) > 0$, where $p(x)$ is an irreducible polynomial, then the degrees of intermediate binary polynomials $r(x)$, $u(x)$, and $v(x)$ in the Montgomery inversion algorithm are always in the range of $[0, deg(p(x))]$, while $deg(s(x))$ is in the range of $[0, deg(p(x)) + 1]$.

The second phase of the algorithm computes the Montgomery inverse (i.e., $r(x) = a(x)^{-1}x^{2n} \bmod p(x)$ given $a(x)x^{2n}$) by applying $2n-k$ repeated multiplication of $r(x)$ by x after the first phase. The Montgomery division algorithm to compute $(b(x)x^n)/(a(x)x^n) \bmod p(x)$ necessitates the reduction of $r(x)$ and $s(x) \bmod p(x)$ when their degrees become n. The second phase of the Montgomery division is $k-n$ repeated division of $r(x)$ by x.

The almost inverse algorithm proposed in [6] is very similar to the Montgomery inversion algorithm; therefore it is not included here.

12.2.3 Comparison of Inversion Algorithms for GF(2^n)

The binary extended Euclidean and Montgomery inversions for GF(2^n) are compared using the approach in Section 12.1.5 and the results are depicted in Figure 12.3. The number of operations (i.e., additions and shifts) from both software and hardware implementation perspectives is given in the figure, where the upper two lines are from the software perspective and the lower ones from the hardware perspective. As can be observed from the figure, the Montgomery inversion algorithm compares favorably against b-EEA for both software and hardware implementations.

12.2.4 Idea of Brent–Kung and Systolic Array Implementations

The b-EEA and the Montgomery inversion algorithm require the time-consuming operation of comparing degrees of two polynomials, that is, $deg(u(x)) > deg(v(x))$, which may dominate the operation speed. To remedy this problem, one can adopt the idea of Brent and Kung [15], which introduces a new variable, δ, to represent the difference between the degrees

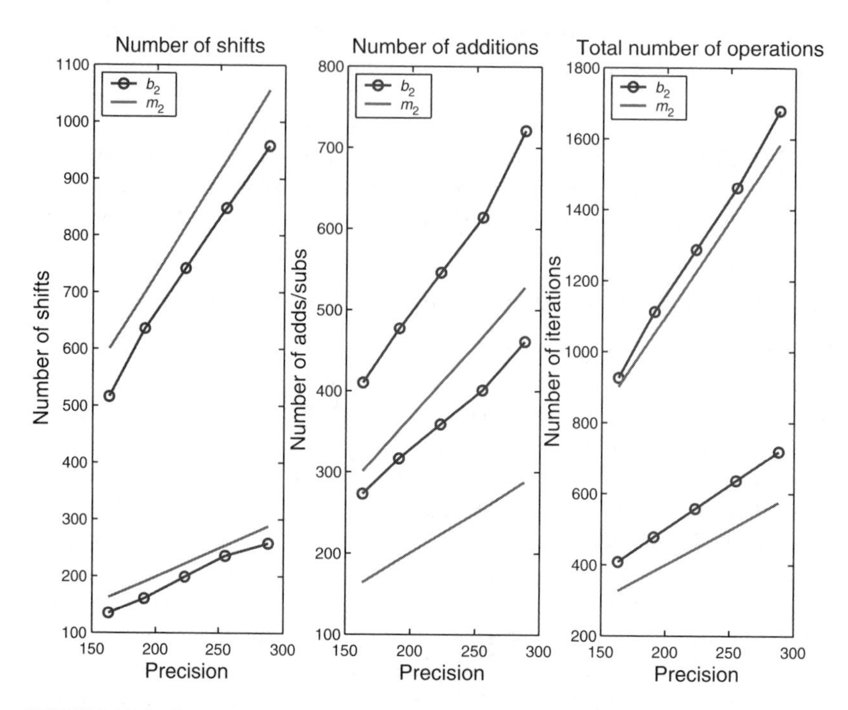

FIGURE 12.3 Comparison of two algorithms in terms of number of operations both from software and hardware point of view.

of $u(x)$ and $v(x)$.* When the need for degree comparison is eliminated, the control circuit of the inversion unit, when implemented in hardware, requires only generation of local signals and hence becomes quite simple to design. In addition, there is no problem of distributing the control signals throughout the circuit. One of the most efficient implementation techniques for VLSI circuits is the systolic arrays [23] due to their attractive features such as regularity, modularity, and concurrency. The systolic arrays yield high-throughput (i.e., the number of inversions per clock cycle) when there are many consecutive inversion operations to be calculated.

In the following, we present three algorithms suitable for systolic array implementations.

12.2.4.1 Stein's EEA

Proposed by Stein [24] and improved and extended to division by Wu et al. [25], Stein's algorithm is the most efficient variant of b-EEA for systolic array implementation. Unlike the algorithms discussed so far, Stein's algorithm executes in a loop with a fixed number of iterations, $2n-1$, where n is the degree of the irreducible polynomial. This property is especially important for two-dimensional systolic array implementation, since the number of rows in the array is equal to the number of iterations. Moreover, Stein's algorithm uses only right-shift operation on variables $r(x)$ and $s(x)$, therefore its adaptation to division is straightforward.* Stein's algorithm is depicted in Algorithm 6.

In the algorithm, since $u(x) := p(x)$ and $v(x) := a(x)$ initially, we have $deg(v(x)) \leq deg(u(x)) - 1$ in the beginning, and hence, δ is initialized to be -1. Therefore, a negative value of δ indicates $deg(u(x)) > deg(v(x))$. In a typical hardware implementation, the variable δ maps onto a simple counter. A systolic array implementation of Stein's algorithm is shown in Figure 12.4 for $n = 3$.

In the figure, a row of cells is responsible for performing an iteration of the algorithm; thus the superscripts represent the iteration number. In each iteration, the variables v, s, u, and r are updated based on three control signals, ctr_0, ctr_1, and ctr_2, all of which can be generated using the least significant bits of these variables and the counter δ by the rightmost control cell in the figure, as in the following equations:

Algorithm 6. Stein's Euclidean Algorithm for GF(2^n) Inversion

Input: $a(x)$ and $p(x)$ irreducible polynomial, where $deg(a(x)) < deg(p(x))$

Output: $r(x)$, where $r = a(x)^{-1}(\bmod p(x))$ and $deg(r(x)) < deg(p(x))$

1: $u(x) := p(x), v(x) := a(x), r(x) := 0, s(x) := 1, \text{and } \delta := -1$
2: **for** i from 1 to $2n$ **do**
3: **if** $v_0 = 1$ **then**
4: **if** $(\delta < 0)$ **then** $(u(x), v(x), r(x), s(x)) \leftarrow$ $(v(x), u(x) + v(x), s(x), r(x) + s(x))$, $\delta := -\delta$
5: **else** $v(x) := v(x) + u(x), s(x) := s(x) + r(x)$
6: **end if**
7: $v(x) := v(x)/x, s(x) := (s(x) + s_0 \cdot p(x))/x, \delta := \delta - 1$
8: **end for**
9: **return** $r(x)$

$$ctr_0 := v_0^i, \tag{12.3}$$

$$ctr_1 := s_0^i + v_0^i \cdot r_0^i, \tag{12.4}$$

$$ctr_2 := (v_0^i = 1) \text{ and } (\delta^i < 0). \tag{12.5}$$

The counter δ is updated as follows:

* Left-Shift operations may increase the degrees of $r(x)$ and $s(x)$ beyond n.

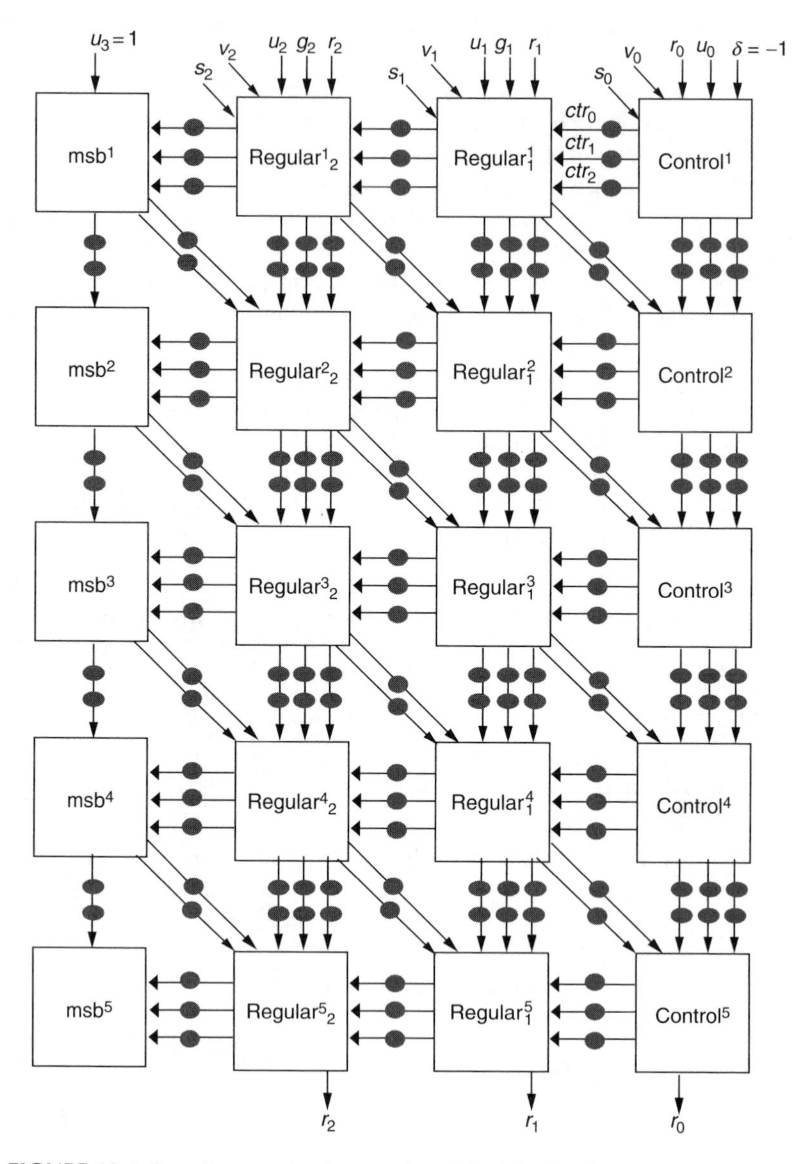

FIGURE 12.4 Systolic array implementation of Stein's algorithm.

$$\delta^{i+1} := \begin{cases} \delta^i - 1 \text{ if } & ctr_2 = 0, \\ -\delta^i - 1 & \text{otherwise.} \end{cases}$$

The control cell in Figure 12.4 is responsible for generating the three control signals and updating δ. The updates of v, s, u, and r are performed in every iteration as follows:

$$v(x) := (v(x) + ctr_0 \cdot u(x))/x, \quad (12.6)$$

$$s(x) := (s(x) + ctr_0 \cdot r(x) + ctr_1 \cdot g(x))/x, \quad (12.7)$$

$$u(x) := \overline{ctr_2} \cdot u(x) + ctr_2 \cdot v(x), \quad (12.8)$$

$$r(x) := \overline{ctr_2} \cdot r(x) + ctr_2 \cdot s(x). \quad (12.9)$$

The jth cell in the ith row performs necessary updates on the jth bit of four variables as follows:

$$v_{j-1}^{i+1} := v_j^i + ctr_0 \cdot u_j^i, \quad (12.10)$$

$$s_{j-1}^{i+1} := s_j^i + ctr_0 \cdot r_j^i + ctr_1 \cdot g_j, \quad (12.11)$$

$$u_j^{i+1} := \overline{ctr_2} \cdot u_j^i + ctr_2 \cdot v_j^i, \quad (12.12)$$

$$r_j^{i+1} := \overline{ctr_2} \cdot r_j^i + ctr_2 \cdot s_j^i. \quad (12.13)$$

For the msb (most significant bit) cell the equations become simpler

$$v_{m-1}^{i+1} := ctr_0 \cdot u_m^i, \quad (12.14)$$

$$s_{m-1}^{i+1} := ctr_1 \cdot g_m = ctr_1, \quad (12.15)$$

$$u_m^{i+1} := \overline{ctr_2} \cdot u_m^i, \quad (12.16)$$

$$r_m^{i+1} := 0. \quad (12.17)$$

As can be observed from the equations, the execution proceeds from right to left and top to bottom. The control signals are conveyed from a cell to the next cell in its left through flip-flops, which are represented as circles on connections in Figure 12.4. Therefore, each circle in the figure indicates one clock cycle delay in the computations. There are two sequential flip-flops in each connection between two cells in a column in systolic array. The values generated in jth cell in row i have to wait for two clock cycles in these flip-flops before they are used by the jth cell in row $i+1$, since the latter also needs the values from $j+1$th cell in row i. Therefore, total latency to compute an inverse in GF(2^n) is $2 \cdot (2n-1) + n = 5n - 2$ clock cycles. After it performs its computation, a cell becomes free for further computation for another inverse operation. At a given time, n consecutive inverse computations execute in the systolic array. The throughput of the systolic array is 1 inverse operation/clock cycle, since it outputs n bits every clock cycle even though these n bits of result do not belong to the same computation.

12.2.4.2 Brunner's EEA

Another inversion algorithm suitable for systolic implementations was proposed by Brunner et al. [26]. Efficient implementation of the algorithm on systolic arrays was presented in [27,28]. It is a left-shift algorithm as opposed to right-shift Stein's algorithm. Brunner's algorithm yields comparable performance to Stein's algorithm when implemented on systolic arrays. However, Brunner's division algorithm that can be easily obtained from inversion algorithm suffers from left-shift operations performed on $s(x)$. As a result of left-shift operations, the degree of $s(x)$ can occasionally become n, necessitating a reduction by $p(x)$.

Algorithm 7. Brunner's Euclidean Algorithm for GF(2^n) Inversion

---

Input: $a(x)$ and $p(x)$ irreducible polynomial, where $deg(a(x)) < deg(p(x))$

Output: $r(x)$, where $r = a(x)^{-1} (\bmod p(x))$ and $deg(r(x)) < deg(p(x))$

1: $u(x) := p(x), v(x) := a(x), r(x) := 0, s(x) := 1$, and $\delta := 0$
2: **for** i from 1 to $2n$ **do**
3: **if** $v_n = 0$ **then**
4:   $v(x) := xv(x), s(x) := xs(x), \delta := \delta + 1$
5: **Else**
6:   **if** $u_n = 1$ **then** $u(x) := u(x) + v(x),\ r(x) := r(x) + s(x)$
7:   $u(x) := xu(x)$
8:   **if** $(\delta = 0)(u(x), v(x), r(x), s(x)) \leftarrow (v(x), u(x), s(x), xr(x)), \delta := 1$
9:   **else** $s(x) := (s(x) + s_0 \cdot p(x))/x,\ \delta := \delta - 1$
10: **end if**
11: **end for**
12: **return** $r(x)$

Montgomery Inversion Algorithm Suitable for Systolic Arrays

The Montgomery inversion suitable for systolic array implementation can be obtained in the same manner resulting in Algorithm 8. The use of counter δ eliminates the need of degree comparison from the algorithm. There is no need for a second phase since the first phase of the algorithm executes exactly $2n$ times yielding $r(x) = a(x)^{-1} x^{2n} \bmod p(x)$. Algorithm 8 provides a comparable performance to those of Stein's and Brunner's algorithms. However, it also suffers from right-shift operations performed on both $r(x)$ and $s(x)$ when it is converted to a division algorithm. Moreover, the Montgomery division algorithm does not compute $(a(x)/b(x))x^n$ but $(a(x)/b(x))x^{2n}$; thus it needs a postprocessing step.

Algorithm 8. Montgomery Inversion Algorithm for GF(2^n) with Brent–Kung Idea

Input: $a(x)$ and $p(x)$ irreducible polynomial, where $deg(a(x)) < deg(p(x))$

Output: $r(x)$, where $r = a(x)^{-1}x^{2n}(\text{mod}\, p(x))$ and $deg(r(x)) < deg(p(x))$

1: $u(x) := p(x)$, $v(x) := a(x)$, $r(x) := 0$, $s(x) := 1$, and $\delta := -1$

2: **for** i from 1 to $2n$ **do**

3: **if** $u_0 = 0$ **then** $u(x) := u(x)/x$, $s(x) := xs(x)$, $\delta := \delta + 1$

4: **else if** $v_0 = 0$ **then** $v(x) := v(x)/x$, $r(x) := xr(x)$, $\delta := \delta - 1$

5: **else if** $\delta < 0$ **then** $u(x) := (u(x) + v(x))/x, r(x) := r(x) + s(x)$, $s(x) := xs(x)$, $\delta := \delta + 1$

6: **else** $v(x) := (v(x) + u(x))/x, s(x) := s(x) + r(x), r(x) := xr(x)$, $\delta := \delta - 1$

7: **end for**

8: **return** $r(x)$

Comparison of Systolic Array Algorithms

Several metrics are used to compare performance of different inversion algorithms for systolic arrays: throughput (inversions per clock cycle), latency (number of clock cycles to compute one inversion), critical path delay, number of flip-flops, total gate counts, and area and time complexities. The three inversion algorithms already mentioned provide the same throughput of 1 inversion/clock cycle and almost the same latency of about $5m$. The time complexities of all three are $O(1)$ because of ring counters used for δ which would dominate the critical path otherwise.* Systolic arrays based on Montgomery and Stein's algorithms have better critical path delay than that of systolic arrays based on Brunner's algorithm. Stein's algorithm usually requires fewer number of flip-flops than others. For division, the best choice is Stein's algorithm since it features only right-shift operations.

Two-dimensional systolic arrays use about $2n$ rows with n cells each; thus their area complexity is $O(n^2)$. This high area complexity is justified for inversion-intensive computations where there are many consecutive inversion operations involved. For cases where there are not many inversion operations and the chip area is limited, one-dimensional systolic arrays are proposed [25]. There are two methods to construct a one-dimensional systolic array from two-dimensional arrays: (i) horizontal projection where all rows are folded into a single row and (ii) vertical projection where all columns are folded into a single column. In the horizontal projection, the latency remains the same but throughput dramatically decreases and the area complexity is $O(n)$. In vertical projection, on the other hand, throughput is higher than that of horizontal projection.

* Note that the control cell features a counter for δ, whose time complexity, $O(n^2)$, dominates some earlier design.

12.2.5 Other Algorithms for Inversion in GF(2^n)

In this section, we briefly mention other inversion algorithms for GF(2^n).

- *Algorithms for Special Binary Extension Fields* GF(2^n). The Itoh–Tsuji algorithm [29] uses Fermat's Little Theorem, which is $a(x)^{2^{n-1}} = 1 \bmod p(x)$, to compute multiplicative inverse as follows:

$$a(x)^{-1} = a(x)^{2^{n-2}} = 1 \bmod p(x).$$

Therefore, multiplicative inverse becomes a special type of exponentiation operation, which consists of repeated multiplication and squaring. The exponentiation can especially be performed very fast when optimal normal basis [30] is used to represent the elements of GF(2^n) since squaring is just a shift in this basis.

- *Unified Inversion Algorithms*. As discussed previously, by effectively changing addition and subtractions in GF(p) to additions in GF(2^n), GF(2^n) inversion algorithms can be obtained from all four GF(p) algorithms proposed in Section 12.1. Therefore, it is possible to design a unified datapath to perform inversion operations in both fields, GF(p) and GF(2^n), as demonstrated in [10,16,19,22].

12.3 CONCLUSION

In this chapter, we investigated binary inversion algorithms proposed for prime GF(p) and binary extension fields GF(2^n) from the perspective of their efficiency in both hardware and software implementations. For arbitrary fields (i.e., fields constructed using random primes and irreducible polynomials), Montgomery inversion algorithms for both fields turn out to be the best choice for software implementations. In hardware, there are more alternatives. For GF(p), the Montgomery inversion and the left-shift algorithms (i.e., Algorithm 3) provide similar performance. The left-shift algorithm requires fewer number of additions while the Montgomery inversion algorithm requires fewer number of shifts and total operations. Therefore, the left-shift algorithm tends to perform better in hardware implementations where shift operations are much less expensive than additions. For GF(2^n), the Montgomery inversion is still better than straightforward b-EEA. However, systolic array implementations based on b-EEA generally outperform the Montgomery inversion on systolic arrays. However, further work is needed for systolic arrays for Montgomery inversion algorithm for a better comparison.

REFERENCES

1. W. Diffie and M.E. Hellman, New directions in cryptography, *IEEE Transactions on Information Theory*, 22:644–654, November 1976.
2. National Institute for Standards and Technology, Digital Signature Standard (DSS), *Federal Register*, 56:169, August 1991.
3. N. Koblitz, Elliptic curve cryptosystems, *Mathematics of Computation*, 48(177):203–209, January 1987.
4. A.J. Menezes, *Elliptic Curve Public Key Cryptosystems*, Kluwer Academic Publishers, Boston, MA, 1993.
5. B.S. Kaliski Jr., The Montgomery inverse and its applications, *IEEE Transactions on Computers*, 44(8):1064–1065, August 1995.
6. R. Schroeppel, H. Orman, S. O'Malley, and O. Spatscheck, "Fast key exchange with elliptic curve systems," In D. Coppersmith, editor, *Advances in Cryptology—CRYPTO '95, Lecture Notes in Computer Science*, No. 973, pp. 43–56, 1995.
7. T. Kobayashi and H. Morita, Fast modular inversion algorithm to match any operand unit, *IEICE Transactions on Fundamentals*, E82-A(5):733–740, May 1999.
8. E. Savaş and Ç.K. Koç, The Montgomery modular inverse—revisited, *IEEE Transactions on Computers*, 49(7):763–766, July 2000.
9. M.A. Hasan, "Efficient computation of multiplicative inverses for cryptographic applications," Technical Report CORR 2001–03, Centre for Applied Cryptographic Research, 2001.
10. E. Savaş and Ç.K. Koç, "Architecture for unified field inversion with applications in elliptic curve cryptography." In Proc. vol. 3, *The 9th IEEE International Conference on Electronics, Circuits and Systems—ICECS 2002*, pp. 1155–1158, Dubrovnik, Croatia, September 2002.
11. Euclid, *Thirteen Books of Euclid's Elements*, Aegean Park Press, Walnut Creek, CA, 1984.
12. E. Berlekamp, *Algebraic Coding Theory*, Dover, New York, 2nd ed., 1956.
13. D.E. Knuth, *The Art of Computer Programming*. Vol. 2, Addison-Wesley, Reading, Mass, 2nd ed., 1981.
14. R. Lórenz, "New algorithm for classical modular inverse." In B.S. Kaliski Jr., Ç.K. Koç, and C. Paar, editors, *Cryptographic Hardware and Embedded Systems*, LNCS, pp. 57–70, Springer-Verlag, Berlin, 2002.
15. R.P. Brent and H.T. Kung, Systolic VLSI arrays for polynomial GCD computation, *IEEE Transactions on Computers*, 47(9):960–970, August 1984.
16. A.F. Tenca and L.A. Tawalbeh, Algorithm for unified modular division in $GF(p)$ and $GF(2^n)$ suitable for cryptographic hardware, *Electronic Letters*, 40(5), 304–306, 4 March 2004.
17. N. Takagi, A modular inversion hardware algorithm with a redundant binary representation, *IEICE Transaction on Information and Systems*, E76–D(8): 863–869, August 1993.
18. E. Savas, "A carry-free Montgomery inversion algorithm," *International Journal of Computer Research*, 13(1), 171–183, 2004.
19. A.A.-A. Gutub, A.F. Tenca, E. Savaş, and Ç.K. Koç, "Scalable and unified hardware to compute montgomery inverse in $GF(p)$ and $GF(2^n)$," In B.S. Kaliski

Jr., Ç.K. Koç and C. Paar, editors, *Cryptographic Hardware and Embedded Systems*, LNCS, pp. 485–500, Springer-Verlag, Berlin, 2002.

20. S.C. Shantz, "From Euclid's GCD to Montgomery multiplication to the great divide." Technical Report SMLI TR–2001–95, Sun Microsystems Laboratory Technical Report, June 2001.
21. M. Brown, D. Hankerson, J. Lopez, and A. Menezes, "Software implementation of the NIST curves over prime fields," Technical Report CORR 2000–56, Centre for Applied Cryptographic Research, 2000.
22. E. Savaş, M. Naseer, A.A.-A. Gutub, and Ç.K. Koç, Efficient unified Montgomery inversion with multibit shifting, *IEE Proceedings—Computers and Digital Techniques*, 152(4), 489–498, July 2005.
23. H.T. Kung, Why systolic arrays?, *Computer*, 15(1), 37–46, January 1982.
24. J. Stein, Computational problems associated with Racah Algebra, *Journal of Computational Physics*, 1:397–405, 1967.
25. C.-H. Wu, C.-M. Wu, M.-D. Shieh, and Y.-T. Hwang, High-speed, low complexity systolic designs of novel iterative division algorithms in GF(2^m), *IEEE Transactions on Computers*, 53(3):375–380, March 2004.
26. H. Brunner, A. Curiger, and M. Hofstette, On computing multiplicative inverses in GF(2^m), *IEEE Transactions on Computers*, 42(8):1010–1015, August 1993.
27. J.-H. Guo and C.-L. Wang, Systolic array implementation of Euclid's algorithm for inversion and division in $GF(2^m)$, *IEEE Transactions on Computers*, 47(10):1161–1167, October 1998.
28. Z. Yan and D.B. Sarwate, High-speed systolic architectures for finite field inversion, *Integration: the VLSI Journal*, 38(3):383–398, January 2005.
29. T. Itoh and S. Tsuji, A fast algorithm for computing multiplicative inverses in $GF(2^m)$ using normal basis, *Information and Computation*, 78, 171–177, 1988.
30. R. Mullin, I. Onyszchuk, S. Vanstone, and R. Wilson, Optimal normal bases in $GF(p^n)$, *Discrete Applied Mathematics*, 22:149–161, 1988/89.

13 Smart Card Technology

Martin Manninger

CONTENTS

Discussing about security, question always arises as to where to implement the security mechanisms? Is a mobile handset that is capable of loading and executing Java applets secure enough as a platform for financial transactions? This depends on the security target of the application and the overall system architecture, but the handset is not the best choice in many cases. To achieve better security on a technical level, we can employ secure hardware such as smart cards. This chapter explains the basics of smart card technology, and it also shows how smart cards can help to establish end-to-end transaction security in wireless environments. This chapter is targeted at readers who have acquired a basic understanding of IT security, cryptography, and wireless technology from the previous chapters.

13.1 TRUSTED COMPUTING BASE

Basically, an information technology system consists of several layers, in which each layer uses services from the next lower level. A simple model of a personal computer shows three layers: application software, operating system, and hardware. This means that the application software depends on the capabilities and the qualities of the operating system, and the operating system depends on the capabilities and the qualities of the hardware. In terms of security, this implies that it is ridiculous to implement a secure application on top of an operating system that is prone to execute malicious code. The malicious software can alter input and output data of the meant-to-be-secure application, at least when the application is communicating with humans. Unfortunately, humans are not capable of performing cryptographic operations with their brains. Hence, they are restricted to cryptographically unprotected communication. In addition, there is no sense in implementing a secure operating system on top of a hardware that allows for unrestricted code changes. Nowadays, all this is rather ancient knowledge. In 1985, the U.S. Department of Defense issued the Trusted Computer System Evaluation Criteria, also called the Orange Book. One key element is the definition of a so-called trusted computing base (TCB):

> The heart of a trusted computer system is the Trusted Computing Base (TCB) which contains all of the elements of the system responsible for supporting the security policy and supporting the isolation of objects (code and data) on which the protection is based. (...) Thus, the TCB includes hardware, firmware, and software critical to protection and must be designed and implemented such that system elements excluded from it need not be trusted to maintain protection.
>
> Department of Defense [1]

In other words, when designing security architecture, we need to start with at least one piece of secure hardware that we can build on. This is the entity that may hold secret keys, perform cryptographic operations, and grant access to certain functions only after a positive authentication. Smart cards are such pieces of secure hardware. They are designed in a way that provides the best possible protection against attackers trying to read stored data or to modify these data or the results of the card's computations. Therefore, smart cards are also counted as pieces of tamper-proof hardware.

13.2 CLASSIFICATION OF SMART CARDS

Giving an exact definition of a smart card is not feasible, because smart card is common speech. The technical term is integrated circuit card (ICC), denominating a (plastic) card containing a tamper-proof integrated circuit.

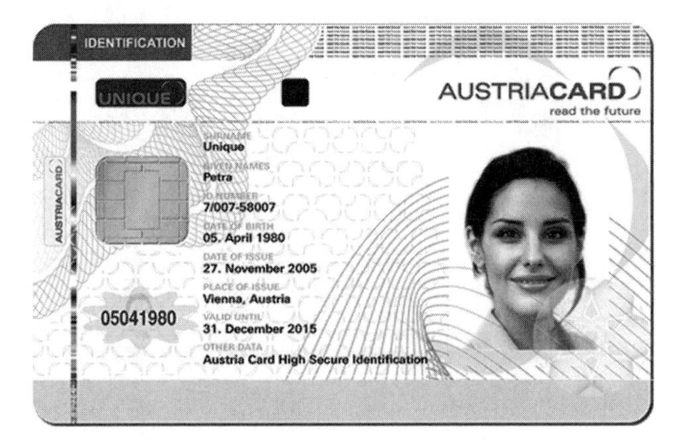

FIGURE 13.1 Integrated circuit card (From Austria Card).

Figure 13.1 shows a typical ICC with a visible electrical contact plate. The card can communicate with the outside world through an interface device (IFD) that connects to this contact plate. The card itself is a passive device that needs power supply from the IFD. The IFD may be a simple card reader that is connected to a background system, or a more complex card terminal that may operate stand-alone and that includes a user interface for interaction with humans.

Obviously, the existence of a contact plate does not imply that the card is particularly smart. Regarding the internal logic of ICCs, we differentiate between memory cards and microprocessor cards.

A memory card consists mainly of a read and write memory, usually an electrically erasable programmable read-only memory (EEPROM), and some hardwired communication and security logic (see Figure 13.2). Typical functions of the security logic of a memory card are an optional write protection of certain memory areas or a primitive authentication mechanism, such as allowing read or write access to the memory only after a password or a personal identification number (PIN) has been entered. Memory cards capable of cryptographic operations with configurable keys have come up in the last two years, but are not yet widespread.

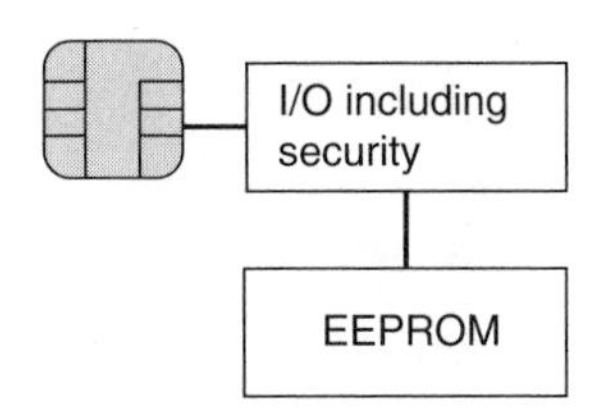

FIGURE 13.2 Block diagram of a memory card.

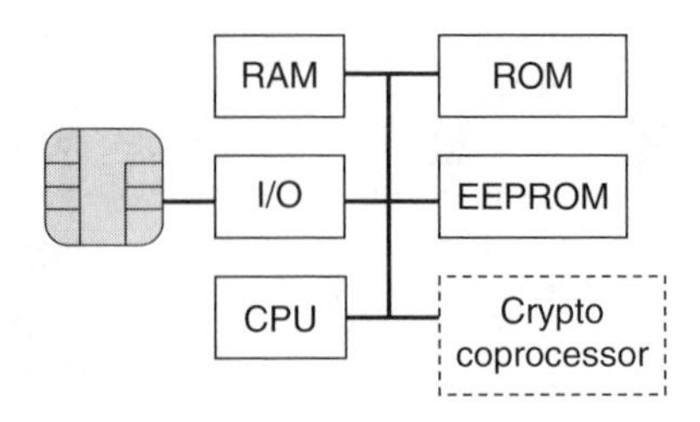

FIGURE 13.3 Block diagram of a microprocessor card.

Complex ICC applications including cryptographic operations are usually realized with microprocessor cards. Figure 13.3 shows the internal structure of such a microprocessor card, containing a central processing unit (CPU), a ROM, an EEPROM, a random access memory (RAM), and an input and output peripheral unit that handles the communication with the outside world. An optional cryptography coprocessor is used to perform the algorithms of asymmetric cryptography, especially the RSA algorithm, with acceptable speed. ICC microprocessors are significantly slower than CPUs of today's desktop computers. They operate at clock frequencies of 5 to 30 MHz. There are also limitations to the memory size resulting directly from semiconductor size restrictions. Typical ICCs offer 64 to 256 KB of ROM, 4 to 256 KB of EEPROM, and 1 to 4 KB of RAM. Employing these components, a piece of software called card operating system (COS) provides services that the application software can use—similar to any other operating system. The ability to execute software has been accepted by many ICC experts as the major criterion for giving a card the attribute smart. Throughout this chapter, the term smart card is used synonymously for the exact term microprocessor card.

The second distinguishing feature of ICCs is the size of the card. The basic standard ISO 7810 defines four different formats, from which only two are widely used. Typical credit cards are produced in the ID-1 format (see also Figure 13.1). The smaller ID-000 format shown in Figure 13.4 is used

FIGURE 13.4 Integrated circuit card in ID-000 format (From Austria Card).

for SIM cards of mobile phones and many other cards built into devices. This is also called "plug-in format." Besides the formats defined in ISO 7810, several other shapes have originated in the past few years. Credit card organizations have come up with fancy roundings, and new applications such as the ICC-based biometric passport require a larger ICC (or the IC alone) to be implemented in a booklet. ICC technology is also implemented in different shapes that do not even resemble a card. One such example is a USB stick.

The third distinguishing feature of ICCs is their communication interface. The first cards were restricted to communication through galvanic contacts (see Figure 13.1). This type of communication has been standardized with ISO 7816-3, and the industry has kept to this standard. Hence cards that are built into devices such as mobile phones also follow this standard. Out of the many existing communication protocols, two distinct protocols have been standardized. $T=0$ is a byte-oriented protocol. It was the first microprocessor card protocol and is still widely used. $T=1$ is a block-oriented protocol. It is more advanced than $T=0$ and offers a clearly layered architecture and good error recognition and recovery techniques. Typical bit rates range from 9600-bits/s to 156,250-bits/s, if the most usual external clock frequency of ~5 MHz is applied [2].

ICCs can directly operate wireless; hence, this is named as contactless communication in the relevant ISO standards that are described in the next section. In any case, the communication between the ICC and the terminal is strictly master–slave oriented. The card receives commands from the terminal and provides the required results.

Figure 13.5 summarizes the classification of ICCs. Additional criteria are the memory sizes of ROM, EEPROM, and RAM, the CPU architecture and speed, and the presence or the absence of an RSA coprocessor.

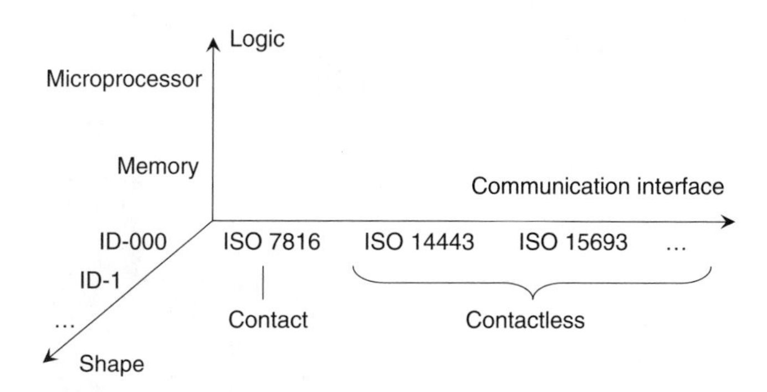

FIGURE 13.5 Classification of integrated circuit cards.

13.3 WIRELESS CARDS

Wireless technology offers a number of benefits for the end-user and the system operator. Wireless systems are in general more convenient to use and they need less maintenance. This is also true with ICCs. Contactless ICCs and contactless card readers are not prone to corrosion of contacts, and more resistant to mechanical and electrical impact, which makes them well suited for outdoor applications. Furthermore, because of their ability to communicate over distance, they allow faster transaction times by easier handling. The card need not be inserted into a terminal but only has to come close enough to establish communication. The card may even remain in the wallet, if the user decides to simply move the wallet close to the terminal.

The technical principle behind today's contactless card systems is to keep the card as a passive element. The card reader is supplying the power through inductive coupling, and the data transfer is in most cases modulated onto the same electromagnetic field. Figure 13.6 shows an example of a contactless card, where the chip and the connected coil needed for the inductive coupling can be clearly seen.

As there are different requirements for contactless cards, depending on the application, a few different standards have been defined already. The first among these was the standard of close-coupled cards according to ISO 10536. These cards never became widespread because of their complex technology, including an additional capacitive coupling for the data transfer. Today, the most widespread contactless cards are the proximity cards, according to ISO 14443, with a typical operating range of 8 to 10 cm. This distance is said to combine easy handling for the user with meeting the demand for an explicit volition. At least the user needs to move his or her card close to a terminal to initiate a transaction. Another standard is named vicinity cards (ISO 15693).

FIGURE 13.6 Contactless integrated circuit card (From Austria Card).

Such cards can operate over a distance of 80 cm and more. They are used for hands-free transactions of humans and as authentication tokens of animals and things. However, because of restriction of the strength of the electro magnetic field, long distance results in a limited power supply for the ICC and in a limited data transmission speed.

In practice, there are not just contact-only and contactless-only cards. Two different combinations are possible. The first option is to put two independent chips into the card, one with contact and the other with contactless communication interface. This is a simple solution and is called a hybrid contact and contactless card. Less development effort is required, but such a card lacks an internal communication between the two chips. This solution is often chosen if independently approved components have to be combined, for example, a contactless memory chip for building access control or ticketing applications and a contact microprocessor chip providing a personal computer sign-on function. The advantage is that the terminals and the background systems that have already been developed for the individual applications can remain unchanged. The second option is to employ special dual interface chips equipped with one contact and one contactless communication interface. This requires the development of special dual interface card operating system (COS), which are much more flexible and allow for complex combined contact and contactless applications. Such an application may even use contact and contactless communication within the same transaction. A good example is a contact-based payment card with a contactless electronic ticketing function for public transport. With a hybrid card solution, the ticket-vending machine would be responsible first for performing the payment transaction through the contacts and then for storing the ticket through the contactless interface. This means that the ticket-vending machine would need both communication interfaces, but there would be another disadvantage. For security reasons, a dual interface card is also preferred, because it can, with a single card command, perform both the payment transaction and the ticket loading in one step. Thus, the consumer is automatically protected from cheating ticket-vending machines.

13.4 NEAR FIELD COMMUNICATION

Near field communication (NFC) is an extension to contactless communication. The new aspect is that this standard specifies communication between active devices also [3]. For practical reasons, passive smart cards that comply with ISO 14443 (Type A) are also compatible with NFC devices. A device with full NFC functionality may either communicate with another active NFC device, or may operate as a contactless terminal in a communication with a smart card, or may behave like a contactless smart card itself. This allows for different devices, such as mobile phones equipped with an additional NFC IC, to include the functionality of contactless cards. As with ISO 14443, the communication distance is limited to ~10 cm. The targeted applications are consumer devices that communicate with each other when a user has started the communication by moving the devices close to each other. The communication may remain fully at the NFC interface, but optionally two hybrid devices may decide to continue their communication over a different interface, for example, Bluetooth.

13.5 TYPICAL WIRELESS CARD APPLICATIONS

There are two categories of wireless card applications: contact cards connected to wireless terminals and contactless cards. Typical contactless card applications include building and parking lot access, time registration, and ticketing applications for public transport, ski lifts, and cultural and sports events. In many cases, the security level of memory cards is sufficient for such applications. Since 2005, classical payment applications such as debit or credit have been offered in a contactless version also. MasterCard has branded this payment technology with the name PayPass. VISA has branded their contactless cards with the name VISA Wave and adopted the PayPass technology for the second version of their contactless payment specifications. Payment applications require a high security level; hence, they are generally implemented on microprocessor cards. In the payment cards sector in 2006, we experience already a strong move toward microprocessor cards that include an RSA coprocessor, opting for dynamic data authentication (DDA), which is the higher one of two different security levels defined in the standard of Europay, MasterCard, and VISA (EMV) [4]. This also influences the contactless payment cards, although RSA operations are relatively slow, and today the main argument for contactless payment is speed.

Today's most important contact cards connected to wireless terminals are the so-called subscriber identity modules, or SIM cards, that are plugged into mobile phones working according to the GSM or the UMTS standards. These cards authenticate themselves when logging into the mobile telephony network, and they derive a cryptographic key that is then used by the mobile handset to encrypt the data stream. Similar to access to mobile telephony networks, WLAN access can also be secured with smart cards. Members of the WLAN Smart Card Consortium [5] have submitted such an Internet Draft named as EAP-Smart Card Protocol (EAP-SC) [6]. Another related Internet Draft is named EAP-Support in Smart card [7].

13.6 SMART CARD OPERATING SYSTEMS

Microprocessor cards are basically one-chip computers that lack only a user interface. Hence their architecture follows that of a conventional computer. An operating system that is specific to the underlying hardware runs this hardware and provides an interface for application programming. A smart

COS is mainly responsible for the file management, maintaining the communication with the IFD, and executing the commands received through the IFD. Unlike a personal computer's operating system, a COS has to meet special requirements. In particular, it must fit into the limited memory and reach proper speed, despite the mediocre computing power of the small microprocessor. As the operating system resides in most cases in a ROM, it must be virtually error-free. Any critical error would require the replacement of all cards issued and the destruction of all ICs produced. Hence smart COSs have to be developed and tested much more carefully than conventional operating systems.

Smart COSs are divided into the categories of monofunctional, multifunctional, and multiapplication operating systems. The major advantage of the last category is that these cards are programmable by the user (meaning the systems integrator, not necessarily the end-user), instead of ordering each new function at the card manufacturer. True multiapplication operating systems implemented on ICs that provide sufficient hardware security to guarantee separated memory areas may even allow new application code to be loaded after the card has been issued. Today, the most widespread multiapplication operating systems are the so-called Java cards that accept Java programs compiled into Java byte code [8].

Still the majority of smart cards issued are of the multifunctional type, which offer different functions that are altogether implemented by the COS developer. Such cards do not allow adding new application program code, but they do allow adding new application data structures including the configuration of the appropriate security mechanisms. The most important aspects of how smart COSs can be used by a security systems engineer according to ISO 7816 are described in Section 13.7 through Section 13.10: file system, cryptographic abilities, access control mechanism, and commands.

13.7 FILE SYSTEM

The file system of microprocessor cards has been standardized with ISO/IEC 7816-4 and further enhanced with later parts of the same standard. The classical tree structure has been adopted for smart cards. There are elementary files (EFs) that contain application data and dedicated files (DFs) that build the nodes of the tree. Unlike a personal computer operating system, the DFs do not necessarily contain a directory of all subordinate files. The root DF is named master file (MF). Figure 13.7 gives an example of a valid smart card file system.

Within the same DF, each file is given a unique file identifier (FID) of 2 byte length. The MF is defined to have the FID = 3F 00 H and to be selected after the card has received a reset signal. An EF may have an additional short FID, which is only 5-bit long and may have a value from 1 to 30. This is used for implicit file selection that makes it possible to apply several card commands directly to specific files, without explicitly selecting a file before.

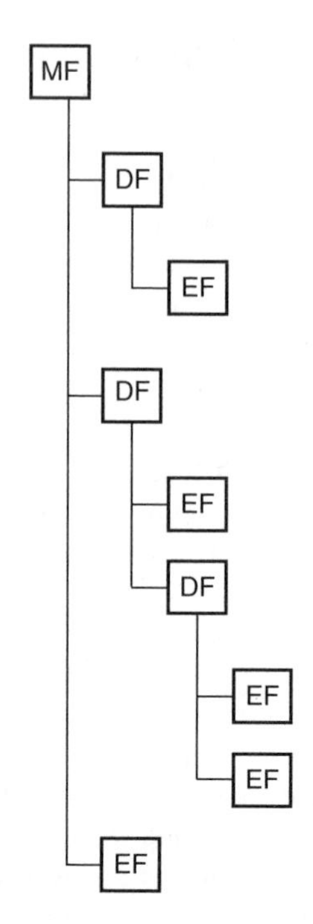

FIGURE 13.7 Example of a smart card file system.

A DF can have a file name in addition to its FID. This name should be unique within the card and may be a text string or a registered application identifier (AID) of 1 to 16 byte length, as defined in ISO/IEC 7816-5. An AID consists of a 5-byte registered identifier (RID) denoting the (juristic) person who has applied to register an application and a proprietary application identifier (PIX) that is up to 11 byte long and chosen by the person who has registered the application.

Four different internal structures are defined for EFs:

- Transparent EFs contain unstructured data.
- Linear fixed EFs are divided into records of fixed length. An index ranging from 1 to FE H is assigned to each record.
- Linear variable EFs are divided into records of variable length. Again, an index ranging from 1 to FE H is assigned to each record.
- Cyclic fixed EFs are divided into records of fixed length, just like linear fixed EFs, but they have a different record numbering scheme that makes it easy to append new records infinitely, in a way that the oldest record is automatically overwritten.

13.8 CRYPTOGRAPHIC ABILITIES

Microprocessor cards are generally capable of symmetric cryptography, namely DES and Triple-DES. In most cases, these algorithms are directly supported by the hardware; this helps to achieve fast execution times. Recent types support AES, but as of today there are not many applications making use of this new standard. Of course, both encryption and decryption are supported. Thus, the card can also secure and check the integrity of (unencrypted) data with the use of message authentication codes (MACs). MACs are cryptographic checksums attached to a (clear text) message. One common method to generate a MAC is to calculate a hash value from the message and then encrypt this hash value. Microprocessor cards are capable of performing a number of hash algorithms. SHA-1, SHA-256, SHA-512, and different versions of RIPE-MD are among these.

As for asymmetric cryptography, RSA is by far the most used algorithm in the smart card world. Because most smart card CPUs are not capable of performing RSA operations within reasonable time, most ICC used for asymmetric cryptography comprise a distinct RSA coprocessor. Note that cards with an RSA coprocessor are on principle also capable of generating an RSA key pair inside the IC. This is beneficial to system security because such an internally generated private key never leaves the card. Only the corresponding public key is exported from the card and distributed further.

In the past few years, DSA based on elliptic curve cryptography (ECC) [9] has been implemented as an alternative for digital signature applications. ECC is assumed to achieve the same security level as RSA at shorter key lengths than RSA. However, the approach to replace RSA by ECC and at the same time to get rid of the need for a coprocessor has not yet succeeded in the industry.

13.9 ACCESS CONTROL

Some data inside a smart card, for example, the secret cryptographic keys, can only be used for internal operations and do not allow any reading. The access to other data and procedures can be configured as follows:

- read application data,
- write application data,
- use cryptographic keys, and
- execute commands.

The card makes the decision whether or not to grant the access based on previous authentications. In general, authentication is the process of verifying the identity or the group membership that a person (or a device) claims to possess. In the area of smart card technology, it is common to handle user authentication and device authentication separately. For the purpose of user authentication, there are basically three options:

- Authentication by possession of a physical item, which could be the smart card itself or a mechanical key, is difficult to check for a device like a smart card.
- Typical examples of knowledge-based authentication are the input of a password or a PIN—the latter which is just a password that is defined as being known by only one person. This is the most common option, as such a comparison is easy to implement in a smart card.
- Authentication by biometrics requires the user to present a physical characteristic of his body to a sensor for capturing and further processing. Practical examples are fingerprints, face, or the iris of the eye. In many cases, the complete processing is too complex to be done by a smart card, but it is possible to have a terminal perform at least a part of it.

For the purpose of device authentication, only one of these options remains. Smart cards or terminals may have knowledge, particularly knowledge of cryptographic keys, and unlike the human user they can also perform cryptographic operations. The basic device authentication method is called cryptographic authentication or challenge/response authentication, because the device that has to authenticate receives a random challenge and encrypts it using a secret key. The cipher text is then transmitted back and can be verified by the communication partner that holds the same secret key (in the case of symmetric cryptography) or the corresponding public key (in the case of asymmetric cryptography).

With such an authentication protocol, the secret information, which is the secret key, is never transmitted and thus cannot be compromised. This is already a major advantage over passwords or PINs that have to be transmitted when used. However, an attacker could still get a large record of pairs of challenges and matching responses, if he continued eavesdropping. Once a challenge is chosen that has been in use earlier, the attacker could then take the proper response from his list. To prevent this kind of replay attack, challenges are generally built as concatenations of random numbers, continuously increasing counter values, and/or time stamps. Thus, each challenge is unique.

There is another improvement in the challenge or response authentication protocol. If two communication partners need to authenticate each other, this could simply be carried out by performing two challenge or response authentications, one after the other. Figure 13.8 shows the more efficient and more secure mutual authentication protocol in an example employing symmetric cryptography. First, communication partner P1 generates a random number N_1 and asks communication partner P2 for another random number N_2.

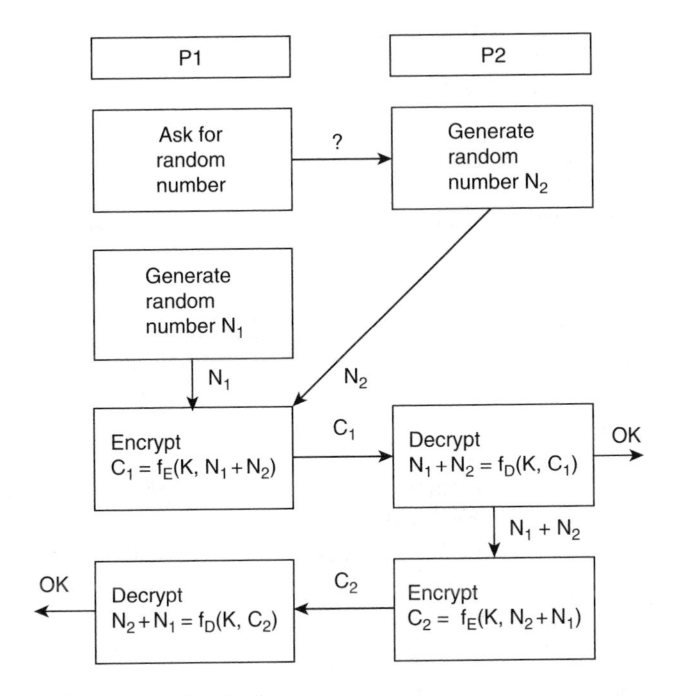

FIGURE 13.8 Mutual authentication.

P1 concatenates the two random numbers to one data package and encrypts it with the key K, resulting in the cipher text C_1. P1 transmits the cipher text C_1 to P2, who decrypts it and verifies that the number N_2 is correct. Then P2 constructs a data package containing the same two random numbers in reverse order and encrypts it with the key K, resulting in the cipher text C_2. Note that with a good cryptographic algorithm, C_2 completely differs from C_1. P2 then transmits the encrypted data package C_2 to A, who in turn decrypts it and verifies that the number N_1 is correct. Doing so, both communication partners have authenticated each other, without giving the attacker a chance of getting to know a clear text and cipher text pair that he might use for cryptanalysis.

In case of asymmetric cryptography, the principle remains the same. The additional benefit is that the secret keys exist only once and that only public keys have to be distributed between the communication partners before a cryptographic authentication.

13.10 COMMANDS

At the application layer, the communication between ICCs and the corresponding card readers or terminals consists of two types of application protocol data units (APDUs):

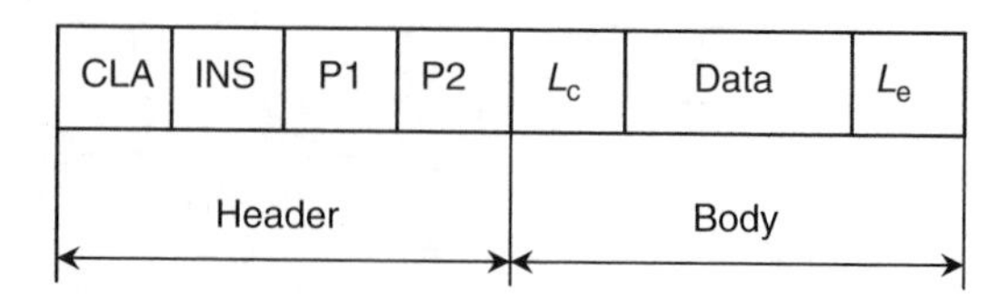

FIGURE 13.9 Command APDU.

- Command APDUs are actively sent from the reader to the card and cause the card to execute one of its internal procedures.
- Response APDUs are sent back from the card to the reader and contain the result of the executed procedure.

Figure 13.9 shows the structure of a command APDU, comprising a mandatory header and a conditional body.

The class byte (CLA) gives a reference to the origin of the command. For example,

- ISO commands have CLA = 0X H,
- GSM commands have CLA = AX H, and
- proprietary commands have CLA = 8X H.

Within these examples, X defines the applied secure messaging format that is explained later in this section.

The instruction byte (INS) specifies one command within a class. The parameter bytes (P1, P2) are used to choose between different options that the command offers. The body contains

- length of the command (L_c),
- data, and
- length of the expected response data (L_e).

According to ISO 7816-4, L_c is not calculated over the whole command but only over the data. Both lengths may be coded in 1 or, depending on the operating system, 3 byte. With this extended length coding, the first byte is used as an escape character and the other two bytes contain the length value.

Figure 13.10 shows the structure of a response APDU, comprising a conditional body and a mandatory status word (two status bytes SW1 and SW2). If the command is processed straightforward, the card responds with the status word 90 00 H, meaning OK. In any other case, various errors or warnings are represented by different values of the status word.

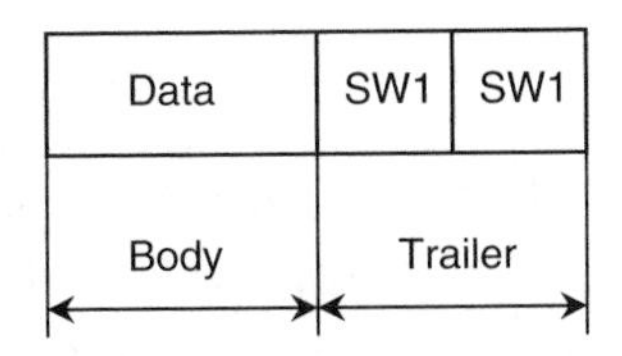

FIGURE 13.10 Response APDU.

Few examples of typical commands help us to understand how smart cards can be used as components of a security system. The first example is the SELECT FILE command that is used to explicitly select one distinct file of the file tree. Figure 13.11 shows the structure of this command and its major options, where H indicates hexadecimal numbers. Files can be generally selected by their FID or by stating a path, which is a concatenation of FIDs starting from the MF or the current DF. DFs can also be selected by their name. This name need not be fully stated in the data of the SELECT FILE command but may be truncated. This is called partial selection and uses the options of selecting first, last, next, or previous file with a name starting with the stated bytes.

The response to the SELECT FILE command may contain information such as FID, file type, file size, record length, and similar. As with any response, a return code indicates error-free processing of the command, or one of the various errors like parameter errors, or that the operating system was unable to find the file.

Another example is the VERIFY command that is used to compare verification data such as PINs with the corresponding reference data stored in the card. Figure 13.12 shows the command's structure and its two options of using either global reference data, for example, a card's master PIN, or

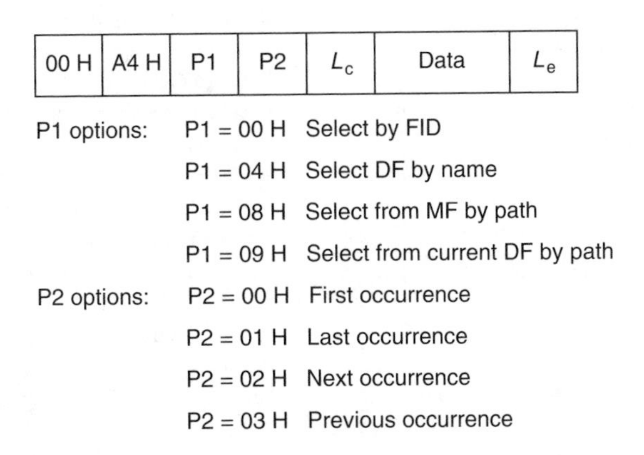

FIGURE 13.11 SELECT FILE command.

00 H	20 H	00 H	P2	L_c	Verification data

P2 options: P2 = 00 H No further information
P2 = 0X H Global reference data
P2 = 8X H Specific reference data

FIGURE 13.12 VERIFY command.

application-specific reference data. The least significant 5-bits of P2 denotes the index of the reference data inside the card. The length L_e is left empty, as the response to the VERIFY command does not contain any data. If the verification fails because of incorrect verification data, the card generally indicates the number of remaining tries in the status word. After the last try, the card blocks the verification of these data. Other important smart card commands include read and write operations, change of reference data, and security operations. Table 13.1 offers a selection of these commands.

TABLE 13.1
Microprocessor Card Commands

INS	Command Name	Purpose
88 H	INTERNAL AUTHENTICATE	Authentication
84 H	GET CHALLENGE	Generate a random number
82 H	EXTERNAL AUTHENTICATE	Authentication
82 H	MUTUAL AUTHENTICATE	Authentication
22 H	MANAGE SECURITY ENVIRONMENT	Activate templates of algorithms and keys for subsequent security operations
2A H	PERFORM SECURITY OPERATION	Authentication, encryption, decryption, hashing, signature, and verification
24 H	CHANGE REFERENCE DATA	Replace reference data stored in the card with new reference data
B0 H	READ BINARY	Read data from transparent EFs
B2 H	READ RECORD	Read data from linear or cyclic EFs
D0 H	WRITE BINARY	Write data to transparent EFs
D2 H	WRITE RECORD	Write data to linear or cyclic EFs
D6 H	UPDATE BINARY	Update data in transparent EFs
DC H	UPDATE RECORD	Update data in linear or cyclic EFs
E2 H	APPEND RECORD	Append a record to linear or cyclic EFs
A2 H	SEARCH RECORD	Search for data within linear or cyclic EFs
CA Hl	GET DATA	Read a specific data object
DA H	PUT DATA	Write a specific data object
C0 H	GET RESPONSE	Fetch response APDUs (needed with the $T = 0$ protocol)

13.11 CRYPTOGRAPHIC AUTHENTICATION AND SECURE MESSAGING

As described in the previous section, user authentication can be performed with the VERIFY command. Cryptographic authentication of devices, as explained earlier in this chapter, can be applied in different ways. One-way authentications may be used either to authenticate the smart card to the outside world or to authenticate the outside world to the smart card. The first procedure is named internal authentication, whereas the second procedure is named external authentication. Two-way authentications are named mutual authentication.

To perform an internal authentication, the terminal simply sends the challenge with the command INTERNAL AUTHENTICATE to the card. The card then performs the encryption of the challenge and responds with the cipher text that can be verified by the terminal or the background system.

To perform an external authentication, the terminal must first send the command GET CHALLENGE to the card. The card responds with a random number that the terminal or the background system encrypts. The terminal then sends the cipher text with the command EXTERNAL AUTHENTICATE to the card, and the card performs the verification.

To perform a mutual authentication, the terminal must first send the command GET CHALLENGE to the card. The card responds with a random number that the terminal or the background system combines with a self-generated challenge and encrypts the result. The terminal then sends the cipher text with the command MUTUAL AUTHENTICATE to the card and the card first performs its own verification and then responds with a different cipher text that can be verified by the terminal or the background system.

The algorithms and the keys used in performing these cryptographic operations may either be stated as parameters of the command or activated beforehand with the command MANAGE SECURITY ENVIRONMENT. This command is also used to activate templates for key agreement, hashing, cryptographic checksums digital signatures, and confidentiality. Many of these functions can be called with the command PERFORM SECURITY OPERATION. This command is able to calculate hash values, MACs, and digital signatures, and to encrypt and decrypt data with both symmetric and asymmetric algorithms, if supported by the COS.

The confidentiality and authenticity of the messages exchanged between the terminal and the card can be ensured with the so-called secure messaging. When using this mechanism, the APDUs are encrypted and secured with MACs. The type of secure messaging is coded in the upper 2-bits of the lower nibble of the class byte. For instance,

- CLA = X0 H indicates that there is no secure messaging used.
- CLA = X4 H indicates proprietary secure messaging.
- CLA = X8 H indicates ISO-compliant secure messaging without authenticated command header.
- CLA = XC H indicates ISO-compliant secure messaging with authenticated command header.

Two more terms are widely used in this context: authentic mode means adding only MAC and combined mode means additionally encrypting the resulting data block. Before the cryptographic algorithms are applied, suitable padding may be needed to make the length of the data block valid.

13.12 SECURITY OF SMART CARDS IN PRACTICE

The security level of smart cards also has to be discussed, as in reality there is no such thing like a 100% tamper-proof device. Every technology that claims to be tamper-proof attracts the effort of attackers until one of them succeeds. The smart card technology has also experienced several serious attacks during the last 15 years. To name only few of the most important attacks, there were

- deletion of EEPROM values by using UV light (1991) [2],
- stopping of the clock frequency and analyzing the RAM with the help of electron beam testers (1993) [2],
- severing of existing connections inside the integrated circuit, establishing new connections, and changing the semiconductor doping with a focused ion beam workstation (1996) [10],
- Bellcore attack, based on the fact that hardware, when performing calculations, often produces incorrect results if environmental conditions are causing stress, was first directed at asymmetric cryptography using algebraic operations and succeeded in calculating the involved secret keys with reasonable effort (1996) [11],
- differential fault analysis (DFA) that transferred the principle of the Bellcore attack to symmetric cryptography (1997) [12],
- simple power analysis (SPA) and more elaborate differential power analysis (DPA), which use the variation of the IC's power consumption, especially during the performance of a cryptographic algorithm, to gain secret information through a statistical analysis (1998) [13],
- Elecromagnetic analysis (EMA) that applies the principle of SPA or DPA to data retrieved from measuring electromagnetic radiation emitted by the integrated circuit (2001) [14], and
- modification of RAM content by the use of light flashes (2002) [15].

The smart card industry has continuously improved the security level of the integrated circuits and the implemented operating systems. An up-to-date smart card comprises protection and active countermeasures against all known attacks. To gain confidence that these measures are effective, both hardware and software need to undergo thorough security evaluations. Independent laboratories analyze the implemented security mechanisms and, after identifying potential weaknesses, try to attack the card. Only if they do not succeed, the product receives a positive evaluation. Thus, the customer does not have to analyze the security level of the product, but can trust in the evaluation result. Depending on the purpose of the card, different evaluation schemes are in place. In general, today's most important scheme is an evaluation according to the Common Criteria [16], where suitable protection profiles for integrated circuits and software such as operating systems and digital signature applications have been defined. Payment organizations such as MasterCard and VISA have set up their own security evaluation schemes that are tailored to the security requirements for payment cards. However, for other purposes such as SIM cards, no mandatory security evaluation has been set in place yet.

The conclusion is that smart cards are practically tamper-proof devices during a limited period, until their security level is no longer state-of-the-art. For applications requiring high security, it is recommended to move on to the next generation of smart cards every 2 or 3 years, depending on the time the card remains in use. For example, if a card based on 4 years old technology were issued with a targeted lifetime of 3 years, at the end of this lifetime the technology would already be 7 years old. If a smart card from 1999 (from the last century) were attacked with today's best methods, the attacker would be likely to succeed.

13.13 SYSTEM SECURITY

Building on this understanding of smart cards as secure hardware and software used for device authentication and user authentication, a systems engineer can design a logically secure information system. To do so, he has to analyze the threats, the security target, and the complete system and its environment. Then he can decide which parts of the system, be it components, data elements, or communication channels, may be regarded as trusted, and which parts need additional security measures.

The definition of a security target helps to clarify the desired security level also. Generally spoken, higher security is more expensive. Thus, there is a trade-off and a point where additional security measures cost more than the value of the covered risk, which makes them inefficient. In the smart card world, the most important parameters are the number of cards, the memory size, and the cryptographic capabilities of the integrated circuit and its operating system. One particularly important issue is how to distribute the cryptographic keys. Secret keys shall never be stored in or transmitted over a medium that is less secure than the smart card. Additionally, the information contained in one smart card shall never be sufficient for an attacker to compromise the whole system. Therefore, there must be a differentiation between individual component keys and master keys that may be kept only by a few components of the highest security level. With the use of a master key and an individual attribute of a chip, such as a hardware serial number, an individual key is derived. This individual key is then used for communication with this particular card only. Several nested key derivations are possible as well, and once a secure communication has been established, a dynamically generated session key can be exchanged and used for further operations on transaction data.

To secure a system that includes user interaction, the following design principles are useful:

- Each component that holds secret or private keys does this only within a smart card or a device of equivalent security level.
- Cryptographic operations with secret or private keys are performed only within a smart card or a device of equivalent security level.
- Components that are likely to get captured by an attacker do not hold master keys.
- Terminals are able to deactivate stolen, copied, or otherwise compromised cards.
- In an off-line system, the terminals are able to check if a card is genuine or a copy.
- Transaction security starts at the user interface.
- Each card is able to verify the authenticity of its user. (This eliminates the need to transmit PINs or other reference data.)
- Each user's card authenticates the user's transactions, preferably by the use of a digital signature.

13.14 INTEGRATED CIRCUIT CARD STANDARDS

Several standards for ICC technology have been mentioned in the previous sections. Table 13.2 gives an overview of the most important international standards for ICCs.

13.15 EXAMPLE: MOBILE PAYMENT SECURED BY A SIM CARD

The subject of mobile electronic payment (m-payment) has been discussed to a great extent, and different approaches have been implemented in commercial payment applications. However, many existing implementations show a security level that is yet to be improved. Examples of common flaws occurring are listed as follows:

TABLE 13.2
Integrated Circuit Card Standards

ISO 7810	2003	Identification cards—Physical characteristics
ISO 7816-1	1998	Identification cards—Integrated circuit cards with contacts—Part 1: Physical characteristics
ISO 7816-2	1999	Identification cards—Integrated circuit cards with contacts—Part 2: Dimensions and locations of contacts
ISO 7816-3	1997	Identification cards—Integrated circuit cards with contacts—Part 3: Electronic signals and transmission protocols
ISO 7816-4	2005	Identification cards—Integrated circuit cards—Part 4: Organization, security, and commands for interchange
ISO 7816-5	2004	Identification cards—Integrated circuit cards—Part 5: Registration of application providers
ISO 7816-6	2004	Identification cards—Integrated circuit cards—Part 6: Interindustry data elements for interchange
ISO 7816-7	1999	Identification cards—Integrated circuit cards with contacts—Part 7: Interindustry commands for structured card query language (SCQL)
ISO 7816-8	2004	Identification cards—Integrated circuit cards—Part 8: Commands for card management
ISO 7816-9	2004	Identification cards—Integrated circuit cards—Part 9: Commands for card management
ISO 7816-10	1999	Identification cards—Integrated circuit cards with contacts—Part 10: Electronic signals and answer to reset for synchronous cards
ISO 7816-11	2004	Identification cards—Integrated circuit cards—Part 11: Personal verification through biometric methods
ISO 7816-12	2005	Identification cards—Integrated circuit cards—Part 12: Cards with contacts—USB electrical interface and operating procedures
ISO 7816-15	2004	Identification cards—Integrated circuit cards—Part 15: Cryptographic information application
ISO 10373	1998–2001	Identification cards—Test methods, Part 1 to Part 7
ISO 10536	1995–2000	Identification cards—Contactless integrated circuit cards—Close-coupled cards, Part 1 to Part 3
ISO 14443	2000–2001	Identification cards—Contactless integrated circuit cards—Proximity cards, Part 1 to Part 4
ISO 15693	2000–2001	Identification cards—Contactless integrated circuit cards—Vicinity cards, Part 1 to Part 3

- There is no end-to-end security between the client (the mobile handset) and the payment server that is finally approving the transaction. The only cryptography involved in the transaction is the standard (GSM) encryption that is securing the air part of the transaction data transmission, but not the (wire-based) data transmission thereafter. (In case of SMS-based payment applications, the transaction data are

vulnerable to any modifications after the SMS has been received by the mobile network operator.) Thus malicious people or pieces of software inside the mobile network operator's organization have many opportunities to launch a successful attack. The goal of transaction data integrity is clearly missed.
- Transaction authentication is solely based on a static PIN that is known to the user and also stored in a database at a server. The goal of nonrepudiation is clearly missed.

If both flaws are combined in a system, the attacker is easily able to sniff PINs and generate faked transactions with these PINs. This leaves room for improvements.

When designing an improved system based on SMS, one can use a smart card as a part of the TCB. For practical reasons, it is the SIM card that is already built into the mobile handset. (Alternative solutions employing a handset that includes a second card reader for another smart card are possible.) In such a system, the SIM card is equipped with a piece of application software (a SIM Toolkit application) that is capable of

- sending and receiving SMS,
- communicating with the user by (indirectly) controlling the display and the keypad of the handset, and
- performing operations of symmetric and asymmetric cryptography.

The smart card holds one symmetric key and one private key of an asymmetric key pair. The corresponding symmetric and public keys, respectively, are held by the payment server. Note that the payment server may be located within the mobile network operator's organization or in a different organization. Involving a different organization has the advantage that this organization can be specialized in payment processing. Therefore, their IT infrastructure is well protected, and they can offer the payment service to all mobile network operators. The consequence of this more complex system is that the cryptographic keys have to be exchanged between the payment organization and the chosen card manufacturer of the mobile network operator before the smart cards are produced (or, in case of on-card key generation of the asymmetric key pair, the public key has to be sent to the payment organization).

Figure 13.13 shows an example of a transaction flow of such an m-payment transaction secured by a smart card. It is assumed that the customer is purchasing goods or services through an Internet connection that secures data integrity and authenticity. However, this does not matter and can be replaced by a purchase at a shop or at a vending machine or even through the mobile phone itself. In any case, the customer makes a purchase request (1) and sends it to the merchant server. The purchase request must

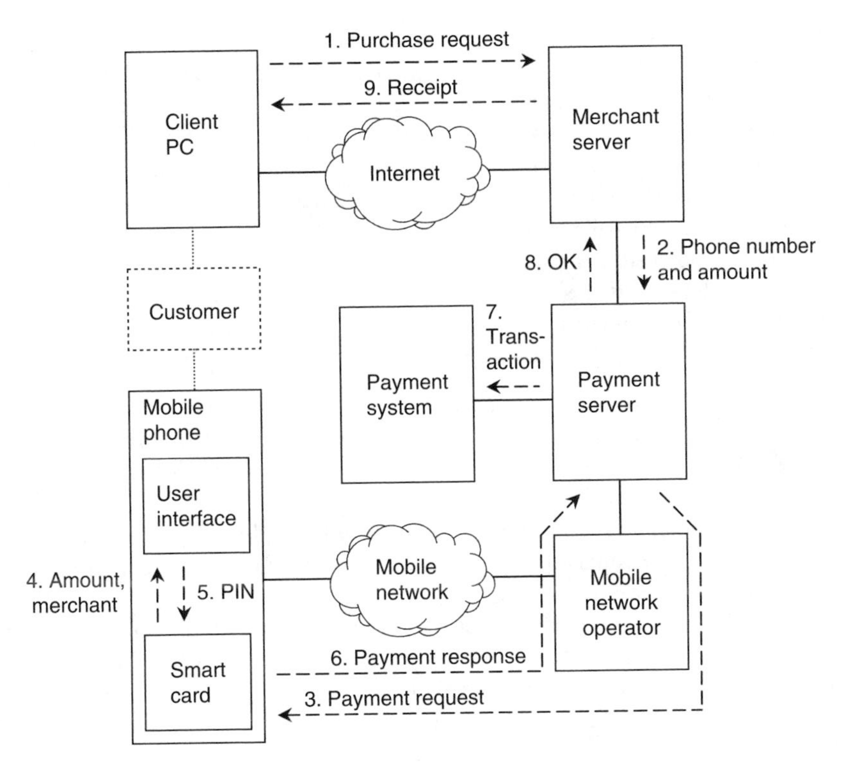

FIGURE 13.13 M-payment transaction secured by a SIM.

include the transaction amount and the customer's mobile phone number. The merchant server sends these data to the payment server over a connection that secures data integrity and authenticity (2). Thus, the payment server can generate a payment request comprising the transaction identification, the transaction amount, the merchant's name and unique identification, and a MAC calculated over these data to secure their integrity and authenticity. (For transaction security, it is unnecessary to encrypt the data; however, privacy concerns can be answered by encrypting the data.) The payment request is then sent to the customer's mobile phone through the mobile network operator who uses an SMS for data transport (3). Inside the customer's mobile phone, the smart card receives the data. The crucial fact is that this whole data transmission needs no additional security measures, as there is an end-to-end security implemented between the payment server and the smart card. After successful verification of the MAC, the card sends the transaction amount, and the merchant's name or unique identification to the user interface to display it to the customer (4). The customer reads this information and confirms the payment by entering his payment PIN. The PIN is directly verified by the smart card (5) that can in turn generate a payment response comprising the transaction identification, its approval, and a digital signature calculated over these data to secure integrity, authenticity, and nonrepudiation. The payment response is sent back to the payment server through the mobile network operator (6). The payment server verifies the digital signature and transmits the necessary transaction data to the customer's chosen payment system (7), which may be a credit card organization. The merchant server is then informed about the successful transaction (8) and can in turn generate a receipt for the customer (9).

An evaluation of this approach shows that all important security requirements of an m-payment system are fulfilled:

- Integrity of the transaction data is secured during each data transmission.
- Transaction data can only be modified at the merchant server and at the payment server. The mobile network operator does not have to take any security measures.
- Customer gets the transaction data as provided by the payment server. (It can be assumed that the customer has trust in the payment server, if he enrols to this payment system.)
- Customer cannot successfully repudiate his transaction confirmation, as it is signed with his private key that exists only in his smart card, and the smart card has verified the PIN that is only known to the customer (and not to any server).

13.16 CONCLUSION

Regarding security, there is no fundamental difference between the classical wire-based Internet and wireless networks. In any case, security must be built on trusted hardware and software. Smart cards are excellent platforms to establish such a TCB, because they can keep secrets and perform cryptographic operations. In practice, the security level of many wireless systems can be improved by making appropriate use of smart card technology.

REFERENCES

1. Department of Defense: Trusted Computer System Evaluation Criteria, 1985, http://www.fas.org/irp/nsa/rainbow/std001.htm
2. W. Rankl and W. Effing: Handbuch der Chipkarten, 4. Auflage, Carl Hanser Verlag, München, 2002.
3. International Organization for Standardization: IT—Telecommunications and Information Exchange between Systems—Near Field Communication—Interface and Protocol, 2004.
4. EMVCo: EMV Integrated Circuit Card Specifications for Payment Systems, Version 4.1, 2004, http://www.emvco.com/cgi_bin/detailspec.pl?id = 5
5. WLAN Smart Card Consortium, http://www.wlansmartcard.org/

6. P. Urien, W. Habraken, D. Flattin, and H. Ganem: EAP Smart Card Protocol (EAP-SC), 2005, http://www.ietf.org/internet-drafts/draft-urien-eap-smartcard-type-03.txt
7. P. Urien and G. Pujolle: EAP-Support in Smartcard, 2006, http://www.ietf.org/internet-drafts/draft-urien-eap-smartcard-10.txt
8. V. Hassler, M. Manninger, M. Gordeev, and C. Müller: *Java Card for E-Payment Applications*, Artech House, Norwood, MA, 2002.
9. V. Hassler: *Security Fundamentals for E-Commerce*, Artech House, Norwood, MA, 2001.
10. R. Anderson and M. Kuhn: *Tamper Resistance—A Cautionary Note*, 2nd USENIX Workshop on Electronic Commerce Proceedings, Nov. 96, pp. 1–11, 1996, http://www.cl.cam.ac.uk/users/rja14/tamper.html
11. D. Boneh, R.A. DeMillo, and R.J. Lipton: *On the Importance of Checking Cryptographic Protocols for Faults*, Lecture Notes in Computer Science, 1294, pp. 37–51, Springer, Berlin, 1997, http://jya.com/smart.pdf
12. E. Biham and A. Shamir: *Differential Fault Analysis of Secret Key Cryptosystems*, Lecture Notes in Computer Science, 1294, pp. 513–525, Springer, Berlin, 1997.
13. P. Kocher, J. Jaffe, and B. Jun: *Differential Power Analysis*, Lecture Notes in Computer Science, 1666, pp. 388–397, Springer, Berlin, 1999.
14. J.-J. Quisquater and D. Samyde: *Electromagnetic Analysis (EMA): Measures and Countermeasures for Smart Cards*, Lecture Notes in Computer Science, 2140, pp. 200–210, Springer, Berlin, 2001.
15. S. Skorobogatov and R. Anderson: *Optical Fault Induction Attacks*, Lecture Notes in Computer Science, 2523, pp. 2–12, Springer, Berlin, 2002, www.cl.cam.ac.uk/~sps32/ches02-optofault.pdf
16. International Organization for Standardization: Information Technology—Security Techniques—Evaluation Criteria for IT Security, Part 1–3, 2005.

Index

A

B

C

E

F

G

H

I

K

L

M

N

O

P